RADIATIVE RECOMBINATION IN SEMICONDUCTING CRYSTALS

IZLUCHATEL'NAYA REKOMBINATSIYA V POLUPROVODNIKOVYKH KRISTALLAKH

ИЗЛУЧАТЕЛЬНАЯ РЕКОМБИНАЦИЯ В ПОЛУПРОВОДНИКОВЫХ КРИСТАЛЛАХ

The Lebedev Physics Institute Series

Editors: Academicians D. V. Skobel'tsyn and N. G. Basov

P. N. Lebedev Physics Institute, Academy of Sciences of the USSR

Volume 30   Physical Optics
Volume 31   Quantum Electronics in Lasers and Masers, Part 1
Volume 32   Plasma Physics
Volume 33   Studies of Nuclear Reactions
Volume 34   Photomesic and Photonuclear Processes
Volume 35   Electronic and Vibrational Spectra of Molecules
Volume 36   Photodisintegration of Nuclei in the Giant Resonance Region
Volume 37   Electrical and Optical Properties of Semiconductors
Volume 38   Wideband Cruciform Radio Telescope Research
Volume 39   Optical Studies in Liquids and Solids
Volume 40   Experimental Physics: Methods and Apparatus
Volume 41   The Nucleon Compton Effect at Low and Medium Energies
Volume 42   Electronics in Experimental Physics
Volume 43   Nonlinear Optics
Volume 44   Nuclear Physics and Interaction of Particles with Matter
Volume 45   Programming and Computer Techniques in Experimental Physics
Volume 46   Cosmic Rays and Nuclear Interactions at High Energies
Volume 47   Radio Astronomy: Instruments and Observations
Volume 48   Surface Properties of Semiconductors and Dynamics of Ionic Crystals
Volume 49   Quantum Electronics and Paramagnetic Resonance
Volume 50   Electroluminescence
Volume 51   Physics of Atomic Collisions
Volume 52   Quantum Electronics in Lasers and Masers, Part 2
Volume 53   Studies in Nuclear Physics
Volume 54   Photomesic and Photonuclear Reactions and Investigation Methods with Synchrotrons
Volume 55   Optical Properties of Metals and Intermolecular Interactions
Volume 56   Physical Processes in Lasers
Volume 57   Theory of Interaction of Elementary Particles at High Energies
Volume 58   Investigations in Nonlinear Optics and Hyperacoustics
Volume 59   Luminescence and Nonlinear Optics
Volume 60   Spectroscopy of Laser Crystals with Ionic Structure
Volume 61   Theory of Plasmas
Volume 62   Methods in Stellar Atmosphere and Interplanetary Plasma Research
Volume 63   Nuclear Reactions and Interaction of Neutrons and Matter
Volume 65   Stellarators
Volume 67   Physical Investigations in Strong Magnetic Fields
Volume 68   Radiative Recombination in Semiconducting Crystals
Volume 70   Group-Theoretical Methods in Physics

In preparation
Volume 64   Primary Cosmic Radiation
Volume 66   Theory of Collective Particle Acceleration and Relativistic Electron Beam Emission
Volume 69   Nuclear Reactions and Accelerators of Charged Particles
Volume 71   Photonuclear and Photomesic Processes
Volume 72   Physical Acoustics and Optics: Molecular Scattering of Light; Propagation of Hypersound; Metal Optics
Volume 73   Microwave—Plasma Interactions
Volume 74   Neutral Current Layers in a Plasma
Volume 75   Optical Properties of Semiconductors

Proceedings (Trudy) of the P. N. Lebedev Physics Institute

Volume 68

RADIATIVE RECOMBINATION IN SEMICONDUCTING CRYSTALS

Edited by
Academician D. V. Skobel'tsyn
Director, P. N. Lebedev Physics Institute
Academy of Sciences of the USSR, Moscow

Translated from Russian by
Albin Tybulewicz
Editor, *Soviet Physics—Semiconductors*

SPRINGER SCIENCE+BUSINESS MEDIA, LLC

Library of Congress Cataloging in Publication Data

Main entry under title:

Radiative recombination in semiconducting crystals.

(Proceedings (Trudy) of the P. N. Lebedev Physics Institute; v. 68)
Translation of Izluchatel'naĭa rekombinatsiĭa v poluprovodnikovykh kirstallakh.
Includes bibliographical references.
1. Semiconductors — Addresses, essays, lectures. 2. Crystal optics — Addresses;
essays, lectures. 3. Luminescence — Addresses, essays, lectures. I. Skobel'tsyn,
Dmitriĭ Vladimirovich, 1892– II. Series: Akademiĭa nauk SSSR. Fizicheskiĭ institut.
Proceedings; v. 68.

QC1.A4114 vol. 68 [QC611.2] 530'.08s [548'.9]   75-17670
ISBN 978-1-4757-6346-1   ISBN 978-1-4757-6344-7 (eBook)
DOI 10.1007/978-1-4757-6344-7

The original Russian text was published by Nauka Press in Moscow in 1973 for the
Academy of Sciences of the USSR as Volume 68 of the Proceedings of the P. N.
Lebedev Physics Institute. This translation is published under an agreement with the
Copyright Agency of the USSR (VAAP).

CONTENTS

Ultraviolet Cathodoluminescence and Photoluminescence of Zinc Sulfide
  Yu. V. Voronov

Abstract   1
Introduction   1

Chapter I

Review of Published Literature   3
1. Optical Properties of Zinc Sulfide   3
2. Ultraviolet Luminescence near Fundamental Absorption Edge of Zinc Sulfide . 10
3. Donor—Acceptor Model of Luminescence Centers 14
4. Efficiency of Cathodoluminescence and Photoluminescence 20

Chapter II

Formulation of the Problem and Experimental Method 24
1. Selection of Investigation Methods and Objects 24
2. Methods Used in Preparation of ZnS Powders and Single Crystals 26
3. Apparatus and Investigation Methods. 26

Chapter III

Thermal and Spectral Properties of Ultraviolet Luminescence of Unactivated ZnS. . 27
1. Efficiency of Cathodoluminescence and Photoluminescence of Unactivated Zinc
      Sulfide 27
2. Spectral Composition of Ultraviolet Luminescence 32
3. Comparison of the Cathodoluminescence of Powders, Single Crystals, and
      Sublimates of Unactivated ZnS. 35
4. Influence of Temperature on the Ultraviolet Luminescence of Unactivated Zinc
      Sulfide 37
5. Quenching of the Edge Cathodoluminescence of ZnS by Infrared Radiation . . . 42

Chapter IV

Influence of Conditions during Synthesis on the Efficiency and Properties of Edge
      Luminescence of ZnS 45
1. Influence of Stoichiometry on the Edge Luminescence of ZnS 45
2. Thermoluminescence of Unactivated Zinc Sulfide Excited by Electron
      Bombardment 51
3. Influence of Quenching on the Efficiency of Edge Luminescence of ZnS Single
      Crystals 54
4. Quenching of the Edge Luminescence of ZnS as a Result of Introduction of
      Luminescence and Quenching Centers. 60

Chapter V

Nature of Ultraviolet Luminescence of Unactivated Zinc Sulfide   66
1.   Various Types of Radiative Transitions   67
2.   Detailed Consideration of the Model of Donor–Acceptor Pairs and
        Discussion of the Polaron Edge Luminescence Model   69
3.   Exciton Mechanism of the Short-Wavelength Bands and the Influence of
        Temperature on Various Types of Luminescence of ZnS   78
4.   Energy Structure of Local Levels in Unactivated Zinc Sulfide   80
Conclusions   84
Literature Cited   86

Ionization Domains in Strong Fields and Motion of
    Luminous Regions in Crystals
        M. V. Fok and E. Yu. L'vova

Abstract   91
Preliminary Experiments   91
Distribution of Luminescence   92
Comparison with Known Types of Instability   93
Possible Nature of the Observed Phenomenon   94
Estimate of the Space Charge   95
Estimate of the Field Intensity in a Double Layer   96
Estimate of the Electron Energy in a Strong Field   98
Formation of a Positive Space Charge in a Double Layer   100
Formation of a Negative Space Charge in a Double Layer   101
Appearance of Ionization Domains   103
Conclusions   104
Literature Cited   105

Investigation of the Ambipolar Diffusion of Free Carriers
    in Inhomogeneously Photoexcited Zinc Sulfide Crystals
        N. N. Grigor'ev and M. V. Fok

Abstract   107
Introduction   107
1.   Energy Transfer over Long Distances   108
2.   Theory of the Luminescence Contour Method   109
3.   Description of Apparatus   116
4.   Method Used in Measurements and Comparison of Luminescence Contours .   118
5.   Shape of Luminescence Contours   120
6.   Influence of the Excitation Rate on the Shape of Luminescence Contours . . .   123
7.   Influence of Long-Wavelength Background Illumination on the Shape of
        Luminescence Contours   125
8.   Theory of Luminescence Contours in an External Alternating Electric Field   127
9.   Experimental Investigation of the Influence of an Electric Field on the Shape
        of Luminescence Contours   134
10.   Estimates of the Transport Properties of Zinc Sulfide   137
Conclusions   141
Literature Cited   142

CONTENTS

Radiative Recombination in Cadmium Telluride Crystals
Zh. R. Panosyan

Abstract... 145
Introduction.. 145

Chapter I

Review of Published Literature..................................... 145
1. Principal Radiative Recombination Mechanisms and Energy Band Structure
    of II-VI Compounds.. 145
2. Edge Luminescence 149

Chapter II

Experimental Method 152
1. Apparatus Used in Determination of Luminescence Spectra 152
2. Determination of the Reflection Spectra 158
3. Preparation of Cadmium Telluride Samples..................... 158

Chapter III

Exciton Luminescence and Absorption in Cadmium Telluride Crystals........ 159
1. General Description of the Photoluminescence Spectra of Cadmium Telluride 159
2. Photoluminescence of Cadmium Telluride Crystals in the Exciton Reflection
    Region 161
3. Free-Exciton Absorption in Cadmium Telluride Crystals 165
4. Comparison of the Free-Exciton Absorption and Luminescence Spectra... 168

Chapter IV

Edge Luminescence and Phonon Effects in the Photoluminescence of Cadmium
    Telluride... 173
1. Dependence of the Edge Luminescence Spectrum on the Impurity
    Composition of Crystals..................................... 173
2. Nature of Radiative Transitions............................. 176
3. Electron—Phonon Interaction................................. 180

Chapter V

Radiative Recombination in Deep Centers in CdTe Crystals............ 184
1. Radiative Recombination between Donors and Deep Acceptors 184
2. Band at 1.1 eV 191
3. Recombination Radiation at $\lambda > 1.2 \mu$ 193
Literature Cited... 195

Investigation of the Energy Band Structure of Semi-
    conductors by Differential Optical Methods
    S. G. Dzhioeva

Abstract... 199
Introduction... 199

Chapter I

Investigations of the Singularities of the Combined Density of Electron States in
    Semiconductors by the Electroreflection Method 202

CONTENTS

1. Experimental Method.   202
2. Electroreflection of n-Type GaAs   205
3. Characteristics of the Reflection of GaAs Crystals with Low
    Free-Carrier Densities   210
4. Exciton Electroreflection of CdTe   214

Chapter II

Investigation of the Thermoreflection of GaAs.   220
1. Experimental Method.   220
2. Results and Discussion   221
Conclusions   224
Literature Cited   225

Luminescence Delay and Determination of Nonradiative
    Relaxation Times
        É. L. Nolle

Abstract.   227
1. Relaxation and Luminescence Kinetics   228
2. Luminescence Delay and Energy Transfer Time from Host to Activator in
    $Y_3Al_5O_{12}:Nd^{3+}$ 230
3. Formation Time of Electron–Hole Drops in Germanium   232
Literature Cited.   233

ULTRAVIOLET CATHODOLUMINESCENCE AND PHOTOLUMINESCENCE OF ZINC SULFIDE*

Yu. V. Voronov

An investigation was made of the properties and nature of the edge and other ultraviolet luminescence bands of zinc sulfide. The excitation conditions (nature of the excitation, its intensity,
temperature of a crystal, intensity of infrared background illumination, etc.), conditions during
synthesis, and composition (stoichiometry, quenching, crystal phase, impurity content) were varied in a wide range. The results were used to determine the energy and spectral characteristics.
The radiative transitions explaining the results were identified. The explanation of the edge
luminescence was based on a donor–acceptor model of radiative transitions at local levels associated with intrinsic lattice defects.

INTRODUCTION

Investigations of the nonequilibrium luminescence of activated crystalline substances have
helped us to understand the general laws governing the luminescence of solids. Experimental
and theoretical studies, a considerable proportion of which were carried out by the Soviet school
of luminescence led by S. I. Vavilov, have established the recombination mechanism of the lu-
minescence and the nature of the luminescence centers in activated ZnS phosphors. Polycrystalline phosphors in which the photoluminescence, cathodoluminescence, and electroluminescence centers are heavy metal atoms (Ag, Cu), are characterized by a high luminescence efficiency in the temperature range ~300°K and are used widely in practice.

Back in 1940 Kröger [1] established that optically excited powders of unactivated zinc
sulfide emitted ultraviolet luminescence at T = 77°K and this luminescence was located near
the fundamental absorption edge. Since at that time the forbidden band width of ZnS was not
known accurately and since luminescence with shorter wavelengths was unknown, the observations were attributed to interband transitions and the observed radiation was called the edge
luminescence.

Later, the luminescence discovered by Kröger was attributed to excitons, but more
recent investigations support the hypothesis that the radiative transitions are associated with
local states in the lattice.

All types of luminescence (interband, exciton, and Kröger) observed near the fundamental
absorption edge of the lattice should be called the edge luminescence. However, we shall adopt
the old but established terminology and we shall apply the term "edge luminescence" only to
the luminescence associated with the structure of ZnS and emitted in the 330-360 nm range.

---

*Thesis for the Degree of Candidate of Physicomathematical Sciences. Defended on October 13,
1969 at the P. N. Lebedev Physics Institute, Academy of Sciences of the USSR, Moscow.

Studies of the properties of the ultraviolet luminescence of zinc sulfide generated by various excitation methods [2-5] and in different objects (powders and single crystals prepared by different technologies) are among the more important tasks in the studies of the luminescence of solids.

There are several factors which make such studies topical:

1) the closeness of the ultraviolet luminescence bands to the fundamental absorption edge of ZnS and the emission of similar low-temperature luminescence by II-VI and other binary compounds;

2) the presence of a fine structure representing the interaction with the host lattice phonons, which is not observed in the luminescence bands of heavy metals (Cu, Ag, Mn) that are typical activators in ZnS phosphors;

3) the absence of well-known impurities which might be regarded as the centers responsible for the observed luminescence and an increase of the intensity of this luminescence with increasing purity, which suggests that the ultraviolet luminescence is the property of the lattice itself, or more probably of its intrinsic defects;

4) the existence of different interpretations of the ultraviolet luminescence (interband recombination, excitons, donor–acceptor pairs of lattice defects);

5) the practical applications of unactivated ZnS, excited by electron bombardment or electric fields, as coherent and noncoherent sources of ultraviolet radiation.

However, the published investigations of the ultraviolet luminescence of ZnS are mainly concerned with its spectral characteristics, which are necessary but insufficient for the determination of its nature and practical applications. Until the investigation reported below, there has been no published information on the absolute efficiency of this luminescence and particularly on its dependence on the excitation conditions and the conditions during synthesis of crystals. It has been assumed that the luminescence is far weaker than the luminescence of typical activators in ZnS crystal phosphors and that it practically disappears on introduction of impurities. It has been known that the ultraviolet luminescence appears only at low temperatures but the thermal quenching has not been investigated. Lack of information on the kinetics of the ultraviolet luminescence of ZnS has been primarily due to experimental difficulties encountered in the optical excitation with mercury discharge lamps, which have been used as the sources in the early studies. In this respect very promising results in investigations of the properties of the ultraviolet luminescence of unactivated ZnS have been obtained by electron bombardment and by comparing the information obtained in this way with that deduced from the photoluminescence.

The first investigations of the ultraviolet cathodoluminescence of polycrystalline unactivated ZnS were started by Levshin's group [2]. Later, intense electron excitation was used by other investigators [3, 6]; however, the immediate aim in these investigations was to achieve a stimulated emission and, therefore, many aspects of the luminescence kinetics were not considered. It should be stressed that the differences between the excitation conditions (nature of the excitation, its intensity, and temperature of a sample) and materials employed in different studies make it difficult to compare the experimental results and to interpret them theoretically. In view of this situation, the present author set himself a task of investigating experimentally the efficiency and kinetics of the edge luminescence under different excitation conditions (studies of the influence of temperature, intensity, wavelength of the exciting light, initial energy of electrons, etc.) using a wide range of objects.

These investigations were carried out on polycrystalline screens (hexagonal and cubic phases) and single crystals of unactivated zinc sulfide.

CHAPTER I

REVIEW OF PUBLISHED LITERATURE

§ 1. Optical Properties of Zinc Sulfide

The importance of zinc sulfide, which is a member of the II–VI group of binary compounds, lies in its practical applications as a highly efficient phosphor. The highest efficiency (close to 1) is obtained when the luminescence centers formed by heavy or rare-earth metals in polycrystalline samples are excited directly by optical means. Therefore, ZnS has been investigated mainly in the form of activated polycrystalline films. At present many studies are being made of unactivated powders, sublimates, and single crystals.

Investigations of the luminescence of semiconducting compounds, particularly those of zinc sulfide, should be carried out bearing in mind other physical properties.

Zinc sulfide crystallizes in the two main modifications, one of which is cubic (sphalerite) and the other hexagonal (wurtzite). These two modifications have the cubic and hexagonal close packings of the S^{2-} anions with a large ionic radius ($r_a = 1.82$ Å). The Zn^{2+} cations with an ionic radius $r_c = 0.87$ Å occupy half the tetrahedral voids in the close-packed structure. Crystallization occurs in such a way that each ion is located at the center of a regular tetrahedron and ions of the opposite sign occupy the vertices of this tetrahedron.

The structure of wurtzite can be regarded as two interpenetrating hexagonal packings; the zinc and sulfur ions are displaced relative to one another along a threefold symmetry axis (111), whereas the structure of sphalerite can be regarded as combination of two cubic lattices of the sulfur and zinc ions which are also interpenetrating and shifted relative to one another by a quarter of the cube diagonal.

Zinc sulfide differs from the other II–VI compounds because it exhibits a polytypism, i.e., it exhibits lattice structures with regular sequences of stacking faults. Zinc sulfide crystals grown by various methods (from the gaseous phase, from solution, and from the melt) have different structures and physical properties.

It is usually assumed that the temperature of the phase transition from the α to the β modification is $\sim 1020 \pm 50$°C, i.e., below this temperature the sphalerite structure is stable and above this temperature the wurtzite form is stable. Usually a ZnS phosphor contains some amount of the unstable phase and this amount depends on the heat-treatment temperature [7]. In fact, studies of the diffuse reflection spectra [8] of phosphors prepared at different temperatures make it possible to follow the transition from the α to the β phase of ZnS. Published information also suggests that the phase transition temperature depends on the medium in which a sample is located [9, 10]. Some investigators [11, 12] have suggested the existence of two stability regions of the cubic modification of ZnS, i.e., they have suggested the existence of two phase transition temperatures (T < 1020°C and T > 1240°C). Therefore, the temperature at which a phosphor has been prepared does not prove that its structure is hexagonal or cubic. In order to understand the physical nature of the processes associated with the electrical conduction, photoconductivity, and luminescence of ZnS, we must know the forbidden band width and the energy structure of levels in this band. The forbidden band width of zinc sulfide can be deduced from the photoconductivity maximum of unactivated ZnS, which can be determined quite accurately. According to Piper [13], who studied ZnS single crystals at T = 293°K, the photoconductivity maximum is located at $\lambda_{max} \approx 337$ nm ($E_g \approx 3.69$ eV).

The fundamental absorption of zinc sulfide is located in this region. However, the quantitative information on the fundamental absorption edge of ZnS, which can be used to determine the forbidden band width directly, is highly contradictory. Therefore, the published experimental estimates of the forbidden band width are very approximate. They give values ranging from

TABLE 1. Principal Energy
States of an Exciton in Hexagonal
Single Crystals

Sample	Exciton maximum		
	E^A, eV	E^B, eV	E^C, eV
ZnS	3.8714	3.8996	3.990
CdS	2.5537	2.5686	2.632
CdSe	1.8258	1.8511	2.259

3.7 to 4.1 eV, but the differences are considerably greater than the difference between the forbidden band widths of the hexagonal and cubic modifications of ZnS and also greater than the narrowing of the forbidden band (by ~0.1 eV) between 77°K and room temperature.

In most cases the absorption spectra have been determined for ZnS crystals only near the long-wavelength edge (in the range $\lambda > 325$ nm); the results have been used to determine the forbidden band width of ZnS. However, doubts have been raised about the validity of the determination of the forbidden band width E_g from the long-wavelength edge and these have arisen from the investigation of the absorption spectrum of ZnS [14, 15] in a wider range of wavelengths. It has been found that there is an unresolved band with k $\sim 10^4$-10^5 cm^{-1} in the region of ~330 nm (T = 293°K) and beyond this band the absorption rises strongly. This band has been attributed to excitons or to some impurities [16] introduced accidentally into ZnS film during preparation.

An absorption band with a maximum at ~325 nm is reported in [17] for ZnS sublimates. The intensity of this band rises and the band becomes narrower and shifts toward shorter wavelengths when the temperature is lowered. It is assumed in [17] that this band is due to the excited states of the lattice (excitons). The existence of exciton states is a characteristic property of the host lattice of this material (Table 1). It is now accepted that exciton states play an important role in the absorption and luminescence of solids. The line structure observed near the long-wavelength absorption edge of some crystals is attributed to the excitation of excitons in the lattices of these crystals.

Investigation of the absorption spectra of hexagonal ZnS single crystals [18] at T = 4.2, 77, and 293°K (Table 2) in the range of thickness from 0.1 to 10 μ has revealed narrow polarized absorption lines in the 3100-3300 Å range on a background of a continuous absorption rising in the direction of shorter wavelengths ($\lambda = 3205$ Å has the $E \perp C$ polarization, $\lambda = 3180$ Å is unpolarized, and $\lambda = 3115$ Å is polarized preferentially in the $E \parallel C$ direction), as well as an absorption step ($\lambda = 3165$ Å with the $E \perp C$ polarization) at T = 4.2°K.

Bearing in mind the positions of the lines in the spectra near the fundamental absorption range and the reproducibility of these lines in the spectra of different types of sample (subli-

TABLE 2. Positions of Maxima
in Absorption Spectra of Hexagonal ZnS Single Crystals at Different Temperatures

T, °K	Positions of maxima, Å		
4.2	3210	3185	3120
77	3220	3190	3130
293	—	3270	3190

mated films and single crystals), as well as the considerable absorption coefficient (10^4-10^5 cm^{-1}), it is concluded in [18] that this absorption is due to direct allowed transitions to exciton states. Absorption maxima in the 200-250 nm range in the spectra of thin sublimated films are reported in [19-21]. These bands are evidently due to the complexity of the energy band structure of zinc sulfide, as pointed out in [22].

The forbidden band width corresponding to the fundamental absorption edge of ZnS is reported to depend on the conditions during preparation of this material, investigation method, and technique used to extrapolate the absorption edge. A theoretical calculation of the energy band structure of ZnS is very difficult even if no allowance is made for various defects in the lattice. The conduction band of zinc sulfide is composed mainly of the S levels of the cations (Zn^{2+}) whereas the valence band is composed of the P levels of the anions (S^{2-}). The energy band structures of hexagonal and cubic zinc sulfide are shown in Fig. 1.

In the sphalerite lattice the valence band is split by the spin–orbit interaction into two sub-bands Γ_8 and Γ_7. In the wurtzite lattice the crystal field splits (because of the axial component) the valence band of sphalerite Γ_8 into two sub-bands Γ_9 and Γ_7 [23].

Recently Birman [24] used a semiempirical variant of the molecular orbital method and calculated the electron structure of II-VI compounds.

Suslina [4] used the theoretical calculations of Birman and Hopfield and the positions of the exciton lines in an accurate evaluation (up to 0.001 eV) of the following important characteristics of the energy structure of hexagonal ZnS at k = 0: the forbidden band width E_g = 3.917 eV, exciton binding energy E_{ex} = 0.049 eV, and splitting energies of the valence band E_1 = 0.03 eV and E_2 = 0.081 eV at T = 4°K.

The hexagonal form of zinc sulfide exhibits a slight anisotropy of the optical properties. The dependence of the absorption edge on the polarization is reported in [25]. Moreover, ZnS exhibits a strong anisotropy of the photocurrent [26].

Zinc sulfide is a semiconductor with mixed ionic-covalent bonds. Since the type of binding governs many of the properties of a material, some work has been done on the bonds in zinc sulfide [27, 28].

The covalent contribution to the binding in ZnS is favored by its tetragonal structure.

If the purely ionic formula for zinc sulfide is expressed in the form $Zn^{2+}S^{2-}$, the purely covalent structure corresponding to a uniform sharing of eight valence electrons by zinc and sulfur should give rise to "formal" charges $Zn^{2-}S^{2+}$.

Estimates of the proportion of the ionic binding in zinc sulfide range from 30 to 60%. Thus, in the first approximation ZnS can be regarded as an ionic compound because the static permittivity ε and the high-frequency permittivity ε_0 are different (ε = 8.3 and ε_0 = 5.07). The Fröhlich relationships between ε and ε_0 and the wave numbers ν_1 and ν_t of the longitudinal and

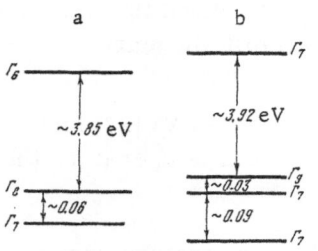

Fig. 1. Energy band structure of sphalerite (a) and wurtzite (b) at the point k = 0 (T = 4.2°K).

TABLE 3.   Effective Masses of Carriers in Some
II-VI Compounds

Effective mass	ZnS [31]	CdS [32]	CdSe [33]
$m_{e\parallel}^{*}$	0.28 ± 0.03	0.205 ± 0.01	0.13 ± 0.01
$m_{e\perp}^{*}$	0.28 ± 0.03	0.205 ± 0.01	0.13 ± 0.10
$m_{h\parallel}^{*}$ (Γ_9)	~1.4	~5	$\geqslant1$
$m_{h\perp}^{*}$ (Γ_9)	0.49 ± 0.06	0.7 ± 0.1	0.45 ± 0.09

transverse optical vibrations are in good agreement:

$$\frac{\nu_l}{\nu_t} = \frac{349 \text{ cm}^{-1}}{274 \text{ cm}^{-1}} = 1.274, \qquad \sqrt{\frac{\varepsilon}{\varepsilon_0}} = 1.280,$$

i.e., the degree of ionicity of zinc sulfide is $\sim 30\%$.

Fok [29] analyzed the effective ionic charge in the crystal lattice of ZnS and reported that $e^{*} = (0.5 \pm 0.1)e$, i.e., it is only half the electronic charge. This is attributed by Fok not only to a considerable proportion of the covalent binding in ZnS ($\sim 60\%$), deduced from the ratio of the squares of the longitudinal and transverse phonons, but also to the high polarizability of the sulfur ion.

The luminescence of zinc sulfide is due to recombination. The luminescence centers can be excited by direct ionization or by the ionization of the host substance. The ionization of the host, which may result from optical or electrical excitation, produces free carriers and an induced conductivity, which exceeds the dark conductivity because the forbidden band width of ZnS is wide. The electron mobility in zinc sulfide, $\sim 120 \text{ cm}^2 \cdot \text{V}^{-1} \cdot \text{sec}^{-1}$ [30], is intermediate between the mobilities in covalent (Ge, Si) and ionic (NaCl, KCl) crystals; the mobility of holes is approximately an order of magnitude lower. The effective mass of free holes is several times larger than the effective mass of electrons [31, 33] and this point governs the ratio of the other properties of ZnS (Table 3).

The forbidden band width and the nature of interband transitions in ZnS phosphors are discussed in [29], where an analysis is made of the experimental investigations up to 1963. It is suggested that the differences between the optical and thermal forbidden band widths ($E_{h\nu}-E_t$), resulting from the local polarization of the crystal lattice of ZnS by free carriers, must be taken into account. We must bear in mind that only the inertial part (heavy ions) of the polarization of a crystal is active in the formation of polaron states. This inertial polarization reduces the energy of free carriers.

Fok [29] analyzed the experimental data on these quantities and showed that it was necessary to control the purity of investigated samples. Extrapolating the data on the edge absorption [34] and the results of Alentsev and Panasyuk [35], Fok concluded that the lower limit was $E_{h\nu} = 3.66$ eV. The experimental points of the dependence $k(\nu)$ fit the curve $k \propto (h\nu - E_{h\nu})^q$, where $q = 2$, which corresponds to indirect allowed transitions [36].

Using the data on the exciton absorption bands (3.87 and 3.78 eV) [37] and bearing in mind the photoconductivity at $E_{h\nu} \sim 4.1$ eV, Fok concluded that at room temperature the optical width of the forbidden band was $E_{h\nu} = 3.9 \pm 0.2$ eV.*

---

*It should be pointed out that allowance is necessary for the temperature dependence of E_g, which was not considered by Fok [29]. The shift observed in the absorption spectra by various authors corresponds to the change by ~ 0.1 eV in E_g between 293 and 77°K.

Fok [29] questioned the experiments of Piper who determined the thermal forbidden band width from the temperature dependence of the current through ZnS (according to Piper, E_t = 3.67 ± 0.1 eV and is identical with the optical forbidden band width). Fok used similar measurements carried out on ZnS powders in the 900-1300°K range and concluded that E_t = 3.2 ± 0.2 eV. This value was refined somewhat by comparing the optical and thermal liberation of carriers from deep traps in ZnS:Cu:Ni and ZnS:Cu:Co flash phosphors [38]. The differences is thus $E_{h\nu} - E_t$ = 0.55 ± 0.15 eV. A similar difference has been reported also for other II-VI compounds. For example, the thermally stimulated current curves and induced photoconductivity spectra were used [39] to show that the optical activation energy of local centers in CdS and CdSe was 1.5-2 times greater than the thermal value. This difference was not observed for CdTe. Thus, Fok concluded that the difference between the thermal and optical widths of the forbidden band of ZnS was due to polaron states of both types of carrier and the difference between the energies of a free carrier and a polaron was ~0.3 eV. This value was compared with a set of four longitudinal and four transverse optical phonons. The possibility of existence of free and polaron states of both carriers, suggested by Fok, was supported later [40] by the results of the study of the electroluminescence spectrum of unactivated ZnS crystals at 293°K. Three bands located at 3.93, 3.67, and 3.45 eV were attributed, respectively, to the radiative recombination of two free carriers, one free carrier with a carrier in a polaron state, and two carriers in polaron states.

The energy of the edge luminescence of unactivated ZnS exceeds the thermal width of the forbidden band width, which is not in agreement with the usual interpretation of this effect. The question of the optical width of the forbidden band of ZnS is closely related to the physical interpretation of long-known edge luminescence of unactivated zinc sulfide.

Apart from the near ultraviolet edge luminescence of the type emitted by ZnS phosphors, many samples of unactivated ZnS exhibit several luminescence bands in the visible region. The intensities and positions of these bands also depend strongly on the method of preparation of ZnS samples (they depend on the materials used in the preparation, ratio of the number of Zn and S ions, ambient temperature, and presence of a flux). Obviously, physical investigations of the luminescence emitted by "self-activated" ZnS should be closely related to chemical investigations of the nature of the reactions that occur during the synthesis of phosphors.

There have been many physicochemical investigations of the luminescence of ZnS phosphors prepared under various conditions. The difficulties encountered in a consistent allowance for all the preparation conditions and in the control of the possible number of various defects in the crystal lattice lead to considerable discrepancies between the interpretations of these luminescence bands.

According to [41], the luminescence spectra depend strongly on the sulfur vapor pressure during the synthesis of ZnS. The dependences of the intensities of these luminescence bands and of the bands reported elsewhere [42] on the ratio of Zn to S in the original charge has served as the basis of the hypothesis that the blue luminescence centers are interstitial excess Zn atoms, i.e., that such phosphors are of the ZnS:Zn type. Other luminescence bands have been attributed to excess sulfur atoms, i.e., to the luminescence of ZnS:S phosphors.

However, the most widely accepted hypothesis is that due to Prener and Weil [43]: according to this hypothesis the luminescence conters are anion sulfur vacancies and cation zinc vacancies located at regular sites in the lattice.

The formation of such vacancies can be regarded as the result of two chemical reactions [44]:

$$Zn^{+2}S^{-2} + Zn \rightleftharpoons \left\{ \begin{array}{cc} Zn^{+2} & S^{-2} \\ \boxed{V_a} & Zn^{+2} \end{array} \right\} + 2e^-,$$

$$Zn^{+2}S^{-2} + S \rightleftharpoons \left\{ \begin{array}{cc} Zn^{+2} & S^{-2} \\ S^{-2} & \boxed{V_c} \end{array} \right\} + 2p^+,$$

where V_a is an anion sulfur vacancy of charge $+2e$ and V_c is a cation zinc vacancy with a double negative charge relative to the normal lattice. The number of vacancies which, in principle, can exist in any ZnS phosphor depends strongly on the reaction temperature and the pressure of sulfur in the vapor phase [45]. The intensity and nature of the luminescence emitted by unactivated ZnS phosphors depends on the presence of halogens (Cl, Br, I) which can be captured during heating or may be present in the flux. Therefore, there is also a point of view that the luminescence of unactivated zinc sulfide is possible only in the presence of these impurities, which are known to be good coactivators and which can improve the efficiency of the luminescence of activated phosphors [46]. However, later experiments have yielded evidence in conflict with this hypothesis.

Similar results were reported in [47, 48] after a high-temperature heat treatment with a control of the absence of any activator admixtures. These results are confirmed in [49] and they support the hypothesis that the blue luminescence of ZnS is due to the presence of zinc vacancies. This is in agreement with the results of a study of sublimated films, which included an investigation of the absorption spectrum. According to this point of view, the influence of chlorine on the intensity of the blue luminescence of ZnS reduces to an improvement of the conditions for an increase in the number of zinc vacancies.

However, some influence of chlorine on the position of the luminescence band attributed to Zn vacancies may indicate a need for considering complex luminescence centers each composed of a Zn vacancy and a Cl atom located in its vicinity [50].

It is shown in [51, 52] that the blue luminescence band of ZnS crystals is not simple but consists of at least four sub-bands.

More complex models of the blue luminescence centers have been suggested in the literature, such as complexes each of which comprises a sulfur vacancy, two zinc atoms at regular sites, and a oxygen atom [53]. This is in agreement with the results in [54, 55], where it is shown that the influence of oxygen and hydrogen, as well as other ions, reduces mainly to a change in the number of zinc and sulfur vacancies.

Thus, there is as yet no agreement on the composition of the centers responsible for the blue luminescence of self-activated zinc sulfide because of the difficulties encountered in ensuring the necessary purity of the samples.

Irrespective of the composition of the blue luminescence centers, they can be described by discrete energy levels in the band model of ZnS phosphors on the basis of purely physical investigation methods (absorption, luminescence, and excitation spectra). Moreover, we must bear in mind that the energy structure of a phosphor depends strongly on the conditions under which it is prepared.

There are several methods for calculating the energy structure of impurity ions in the crystal lattice and these are approximately valid in some extreme cases.

For example, introduction of some impurities (such as Tl) which differ little from the cations being replaced, deforms only slightly the lattices of alkali halide crystals and, therefore, the energy structure of such centers is close to the electronic configuration of free Tl ions. The luminescence may be due to optical transitions within the luminescence centers without ionization. This type of luminescence center was first considered by Seitz [56]. Subsequently, Williams [57] used the model of configuration curves and approximate quantummechanical methods in dealing with such centers.

Klick and Schulman [58] used a semiempirical method for the calculation of the luminescence bands and of the thermal quenching within the luminescence centers due to nonradiative transitions at higher temperatures: this method was applied to KCl:Tl phosphors, and to $ZnSiO_4$:Mn and $CaWO_4$.

TABLE 4. Energy Depths of Levels (E) of Donors
and Acceptors in Different Charge States (Z), and
Radii of Equivalent Bohr Orbits of Carriers ($R_{e,h}$)
in ZnS Lattice

Level	z	E, eV	Orbit radius, Å	
			R_e	R_h
Donor	1	0.1	15	—
	2	0.25	7.5	—
Acceptor	1	0.2—0.35	—	3—9
	2	0.5—1.1	—	1.5—4.5

Donor and acceptor impurities in narrow-gap semiconductors such as Ge and Se represent the other (opposite) case of the structure of energy levels of an impurity atom which interacts strongly with the host lattice.

In this case an electron or a hole is weakly localized near a structure defect and, therefore, it is not very sensitive to the potential in the immediate vicinity of an impurity. The theory [59] describing the energy levels of localized carriers close to the conduction band can be reduced to the model of a hydrogen-like atom if allowance is made for the static permittivity of the medium ε and for the effective masses of the carriers.

The energy depth of a level E_i and the radius of the equivalent Bohr orbit R_i are then described by the formulas

$$E_i = \frac{q^2 m^*}{2\hbar^2 \varepsilon^2} \simeq 13.5 \left(\frac{m^*}{m}\right) \frac{1}{\varepsilon^2} Z,$$

$$R_i = \frac{\hbar^2 \varepsilon}{q m^*} \simeq 0.52 \varepsilon \left(\frac{m^*}{m}\right) \frac{1}{Z}.$$

For example, in the case of silicon we have $\varepsilon = n^2 = 12$; $m_e^* = 0.3m$; the energy depth of a level is $E_i \sim 0.029$ eV and the radius of the Bohr orbit is $R_i \sim 20$ Å, i.e., it amounts to several lattice constants (d $\sim$ 5.42 Å), which justifies this theory.

Strictly speaking, the luminescence centers which occur in zinc sulfide phosphors do not fit any of the theories mentioned above. They represent an intermediate case between typical semiconducting crystals (Ge, Si) and typical ionic crystals (KCl, etc.). Therefore, in general, we must allow for the influence of the crystal field of the lattice and for the internal structure of the luminescence center itself. This makes the problem extremely difficult and only the simplest cases have been solved [24].

However, in a qualitative analysis of shallow trapping levels and of centers which emit photons differing only slightly from the forbidden band width of ZnS (< 0.2 eV), it is preferable to use the effective mass theory. Table 4 gives the results of calculations carried out on zinc sulfide by the present author.[*]

Shallow traps (~ 0.1 eV) are associated with singly charged donors. The presence of such shallow electron traps, which are emptied at T ~50°K at the usual rates of heating, is mentioned in [60, 61]. It should be pointed out that low-temperature traps with an activation energy of ~ 0.1 eV or less in ZnS phosphors have not been reliably established because structure defects

---

[*]In the case of doubly charged donor and acceptor states it is necessary to use not the hydrogen-like model but the helium-like model [63] in which the ionization energy needed to detach the first electron is $E_{i1} = 24.5$ eV and that needed to detach the second electron is $E_{i2} = 54.2$ eV.

in ZnS, associated with the alternation of the wurtzite and sphalerite blocks, have approximately the same activation energy of ~0.05-0.1 eV; such structure defects have been investigated intensively for SiC.

As pointed out by Riehl [62], the tunnel effect has to be allowed for in considering the liberation of carriers from such shallow levels.

Singly charged acceptor traps (liberation of holes) and doubly charged donor traps (liberation of electrons) are active in the ~0.2-0.4 eV range, i.e., in the range where carriers of both signs are liberated and where the strongest de-excitation occurs via impurity centers in ZnS phosphors.

Deep traps (~1 eV) are doubly charged acceptor centers which may be active at high temperatures and responsible for the final stages of the luminescence decay.

§2. Ultraviolet Luminescence near Fundamental Absorption Edge of Zinc Sulfide

In the last decade one of the most important problems in the studies of luminescence, closely related to other problems in solid-state physics, has been the emission of the edge luminescence from undoped semiconductors. Such investigations are particularly important in the case of zinc sulfide, which is used widely in high-efficiency activated phosphors.

This intrinsic luminescence is located in the near ultraviolet and was first reported by Kröger [1] in 1940. He studied zinc sulfide excited optically at 77°K. The ultraviolet luminescence band (330-360 nm), located close to the fundamental absorption edge of unactivated zinc sulfide, was called by him the edge band.

Somewhat later similar luminescence was reported for other compounds (CdS, ZnO, CaS) [64, 66].

The edge luminescence emitted by various unactivated phosphors has certain properties in common.

The main feature of this luminescence, which is located near the fundamental absorption edge (the difference between E_g and the edge $h\nu_{edge}$ is ~0.2 eV), is the equidistant structure of the luminescence spectrum, which is manifested clearly at low temperatures (Fig. 2). When the temperature is raised, the total intensity of this luminescence decreases strongly. Moreover, the spectrum becomes broader. The fine structure may be attributed to the phonon replicas resulting from the modulation of the luminescence by the frequency of longitudinal optical

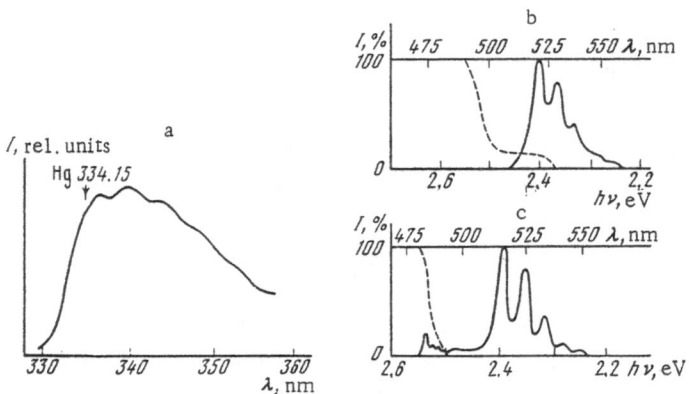

Fig. 2. Low-temperature edge luminescence spectra of zinc sulfide [62] and cadmium sulfide [64]: a) ZnS at 77°K; b) CdS at 77°K; c) CdS at 4.2°K (the dashed curves represent the absorption spectra).

phonons. In fact, the positions of the maxima of the luminescence spectrum correspond accurately to the energies of longitudinal optical phonons in various crystals which have been found by other independent (Raman scattering, infrared absorption) methods.

The theoretical paper of Hopfield [67] can be regarded as completing the first stage of the investigations of the edge luminescence. Hopfield made a thorough analysis of the edge luminescence spectra. A comparison of the results of theoretical calculations of the ratio of the intensities of the phonon replica peaks with the experimental results for CdS demonstrated that the spectrum, which usually had 5-6 clear lines, could be described by the Poisson curve.

Pekar [68] and Krivoglas [69] considered impurity centers interacting with lattice vibrations and derived a relationship between the integrated luminescence band intensities and their numbers in the electronic-vibrational series:

$$I_n = I_0 \frac{\overline{N}^n}{n!},$$

where I_0 and I_n are, respectively, the integrated intensities of the zero-phonon line and of the line with a number n; $\overline{N}$ is the number of phonons emitted in a recombination event. According to Hopfield, the average number of phonons $\overline{N}$ is

$$\overline{N} = \left(\frac{\bar{e}^2}{R} \right) \left(\frac{1}{\hbar \omega_0} \right) \frac{1}{\sqrt{2\pi}} \left(\frac{1}{n_r^2} - \frac{1}{\varepsilon_0} \right),$$

where R is the orbit radius; $\hbar \omega_0$ is the energy of a longitudinal optical phonon; ε_0 is the high-frequency permittivity; n_r is the refractive index.

A good agreement between the experimental distribution of the peaks in the edge luminescence spectrum and the theoretical calculations was obtained for cadmium sulfide [64] on the assumption that $\overline{N} = 0.87$.

Later experimental investigations carried out using spectroscopic instruments with a better resolution than that employed by Kröger made it possible to determine the width of the phonon replica peaks (~ 10 Å for CdS). It was found that this width exceeded considerably the theoretical estimates obtained for the thermal broadening of the lines (~ 2 kT) and for broadening due to the dispersion of the optical phonons (~ 2 Å). This forced Hopfield to postulate the existence of other causes of the broadening of the phonon spectrum, which were not included in the model of the centers employed by him. This additional broadening was due to the interaction with the acoustic phonons or due to the influence of lattice defects on the binding energy of trapped carriers.

More detailed investigations of the edge luminescence spectra were again concentrated on cadmium sulfide and were carried out at low temperatures between 77 and 4°K [70-75]. Two series of the edge luminescence were found and the ratio of the intensities in the two series depended on the preparation method and on the temperature at which measurements were carried out. The intensity of the short-wavelength series increased when the temperature was raised.

The presence of two series of maxima in the luminescence spectra of these crystals was attributed to the formation of two types of defect (Cd and S). The existence of two series (Fig. 3) in the edge luminescence of zinc sulfide was attributed in [76] to the presence of two different donors.

The earliest information on the fine structure of the edge luminescence bands of cubic ZnS crystals at 17-77°K was reported in [77] (Fig. 4a). Spectroscopic investigations and thermoluminescence experiments indicated that the optical width of the forbidden band of cubic zinc sulfide was 3.84 eV (at T = 10°K). The existence of bands with even shorter wavelengths in the

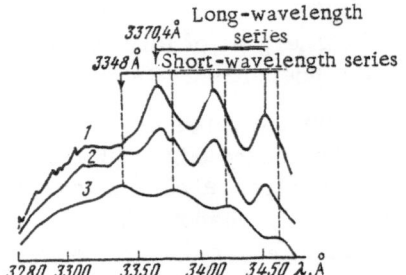

Fig. 3. Influence of temperature on the
edge luminescence spectra of ZnS crystals:
1) 4.2°K; 2) 4.2°K < T < 77°K; 3) 77°K.

330-327 nm range was also pointed out in [77] (Fig. 4b). However, the interpretation adopted in [77] for the edge luminescence bands of ZnS (it was assumed that the luminescence was due to the recombination of electrons localized at chlorine levels with valence-band electrons) has not found acceptance.

In the early investigations the edge luminescence was attributed to the interband transitions [78], and then to excitons [79]. Recently this luminescence was attributed to localized states [2, 80-82].

The Schön model [145] has recently found the widest acceptance and it was used by Williams [81, 83] to attribute the edge luminescence to the radiative recombination of a free hole with an electron trapped by a level due to a dipole formed by oppositely charged lattice defects. This associative model was used by the present author as a working hypothesis and was refined in [84].

The assumption of the participation of lattice defects in the edge luminescence was used by many workers. For example, Shalimova et al. [85] suggested that a donor-acceptor pair should be composed of an excess Zn atom and an atom of S, both located at irregular sites in the crystal lattice of ZnS. However, if the donor-acceptor model is adopted, it is preferred to postulate the existence of Schottky defects, i.e., pairs of sites. Naturally, Schottky defects give rise to smaller deviations from the regularity of the lattice and this reduces the difference

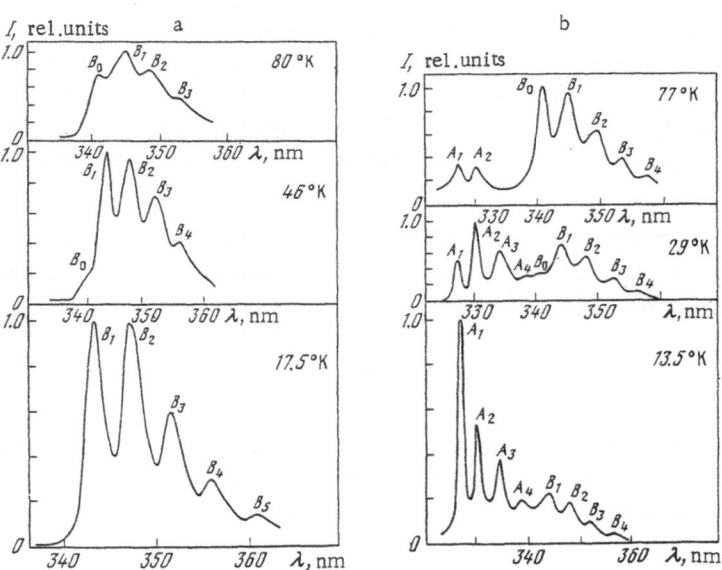

Fig. 4. Edge luminescence spectra of ZnS single crystals recorded at different temperatures: a) type I; b) type II (λ_{exc} = 313 nm).

between the quanta corresponding to the lattice absorption edge and the quanta corresponding to the head (zero-phonon) line of the edge luminescence of ZnS.

A direct proof of the donor—acceptor model was obtained by Gross and Suslina [76] in an investigation of the dependence of the fine structure of the edge luminescence spectra of ZnS on the excitation intensity and in a study of changes in the spectra which occurred during decay (Fig. 5). They found that during the decay the phonon peaks shifted by ~0.01 eV toward longer wavelengths. This followed directly from the Williams model of the centers [83, 86] for pairs with different distances between the donor and acceptor and characterized by different transition probabilities.

A confirmation of the donor—acceptor model of the edge-luminescence centers was also provided by the temperature dependence of the degree of polarization of the edge luminescence of CdS [73]. It was concluded there that the donors and acceptors were separated by just one or two lattice constants.

However, it should be stressed that there is as yet no agreed view on the nature of the centers responsible for the edge luminescence, which depends considerably on the purity of a sample (conditions during synthesis, which influence the stoichiometry, and introduction of additional luminescent impurities, which reduce considerably the luminescence intensity).

Since the edge luminescence is excited readily by optical excitation in the fundamental absorption band, it follows it should also appear when the host lattice is excited by nonoptical means. In fact, the excitation by electron bombardment is even more convenient than the optical excitation: firstly, it avoids difficulties resulting from the overlap of the edge luminescence spectra of ZnS with the light which usually originates from a mercury lamp and is reflected by the surface; secondly, much higher volume densities of nonequilibrium carriers can be achieved by electron bombardment than by optical excitation. This is due to the shallowness of the penetration of electrons into solids and the ability to focus electrons; it is also due to the availability of suitable efficient thermionic cathodes.

Elementary calculations show that the volume density of carriers generated by a $\sim 1\,A/cm^2$ beam of ~ 20 keV electrons is $\sim 10^{26}$ pairs $\cdot cm^{-3} \cdot sec^{-1}$. However, even if $j \approx 10^{-6}\,A/cm^2$, the intensity of the electron-bombardment-induced excitation is considerably greater than the intensity of the conventional optical excitation.

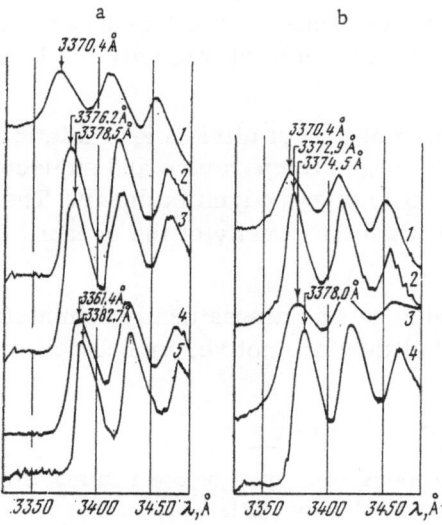

Fig. 5. Results of an investigation of the fine structure of the edge luminescence of ZnS at 4.2°K [76]. a) Edge luminescence spectrum (1) and afterglow spectra: 2) excited in the fundamental absorption region and recorded after $\tau_1 \geq (1-3) \times 10^{-3}$ sec; 3) same as 2, but for $\tau_2 \geq (4-7) \times 10^{-2}$ sec; 4) excitation with long wavelengths, recorded after τ_1; 5) same as 4, but after τ_2. b) Influence of the intensity of exciting light on the edge luminescence: 1) maximum excitation with light corresponding to the fundamental absorption region; 2) same as 1 but minimum excitation; 3) maximum excitation with long-wavelength radiation; 4) same as 3 but minimum excitation $(I_{min} = I/40)$.

It should be stressed that the efficiency of the edge luminescence emitted by, for example, GaP is a superlinear function of the excitation intensity in the weak excitation range [87]. This observation is used as the basis of the attribution of the thermal quenching to the liberation of carriers from localized centers (and not to some internal effect).

It has recently been shown [6] that when the excitation intensity is increased considerably (by the use of electron pulses), the edge luminescence of ZnS is replaced with shorter luminescence bands. This pulsed excitation is used mainly to exclude the possibility of the surface heating by the electron beam because under continuous conditions the excitation of powders ($j \sim 10^{-6}$ A/cm^2) and thin screens ($j \sim 10^{-4}$ A/cm^2) results in the heating of the surface by tens of degrees [88].

The emission of several different luminescence bands may be due to differences between the thermal conditions and the associated differences between the forbidden band widths or it may be due to differences between the crystal structures of the samples.

The need for a careful control of the phase composition of polycrystalline ZnS is stressed in [85].

This point is also made by Klein [89]. He studied hexagonal single crystals of ZnS at 4.2°K and, using a high-resolution spectrograph, found a fine structure of the luminescence band in the $\sim 321\text{-}323$ nm range, which shifted to $\lambda \sim 330$ nm when an admixture of the cubic phase was present. The line with the shortest wavelength (there were about ten such lines) was attributed to the recombination of free excitons and the other lines were attributed to the recombination of bound excitons.

The free-exciton luminescence of CdS was investigated in [90], where a study was made of the influence of temperature on the spectra recorded at low electron excitation levels (V = 60 keV, $j \leq 5$ A/cm^2). Later papers [3, 6] were concerned mainly with the stimulated emission due to excitons and theoretical calculations indicated that the population inversion could only be achieved for bound excitons [91].

The luminescence of pure zinc sulfide crystals observed near the fundamental absorption edge has been attracting considerable interest because of the use of the transitions involved in ultraviolet lasers. Theoretical estimates [92] have shown that high electron-beam excitation densities are needed for the laser action in ZnS.

In this situation somewhat unexpectedly Hurwitz [93] reported building of an efficient semiconductor laser made of a pure zinc sulfide crystal and excited with electron-beam pulses of relatively low intensity. The laser emitted in the ultraviolet range ($\lambda \sim 324.5\text{-}330$ nm) at liquid nitrogen and helium temperatures when the electron-beam current was only ~ 0.1 A/cm^2 (V = 28-40 keV).

In a later paper [94] it was reported that stimulated emission of ultraviolet radiation was observed when ZnS single crystals were cooled to liquid nitrogen temperature and subjected to two-photon pumping with the second harmonic ($\lambda \approx 0.503\,\mu$) of a neodymium laser. The emission wavelength was $\sim 0.33\,\mu$ and the line width was ~ 15 Å when the density of the exciting optical radiation was ~ 150 MW/cm^2.

These investigations of the effects of high excitation densities are naturally stimulating practical interest in the ultraviolet luminescence of ZnS but they are not very much concerned with the nature of this luminescence.

§ 3. Donor—Acceptor Model of Luminescence Centers

The idea of donor—acceptor complexes as one of the possible types of the radiative recombination center in crystal phosphors was introduced by Williams [83].

The hypothesis of the formation of donor—acceptor pairs has been used by many workers in explaining a large number of various radiative transitions in typical phosphors and in elemental semiconductors (Ge, Si).

The Coulomb attraction between oppositely charged donors and acceptors (which can be impurities or lattice defects) ensures that the components of a donor—acceptor pair tend to occupy the nearest lattice sites. In this situation a small but finite proportion of donors and acceptors are separated by such short distances that the wave functions of the localized carriers overlap so that the probability of direct radiative transitions of an electron from a donor to a hole in an acceptor is higher than the probability of the liberation followed by recombination.

The maximum distance between a donor and acceptor necessary for this to occur, i.e., the maximum distance for a tunnel transition between bound carriers without emergence into an allowed band, can be estimated roughly as being equal to the sum of the radii of the localized-carrier orbits.

The general questions in the theory of donor—acceptor pairs can be formulated as follows: 1) Is the number of such pairs, which are present in any phosphor, sufficient to become the main or one of the main radiative recombination channels? and 2) Can one adduce decisive or weighty arguments in support of this model for a specific experimentally observed luminescence band?

Unfortunately, it is very difficult to answer these questions directly but as an approximation we can consider the distribution of the number of donor—acceptor pairs in accordance with their internal separation and we can discuss the model of a center and some calculations of the energy structure and probabilities of radiative transitions. We can also consider some results of the physical investigations of the spectral and kinetic characteristics of the luminescence of ZnS using the donor—acceptor model.

Distribution Function of the Number of Pairs Separated by a Given Internal Distance. The distribution of the number of pairs as a function of the internal distance between the donor and acceptor is a nonequilibrium property under normal temperatures but can be regarded as an equilibrium characteristic at a temperature during the preparation of a crystal below which diffusion stops.

In elemental semiconductors donors are usually located at interstices and acceptors at regular sites. In semiconductor compounds and phosphor crystals an acceptor and a donor are usually located at regular sites.

A random distribution of charged ions allowing for their interaction has been considered for liquid electrolytes, for which this distribution is continuous. The distribution in crystals is at least partly discrete; nevertheless, the theory of a continuous distribution can be adapted to discrete systems [95].

The Debye and Hückel theory [96] is concerned with the interaction between ions in the weak binding approximation: $ZeV(R) < kT$ (Z is the charge of an ion and $V(R)$ is the potential at a given point); this theory shows that the Poisson distribution is valid. However, there is no pair distribution in this theory because the effect of many ions and not of the pair interaction predominates.

Fowler and Guggenheim [97] developed this theory somewhat and obtained the following distribution function for pairs:

$$G(R) = 4\pi R^2 N(R)\left[1 - \int_{R_0}^{R} G(R')\,dR'\right],$$

where

$$\int_{R_0}^{\infty} G(R)\,dR = 1.$$

If it is assumed that there is no interaction between the charges, a donor and an acceptor can occupy any position, and the distribution has the Poisson form, i.e., it is a random temperature-independent distribution

$$G_0(R) = 4\pi R^2 N \exp\left(-\frac{4\pi R^3 N}{3}\right).$$

However, in this case a certain number of donors and acceptors is separated by sufficiently short distances and the number of pairs formed by these centers depends on the concentration.

However, if we allow for the Coulomb attraction between the ions, $V = Z_i Z_j e^2/R\varepsilon$, the distribution function of pairs becomes

$$G_C(R) = 4\pi R^2 NC \exp\left(\frac{Z_i Z_j e^2}{\varepsilon R kT}\right)\left[1 - \int_{R_0}^{R} G(R')\exp\left(\frac{Z_j^2 e^2}{\varepsilon kT}\right)\frac{dR}{C}\right],$$

where C tends to 1 at high concentrations; $\int_{R_0}^{R} G(R)\,dR$ represents the closely associated pairs, whereas $\int_{R_C}^{\infty} G(R)\,dR$ represents pairs separated by large internal distances, i.e., the unassociated pairs (Fig. 6).

At high concentrations the minimum disappears and the difference between these two parts of the distribution becomes less important.

The observed luminescence bands have been interpreted by different workers using the concept of donor–acceptor pairs with different internal distances. In the early investigations Williams assumed that a donor and an acceptor are separated by between one and three lattice constants; subsequently, he analyzed data on the activated ZnS phosphors and reached the conclusion that the internal separation in such pairs was greater (5-7 lattice constants).

Recently many investigators have discussed donor–acceptor pairs separated by larger internal distances. The spectral properties and the probabilities of radiative transitions are very different for donor–acceptor pairs with different internal separations. This will be con-

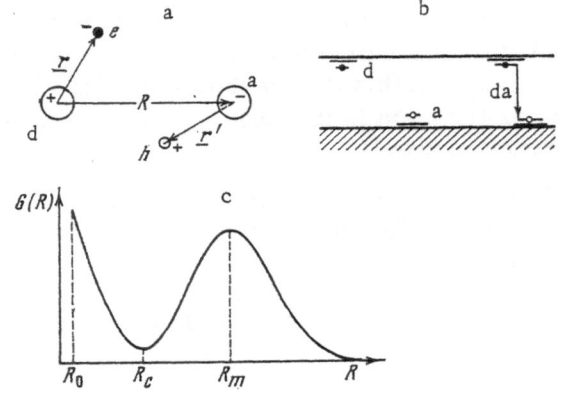

Fig. 6. Model used in calculations of electronic states (a), model of a donor–acceptor center (b), and dependence of the distribution of ion pairs on the separation between the donor and acceptor (c).

sidered and used to refine the donor–acceptor model of the edge-luminescence centers in zinc sulfide.

Electronic States and Radiative Transitions. Each pair may be 1) in the ground state of a semiconductor with equal concentrations of donors and acceptors in the form of ionized defects or 2) it may exist in an uncompensated or compensated semiconductor into which charges of the same sign are injected and carriers may exist in the ground state in each pair. Alternately, 3) carriers of both signs may exist in the ground state under photoexcitation or injection conditions.

In the last case neutral excited pairs can be regarded as an exciton which includes two localized carriers (a hole at an acceptor and an electron at a donor).

A suitable description is adopted depending on the relative force of the interaction (I_{+-}) between two electronic particles and of the interaction between each particle and a corresponding ion (I_{d+}, I_{a-}). Radiative transitions in donor–acceptor pairs include those occurring between all three states. The transitions between the first and last states represent the formation and annihilation of a bound exciton or an electronic transition between a donor and an acceptor.

A theoretical investigation of the energy levels of a donor–acceptor pair was carried out by Williams [83] in the effective mass approximation and he used a method analogous to the Heitler–London approach to the H_2 molecule.

The model used in the calculations of electron states and a donor–acceptor model of a luminescence center are shown in Fig. 6. The nucleic of a donor–acceptor pair are approximated by point charges in a medium of permittivity ε. The coordinates of an electron and a hole are generally described by a two-particle function

$$\psi(-, +) = \psi_d(-) \psi_a(+).$$

The state of a luminescence center can be described as a state of a perfect crystal disturbed by a defect (a donor or an acceptor):

$$\left[\frac{\hbar^2}{2m_{e,h}^*} \Delta - \frac{e^2}{\varepsilon R} \right] \mathcal{F}_{e,h}(R) = E_{d,a} \mathcal{F}(R),$$

where $e^2/\varepsilon R$ is a perturbation due to the Coulomb interaction. The eigenvalue of the energy is similar to the terms of the hydrogen atom:

$$E = - \frac{m^* e^4}{2\hbar^2 \varepsilon^2 \eta^2},$$

where $\eta = 1$, 2, and 3, provided m is replaced by m*, and allowance is made for ε.

The Williams theory of the electronic states of donor–acceptor pairs is based on the distinguishability of an electron and a hole so that the two-particle wave function is due to one donor level and one acceptor level.

The total energy of such a system can be represented by a sum of the Coulomb interactions between point charged particles in accordance with the scheme in Fig. 6a:

$$I = I_{a-} + I_{d+} + I_{+-} - \frac{e^2}{\varepsilon R}.$$

Thus, because of the association, the donor and acceptor levels shift toward the conduction and valence bands (compared with the positions of the levels of isolated donors and acceptors), and the magnitude of the shift decreases with increasing distance between the donor and acceptor.

TABLE 5. Dependence of Energy Parameters on Number of Donor –
Acceptor (d-a) Pair

Number of d-a pair	R, Å	$E_g - E_{exc}$, eV	$E_g - E_{lum}$, eV	Number of d-a pair	R, Å	$E_g - E_{exc}$, eV	$E_g - E_{lum}$, eV
1	3.84	0.150	0.193	10	12.14	0.333	0.539
2	5.43	0.207	0.299	20	17.60	0.356	0.613
3	6.65	0.252	0.366	∞	∞	0.394	0.776
5	8.59	0.293	0.446				

Williams carried out a theoretical analysis and made numerical calculations for the zinc sulfide structure without allowance for the polarity of the lattice and allowing for its polarization. In the latter case there was a small Stokes shift between the absorption and luminescence spectra of donor–acceptor pairs.

The main results are given in Table 5.

It is clear from Table 5 that the association energy for the pairs with the shortest internal separation ($R = 3.84$ Å) is 0.5 eV and the difference between the luminescence photons for the first and second pair is ~ 0.1 eV. The association energy and the difference between the luminescence photons emitted by two neighboring pairs (n and n + 1) gradually decreases with increasing number (n) of a pair. This dependence of the scatter of the energy parameters on the pair number can be used to find more accurately the average distance between active pairs in the interpretation of the experimental results.

Application of the Theory of Donor–Acceptor Pairs. The important role played by donor–acceptor pairs in the luminescence of crystals and semiconductors has been understood only in the last decade.

The appearance of different types of luminescence after introduction of the same activator naturally suggests the possibility of formation of complex luminescence centers composed of more than one impurity atom. Gurvich's review [98] discusses the published information on the origin of various luminescence bands.

The existence of donor–acceptor pairs has been deduced from electrical properties of lithium-doped germanium [99].

The model of donor–acceptor centers is used in [100] to explain the luminescence of ZnS:Cu:Cl phosphors. It is shown that copper, which forms at the zinc sites as a result of the radioactive decay of Zn^{65}, makes no contribution to the luminescence of the type observed for copper which occupies zinc sites as a result of a chemical treatment. Moreover, the radioactive decay produces divalent copper, which does not luminesce in the Cu^{2+} form. It is also pointed out that the luminescence of Cu may be excited at low temperatures and this may not be accompanied by the photoconductivity (such a behavior is attributed to the existence of an excited state of Cl near an allowed band).

A more direct proof of radiative transitions in donor–acceptor pairs is provided in [101] for ZnS:Cu:Ag with Ga in In donors. The positions of the long-wavelength luminescence bands of these phosphors depend on the donor and acceptor, which follows directly from the model of associated centers, i.e., the luminescence may be attributed to transitions between the ground states of a donor and an acceptor. The dependence of the luminescence intensities on the temperature and on the donor and acceptor concentrations are in agreement with this interpretation of the transitions.

In recent years several workers have reported that the activator luminescence bands are not simple, as manifested in the course of the luminescence decay (ZnS : Cu : Cl) [102], and that the luminescence band profile changes in the presence of background infrared radiation (ZnS : Cu [103]). This complexity of the luminescence bands is due to the scatter of the parameters of the luminescence centers resulting from the different degrees of association of the donor-acceptor pairs.

It has also been reported [104] that the blue luminescence band exhibited by ZnS crystals is not simple. Riehl [62] analyzed the afterglow of ZnS phosphors ($\tau \sim 10\text{--}1000$ sec) at low temperatures ($\sim 4°K$) and concluded that the thermal liberation of carriers even from shallow traps (~ 0.05 eV) cannot explain the observed prolonged afterglow. The shift of the luminescence spectrum in the direction of longer wavelengths during the afterglow and the high efficiency of the afterglow process can be explained by the tunneling of electrons from traps to luminescence centers. The tunneling distances may even exceed the average random separation between a donor and an acceptor (~ 100 Å).

Gross [105] was the first to report the luminescence spectra of GaP at low temperatures; he attributed various lines to local levels and postulated the existence of bound excitons.

However, Hopfield et al. [106] attributed the spectrum of GaP to donor-acceptor pairs and they calculated the transition energies for a series of lines from a distribution function of the internal separation in donor-acceptor pairs.

The theory discusses donor-acceptor pairs with particular internal separation corresponding to a maximum in the luminescence spectrum. As this separation varies, so do the probability of radiative recombination $\beta = \beta_0 \exp(-2r/R_H^*)$, effective capture cross section σ, activation energy E, and total number of donor-acceptor pairs N(R). These changes alter the luminescence spectrum when the temperature and the excitation intensity are varied. This can be explained qualitatively as follows. Each line emitted by a donor-acceptor pair with a definite discrete internal separation is very narrow (if broadening is ignored) so that the width of the unresolved luminescence band of GaP represents a distribution of the contributions of different donor-acceptor pairs to the combined luminescence and the energy of each line is given by the expression

$$E(r) = E_g - (E_d + E_a) + \frac{e^2}{\varepsilon R},$$

where E_g is the forbidden band width; E_d and E_a are the binding energies of a donor and an acceptor; ε is the static permittivity; R is the distance between the donor and acceptor.

In the case of high values of R the energy intervals between the neighboring lines become negligible and this gives rise to a wide unresolved edge luminescence spectrum of GaP.

Maeda [70] considered two possible models of a donor-acceptor center allowing for the order of capture of the carriers, i.e., distinguishing the cases when an electron or a hole was the first to be captured, and he compared the results with the experimentally determined temperature dependence of the intensity of the luminescence band of GaP doped with S donors and Si acceptors.

Two cases can indeed be distinguished: in one case an electron is captured first by a donor and in the other a hole is captured first by an acceptor. The capture probability depends on the densities of free carriers and on the capture cross sections of the center in question. It is difficult to see how these quantities could be of the same order of magnitude for electrons and holes in a real crystal. Therefore, the empty centers capture preferentially one particular type of carrier. After the capture of two carriers by a center, the activation energy for the thermal liberation of a carrier from each defect is the same as for an isolated impurity. Consequently, the probability of the thermal liberation of a carrier which is captured first is much greater than the probability of the capture of a second carrier.

It is assumed that the mechanism of the thermal quenching of the luminescence involves the thermal liberation of one of the captured carriers followed by a nonradiative recombination center. Therefore, the behavior of the quenching depends strongly on which carrier is captured first.

This brief review shows that many physical studies have confirmed the donor—acceptor model of the luminescence centers. In any case, many characteristics of the luminescence suggest that both recombining carriers are in the bound state.

It should be pointed out that the models of donor—acceptor centers seem to be most promising in the explanation of the multiplicity of the luminescence bands emitted by II-VI compounds. However, these models must be developed further and a more careful analysis is needed of new and more refined experimental studies of the spectral and kinetic characteristics of the observed luminescence bands.

§ 4. Efficiency of Cathodoluminescence and Photoluminescence

The magnitude and mechanism of nonradiative energy losses in crystalline substances depend strongly on the excitation method and intensity, composition of the substance, temperature, etc.

In this section we shall consider briefly the main published information on the various sources of energy losses which accompany the photoluminescence and cathodoluminescence of ZnS phosphors.

In spite of the presence of nonradiative recombination centers (dislocations, residual quenching impurities, grain boundaries, and other imperfections), the quantum efficiency of modern ZnS phosphors in the case of direct optical excitation of such activators as Cu or Ag (provided the activator concentration is sufficiently high, reaching $\sim 10^{17}$-10^{18} cm^{-3}), can be close to unity ($\sim$60-90%).

The kinetics of the recombination interaction between the luminescence and nonradiative centers governs the dependence of the luminescence efficiency on the excitation intensity and temperature, as discussed in detail in the monographs of Fok [107] and Antonov-Romanovskii [108].

The energy efficiency of the cathodoluminescence, defined as the ratio of the luminescence energy to the total energy of the electron beam reaching a phosphor, is considerably lower [109].

This can be explained by the various intermediate stages in the conversion of the kinetic energy of high-energy electrons (usually 10-20 keV) into the photon energy ($\sim$2-3 eV) and at each of these stages there may be nonradiative energy losses [110]. Nevertheless, the energy efficiency of the best crystal phosphors is 20-25% under electron excitation conditions [111, 112], i.e., the ratio of the emitted photons to the number of the primary electrons is of the order of several thousands.

In the case crystal phosphors with one type of luminescence center, the cathodoluminescence and photoluminescence spectra are usually practically identical, which is a demonstration of the common nature of the final stages, pointed out many years ago by Lenard [113].

The experimentally determined dependence of the luminescence efficiency on the electron excitation conditions has been described by the empirical formula

$$\eta = \frac{L}{jV} = \frac{f(j)(V - V_0)^n}{jV},$$

where f(j) is a linear function of j at low excitation densities; V is the accelerating voltage; V_0 is the "dead" potential. The exponent n ranges from 1 to 3; it is different for different crystal phosphors and depends on the acceleration voltage.

According to the modern theory of the luminescence processes, which distinguishes three stages (absorption, internal conversion, and finally excitation of luminescence centers [107]), the cathodoluminescence involves the following stages [110]:

1) penetration of the primary electrons into a phosphor (some of these electrons are scattered elastically and inelastically out of the phosphor);

2) formation of a secondary-electron plasma in the conduction band (and a corresponding hole plasma in the valence band) due to the ionization losses of the energy experience by the primary electrons inside the phosphor (some of these electrons may escape from the surface in the form of secondaries);

3) transfer of the excitation to the luminescence centers and their transition to the original unexcited state with the excess energy being emitted in the form of luminescence photons or of heat (losses).

Following this scheme, the overall cathodoluminescence efficiency can be represented, in the first approximation, by the product $\eta_{cl} = \eta_1 \eta_2 \eta_3 \eta_4$ [112, 114]. Here, η_1 represents the losses in the absorption of the energy (secondary emission and reflection of electrons, over-coming of retarding fields, etc.); η_2 represents the Stokes losses in the conversion of fast secondary and other electrons into slow carriers with an energy close to that corresponding to a thermal equilibrium in the allowed bands; η_3 represents the Stokes losses in the capture of a carrier by the level of a center and thermalization of the excited levels in the center it-self; η_4 are the energy losses in the recombination stage, i.e., the ratio of the radiative and nonradiative recombination events.

Naturally, the losses during the first two stages are governed by the specific nature of the electron excitation.

Similar energy losses in the optical excitation (due to reflection and partial absorption of light) of thick layers are usually small ($\sim$10-20%); they are relatively easy to measure and they can be ignored in the definition of the luminescence efficiency which is regarded as the ratio of the emitted to the absorbed energy [115].

In the electron excitation case an exact allowance for the absorbed energy is much more difficult. According to Garlick [116], the energy losses due to the reflection and scattering may exceed half the incident electron energy. Since the energy of the secondary electrons is low and the distribution of reflected electrons is wide (the average energy of these electrons does not exceed half the initial energy), these losses are not too serious [117, 118].

On the other hand, the conditions which ensure the removal of the charge brought in by the electron beam may affect strongly the efficiency of thick phosphor layers [119], which may have a low conductivity. In this case the kinetic energy of the electrons reaching a phosphor may be considerably less than the energy corresponding to the acceleration voltage because of the formation of a negative charge on the surface and the appearance of a retarding field. This effect can be avoided by depositing thin layers on metal substrates or protecting a phos-phor with a conducting coating (usually one of aluminum) whose small thickness ($\leq 0.05\mu$) does not interfere with the passage of electrons [120]. Aluminization of phosphor screens raises the luminescence intensity by a factor of 2-2.5. The occurrence of charging can be deduced also from the dependence of the luminescence efficiency of single crystals on the angle of in-cidence of the primary electrons, because an oblique incidence increases the secondary-emis-sion coefficient and reduces correspondingly the retarding field. In particular, in the case of $Al_2O_3 : Cr^{3+}$ single crystals the luminescence efficiency is found to depend strongly on the angle of incidence of the primary electrons [121]. When this angle is ~60%, the luminescence efficiency is about 10 times higher than in the case of the normal incidence of electrons and it approaches the efficiency of aluminized crystals.

The occurrence of charging is also indicated by the sublinear dependence of the luminescence brightness on the acceleration voltage when a certain critical value of the voltage is exceeded; this critical value is governed by the dynatron characteristic of the phosphor and it corresponds to a secondary emission coefficient $\sigma_{sec} = 1$ (for ZnS, ~10 keV) [109].

Thus, the currently available experimental methods make it possible to demonstrate charging directly as well as to take the necessary measures to prevent its occurrence.

In the range of the usual primary electron energies (~10-20 keV), the energy losses due to bremsstrahlung [122], direct formation of lattice defects, and absorption in the inactive surface layer of the phosphor are all small.

The existence of an inactive ("dead") layer, where the concentration of nonradiative recombination centers is high, was discovered for phosphors by Lenard. He deduced the existence of such a layer from a threshold energy, ~1-3 keV for ZnS, below which there was practically no luminescence. The thickness of the dead layer, ~10^{-5}-10^{-6} cm, can be estimated directly from this energy and from the data on the penetration of electrons into solids [123].

Theoretical estimates and experimental results indicate that the dependence of the depth of penetration R_p of the primary electrons on their energy E_0 is of the form $R_p = CE_0^n$, where C is inversely proportional to the density of the investigated material and $1 < n < 2$. In the range of moderate electron energies given above, the exponent n is close to 3/2.

Figure 7a shows the experimentally determined dependence of the distribution of the energy losses with depth in a ZnS phosphor [112].

On the one hand, the curves in Fig. 7a are governed by the increase in the specific energy losses of a single primary electron due to the ionization of the phosphor when the energy of this electron decreases:

$$\frac{dE}{dR_e} = \frac{A}{E} \log \frac{E}{\bar{J}_0},$$

where $\bar{J}_0$ is the average ionization energy of the lattice; A is a constant proportional to the density of the investigated material; R_e is the average range of electrons. On the other hand, the inelastic and elastic scattering of electrons causes a considerable deviation from the initial path (Fig. 7b [124]). Consequently, the number of electrons which can cause ionization decreases well before the depth corresponding to the total range of electrons and the curves of dE/dx have a maximum at $x < R_0$, where R_0 is the total range of electrons.

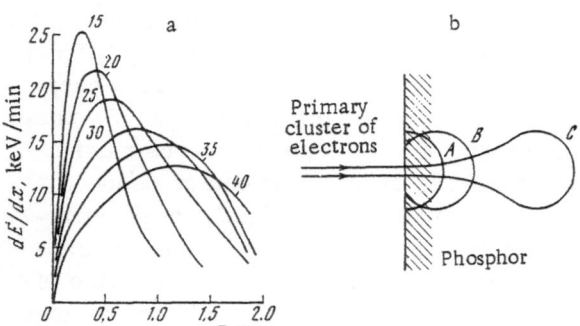

Fig. 7. Distribution of the specific losses experienced by electrons (a) and penetration of exciting electrons into a phosphor (b). The numbers by the curves in Fig. 7a represent the initial electron energy (keV); A, B, and C correspond to increasing energies of primary electrons.

This initial distribution of the absorbed energy with depth in a phosphor is very important for the determination of the volume density of the electron excitation, which influences strongly the ratio of the radiative and nonradiative transitions during the recombination stage.

Under photoexcitation conditions (λ_{exc} = 313 nm), the absorption coefficient of ZnS is $k \approx 1 \times 10^4$ cm^{-1} so that the volume densities of the electron and optical excitation (for the same surface energy) differ by more than an order of magnitude. However, the surface energy density, typical of the optical excitation by mercury lamps, is $\sim 10^{-4}$ W/cm^2, whereas the use of modern cathodes may ensure that the electron-beam density is considerably greater than ~ 1 μA/cm^2, i.e., if V = 20 keV, the volume density of the electron excitation can be at least two orders of magnitude higher than the volume density of the optical excitation.

When these high electron-beam densities are used, several effects tend to reduce the luminescence efficiency and they are known collectively as the cathodoluminescence saturation. These effects include: 1) heating accompanied by thermal quenching (the effect of heating can be deduced, for example, from a change in the wdith of the luminescence spectrum of Zn$_2$SiO$_4$: Mn and from the thermal radiation emitted by such phosphors [88]); 2) de-exciting action of the incident electrons [125], which is manifested by the liberation of carriers localized at centers and is analogous to the de-exciting action of incident light [108]; 3) saturation of the centers with forbidden internal transitions and correspondingly long lifetimes (Zn$_2$SiO$_4$:Mn) [114, 126]; 4) three-particle nonradiative recombination, which — according to Popov [127] — is important for currents $\geq 10^{-2}$ A/cm^2; 5) "burnup" of crystal phosphors accompanied by various changes in their energy structure.

At lower current densities ($\sim 10^{-6}$ A/cm^2) the dependence of the luminescence brightness on the current is frequently linear. Under optical excitation conditions corresponding to even lower excitation densities a superlinear dependence $L = CE_0 + CE_0^{3/2}$ is frequently observed for activated phosphors under external quenching conditions and this dependence is in agreement with the theoretical calculations concerned with external quenching of the second kind [107].

The external thermal quenching may be affected strongly by changes in the excitation intensity.

Thus, the efficiency of the recombination stage of the recombination stage of cathodoluminescence may be higher or lower than the corresponding efficiency of photoluminescence.

Different behavior of the thermal quenching and differences between the rise and decay times as well as spectral composition of the luminescence of phosphors with two types of center are also primarily due to the differences in the volume excitation density in the case of electron bombardment and illumination.

The Stokes losses η_3 and the conditions for emergence of luminescence are governed by the energy and geometric structure of a sample and, in the first approximation, are independent of the excitation method.

In the case of activated phosphors the quantity $\eta_3 = h\nu/E_g$ represents 50-70% [117]; the energy lost in the emergence of the luminescence characterized by a low absorption coefficient does not exceed 10-15%.

When all these energy losses are eliminated and the volume excitation densities are made similar, it is found that the photoluminescence and cathodoluminescence efficiencies are in the ratio of about 3. This is in satisfactory agreement with the ratio of the average energy needed to form one electron-hole pair to the forbidden band width of a given compound (for example, in the case of cadmium sulfide we have $E_g \approx 2.6$ eV and $\bar{\varepsilon}_{eh} = 7.5$ eV [128]). No direct data are available on $\bar{\varepsilon}_{eh}$, but a comparison of the ratio of the cathodoluminescence and photoluminescence efficiencies [129] shows that $\bar{\varepsilon}_{eh} \approx 12$ eV, i.e., it is also approximately equal to the trebled forbidden band width (E_g).

This ratio has a theoretical justification given by Popov [112], who discussed the final stage of the cascade ionization energy losses experienced by primary electrons in solids.

The losses are due to the formation of a fairly wide distribution of secondary carriers of both signs, which extends along the energy scale from 0 to $1.5E_g$ and the kinetic energy of these secondary carriers is insufficient to form additional electron–hole pairs.

This excess energy is dissipated in $\sim 10^{-12}$ sec because of the interaction with the lattice phonons, i.e., the thermalization of slow carriers occurs in a time which is much shorter than the radiative transition time ($\sim 10^{-8}$ sec).

The ionization threshold ($\sim E_{th} > E_g$) arises because it is necessary to satisfy the law of conservation of momentum as well as of energy.

Popov assumed that the distribution of the carrier energies corresponds initially to the density of states in the bands ($\propto \sqrt{E}$) and he found that the maximum cathodoluminescence efficiency limited by these thermalization losses is $\sim 30\%$.

Thus, the thermalization losses are obviously the main cause of the difference between the cathodoluminescence and photoluminescence efficiencies and they have to be allowed for in comparisons of these efficiencies in the same phosphors.

In estimating the number of recombinations per one electron–hole pair we would use the experimental data on the quantum efficiency of the photoluminescence and in the case of cathodoluminescence we should multiply the experimental efficiencies by $3E_g/h\nu_{lum}$.

The Stokes losses in the ultraviolet photoluminescence ($\lambda_{exc} = 313$ nm) do not exceed $\sim 5\text{-}10\%$ and in the cathodoluminescence they have to be multiplied by $3E_g/h\nu_{rad} \approx 3$.

Apart from allowance for the thermalization losses of the energy in a comparison of the radiative recombination efficiencies of the ultraviolet cathodoluminescence and photoluminescence of zinc sulfide, one must take measures to eliminate the additional energy due to electron excitation, as discussed above.

CHAPTER II

FORMULATION OF THE PROBLEM AND EXPERIMENTAL METHOD

§ 1. Selection of Investigation Methods and Objects

The published investigations of the properties of ultraviolet luminescence of ZnS have been fairly narrow in scope so that the available experimental data are extensive but difficult to compare. In fact, the considerable differences between the luminescence spectra reported by many workers may be attributed to: 1) differences between the objects themselves (purity, stoichiometry, concentration of defects, crystal structure, etc.); 2) differences between the excitation methods (photons or electrons of different energies, electric fields); 3) different excitation intensities and temperatures during measurements; 4) combined influence of all these factors.

The disagreements between the reported luminescence spectra have created an impression that the ultraviolet luminescence emitted by a given object is unique and that it is not possible to obtain reproducible results. The absence of even qualitative data on the efficiency of the ultraviolet luminescence of ZnS has suggested that the efficiency is very low and that the luminescence is due to a secondary radiative recombination channel. The kinetics of the thermal quenching has hardly been investigated and the dependences of the efficiency on the exictation intensity have been obtained only for high densities of electron beams.

Thus, it has been difficult to compare the properties of luminescence bands located near the fundamental absorption edge of ZnS itself and then compare these properties with the results of investigations of other II-VI compounds (particularly CdS).

In view of this the author carried out an experimental study of the efficiency and spectra of ultraviolet cathodoluminescence and photoluminescence of zinc sulfide as a function of the excitation conditions (temperature, excitation method and intensity, additional infrared illumination, etc.) and of the conditions during synthesis and composition (powders fired at various temperatures, stoichiometry, presence of activators such as Ag, Sm, etc., single crystals in which the thermodynamic equilibrium conditions are frozen-in by slow cooling and by quenching).

The wide range of the preparation conditions and the use of intensive electron excitation in the temperature range from 17°K to ~600°K made it possible to: 1) establish the emission (or lack of it) of ultraviolet luminescence from objects of various kinds; 2) compare the luminescence spectra and follow gradual changes under strong luminescence quenching conditions and thus separate the various types of ultraviolet luminescence; 3) determine the conditions during synthesis and excitation which ensure the maximum efficiency of the edge luminescence of ZnS, estimate the optimal efficiency, and compare it with the efficiency of activated crystal phosphors; 4) determine the relationship between the radiative recombination in the ultraviolet and visible parts of the spectra of unactivated ZnS; 5) compare the properties of the ultraviolet luminescence with the published information (both that relating to ZnS and to other II-VI compounds); 6) select most suitable models of radiative transitions and to refine some of their characteristics.

In carrying out this investigation a special attention was paid to the excitation method, and suitable apparatus and techniques were developed. The use of electron beams of moderate energy (5-20 keV, $j \approx 10^{-6}$ A/cm^2) made it possible to:

a) ensure a sufficiently high luminescence intensity up to 10^{16} photons/cm^2, which was quite sufficient for the measurement of the spectra of different samples at different temperatures; vary excitation power density from ~150 μW/cm^2 to ~0.1 W/cm^2; avoid, at these excitation power densities, heating and decomposition of the samples; use low electron-beam current densities of the electron-beam and optical excitations, which was essential in comparisons of the kinetics;

b) eliminate light from the excitation source, which caused considerable experimental difficulties when mercury lamps were used (it was necessary to eliminate the peaks and background of the reflected exciting light of wavelengths in the near ultraviolet).

Recording of the luminescence spectra in the ultraviolet region required the use of the quartz optics and of calibration of the spectral sensitivity of the detector (a photomultiplier) with the aid of a phosphor characterized by a constant quantum efficiency.

In contrast to the majority of other investigations, the luminescence was recorded not by the photometric but by the photoelectric recording method, which made it possible to determine accurately the ratio of the intensities of the components of individual luminescence lines and to compare the intensities in different parts of the spectrum.

Considerable care was taken to ensure the identity of the excitation conditions of different objects. For this purpose, comparisons were made of the luminescence of thin polycrystalline screens deposited not only on metal substrates but also on single-crystal surfaces. This revealed the identity of or difference between the specific nonradiative losses of the electron energy (depending on the conditions of heat and charge removal) in single crystals and powders, and the identity of or difference between the conditions of the emergence of luminescence from various objects. Variation of the acceleration voltage from 5 to 20 keV altered the depth of penetration of electrons from ~0.1 to ~1 μ, which was important in studies of the luminescence characterized by a high reabsorption coefficient (10^3-10^4 cm^{-1}) and of phosphors with an inhomogeneous (across the thickness) energy band structure. In view of the known strong temperature dependence of the edge luminescence and the absence of suitable quantitative data, it was decided to determine the influence of temperature on the luminescence in a wide range from 17 to 600°K. Most of the measurements were carried out at 77°K, which corresponded – in the case of the ultraviolet edge luminescence – to the thermal conditions obtaining at 293°K in ordinary activated ZnS:Cu and ZnS:Ag crystal phosphors.

Apart from the task of separation of the different types of ultraviolet luminescence of unactivated zinc sulfide on the basis of the physical properties (particularly thermal properties) and of the conditions of appearance of such luminescence, the investigation was also aimed to obtain more precise information on the nature and characteristics of the various types of luminescence and on the model of the luminescence centers. This information was then compared with the published data on ZnS and on other II-VI compounds.

§2. Methods Used in Preparation of ZnS Powders and Single Crystals

Phosphors were prepared from amorphous ZnS of the "phosphor" purity grade. The original charge was treated in a stream of H_2S for 2 h at 400°C and for 1.5 h at 1200°C. The gas was first purified by removing moisture, oxygen, and oxygen-bearing impurities. The purity of phosphors prepared in this way was deduced from the absence of the visible luminescence at T = 293°K (when excited with a PRK-4 lamp), absence of the oxygen maximum in the thermoluminescence curves, and absence of significant afterglow. Batches of the investigated samples were prepared by firing the powders in evacuated sealed quartz ampoules at T = 1110°C applied for t = 35 min. Phosphors with an excess of Zn or S were prepared by introducing these impurities as elements before firing and adding them to the original charge in amounts of ~5, 10, and 30% of ZnS. During the subsequent firing at T = 1200°C the vapor pressures of these elements were ~10, 20, and 60 atm.

Thus, by heating ZnS in sealed ampoules in the presence of excess sulfur or zinc we altered the ratio of the anion and cation vacancies. The investigated "pure" zinc sulfide phosphors exhibited a strong ultraviolet luminescence at T = 77°K.

Activated phosphors were prepared by adding a rare-earth element (Sm) or Ag (in the form of a solution of silver nitrate). The activator concentrations ranged from 10^{-7} to 10^{-2} g/g (N g of the metal per 1 g of ZnS). The phosphors were prepared without a flux or using 4 wt.% of $MgCl_2$ as a flux.

Single crystals were synthesized from ZnS subjected to an oxygen-removal treatment. The temperature of the treatment of the original charge varied from 1100 to 1200°C; this temperature was applied for 2 h and cooling took ~30 min. Single crystals of unactivated ZnS were grown by the sublimation of the ZnS charge in a closed ampoule which was pulled through a high-temperature zone (T ~1380-1400°C) in a furnace at a rate of ~0.18 mm/h. Usually the crystals were cooled together with the furnace (~12 h). In the quenching experiments a tube containing ampoules enclosing ZnS single crystals was taken out rapidly from the furnace and cooled in water (T = 293°K) in ~15 min.

§3. Apparatus and Investigation Methods

The properties of phosphors were studied using a PRS unit [130], which was an image converter suitable for the investigation of the efficiency and spectral composition of the luminescence emitted by screens under transmission conditions and by powders and single crystals under reflection conditions [31], when excited with an electron beam. The apparatus was modified considerably. Improvements were made in the vacuum system and the following attachments were built for measuring the luminescence characteristics:

1) a temperature attachment for making measurements from −186°C to + 350°C [132];

2) an attachment for studies of the angular characteristics of the cathodoluminescence [121].

The measurement methods were described in [2, 80, 84, 112]. The ultraviolet luminescence spectra were recorded with an FÉU-18 photomultiplier coupled to a ZMR-3 monochromator. The photoexcitation source was a PRK-7 or a DVS-200 lamp. The excitation line was selected with a Zeiss monochromator. The luminescence was extracted by a special optical system which included quartz fibers.

CHAPTER III

THERMAL AND SPECTRAL PROPERTIES OF ULTRAVIOLET LUMINESCENCE OF UNACTIVATED ZnS

§ 1. Efficiency of Cathodoluminescence and Photoluminescence of Unactivated Zinc Sulfide

According to Kröger [1], the luminescence spectrum of unactivated ZnS recorded at 77°K is a structured band in the region of 330-360 nm, known as the edge band. In the case of our samples this band is the main radiative recombination channel at low temperatures; the intensity of the visible luminescence recorded at such temperatures does not exceed 1-2% of the ultraviolet luminescence. This ratio of the intensities can be seen in the general spectrum shown in Fig. 8a, which is plotted on a logarithmic scale. We shall now consider the influence of the excitation conditions on the edge luminescence efficiency; the results of more detailed studies of the spectra will be considered later.

The excitation conditions corresponding to the maximum intensity of the edge luminescence of ZnS can be deduced from the excitation spectrum (Fig. 8b). This spectrum (curve 1) was determined separately for the three principal maxima (334.5, 338, and 342 nm), usually observed in the edge luminescence of hexagonal ZnS (curve 2). This study made it possible to attribute

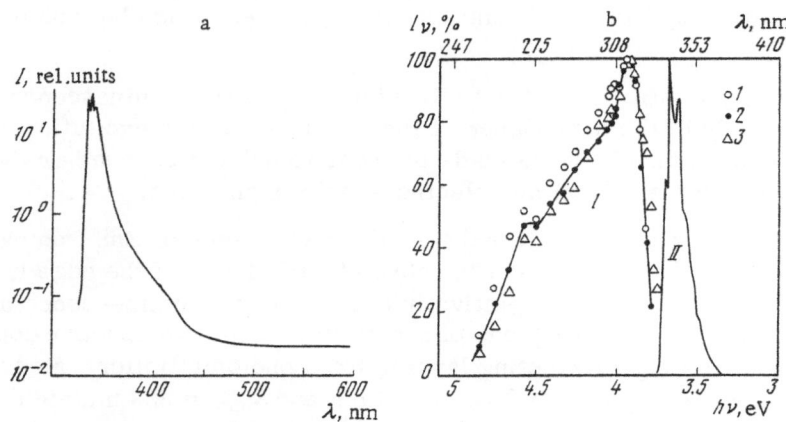

Fig. 8. Edge luminescence and excitation spectra of hexagonal zinc sulfide: a) luminescence spectrum (on a logarithmic scale); b) normalized excitation spectrum (curve I) for individual lines at λ = 334.5, 338, and 342 nm (points 1-3, respectively) and for the whole luminescence spectrum (curve II).

all three transitions to the same type of luminescence, i.e., in the first approximation the luminescence spectrum was independent of the wavelength of exciting light.*

The main maximum in the excitation spectrum is located in the region of 316 nm (~ 3.92 eV), i.e., within the limits of the experimental error (~ 0.03 eV) this maximum coincides with E_g (optical width of the forbidden band) of hexagonal ZnS (~ 3.917 eV) [4].

Consequently, the edge luminescence of unactivated zinc sulfide is excited efficiently by generating free nonequilibrium carriers using various types of excitation (photons with $h\nu_{exc} \geq E_g$, electrons, electric field).

There are no excitation maxima in the range $h\nu_{exc} < E_g$ and, in particular, there are no such maxima at the positions of the exciton absorption maxima in the region of ~ 321 nm. When the wavelength of the exciting light is increased, the excitation intensity falls ($k_{313} \sim 10^4$ cm^{-1} decreases to $k_{340} \sim 10$-20 cm^{-1}) and the edge luminescence efficiency drops sharply. The excitation spectrum extends practically to the short-wavelength edge of the luminescence spectrum and this is evidence of the possibility of direct excitation of the centers.

The spectral range of the exciting light was ~ 3 nm, so that we could not investigate the excitation spectrum in the range $\lambda_{exc} > 325$ nm or consider in greater detail its fine structure.†

The fall of the excitation efficiency in the $h\nu_{exc} > E_g$ range was due to an extremely high value of the absorption coefficient ($k \geq 10^5$ cm^{-1}) and a correspondingly shallow penetration of the exciting light. The strong fall of the efficiency under surface excitation conditions ($d \leq 0.1\,\mu$) is well known in cathodoluminescence (it is due to the presence of the dead Lenard layer) and it was confirmed by our investigation of the ultraviolet luminescence excited by bombardment with ~ 5 keV electrons (whose depth of penetration was approximately the same as that of light, $d \sim 0.1\,\mu$), instead of the usual ~ 15-20 keV electrons.

The structure of the excitation spectrum observed in the $h\nu_{exc} > 4$ eV range is governed by the dependence of the absorption coefficient on the photon energy $k = f(h\nu_{exc})$, i.e., it is related to the structure of the corresponding energy bands of ZnS. These results are in quantitative agreement with those reported by Arapova [21] on the absorption coefficient of ZnS (she reported an increase of $k \sim 10^5$ cm^{-1} at $\lambda \sim 300$ nm to $k \sim 4 \times 10^5$ cm^{-1} at 220 nm).

The inhomogeneous structure of the surface layers was deduced from the thermoluminescence curves obtained using different initial energies of an exciting electron beam (see Chap. IV).

A comparison of the efficiency of the ultraviolet edge photoluminescence and cathodoluminescence and a study of the dependence of the efficiency on the excitation conditions (current density j and temperature T) were made in order to determine whether the luminescence kinetics was the same for the optical and electron-beam excitations.

The investigation demonstrated that the edge luminescence of ZnS was excited very efficiently by a mercury line at 313 nm and the absolute efficiency of the edge luminescence was found for this line. We compared the relative intensities of the luminescence and of the light reflected from MgO for the same positions of a mercury lamp, two spectroscopic instruments, a sample (ZnS phosphor or MgO reflecting layer), and a photomultiplier. Measurements of the intensities in these two cases, made at $\lambda_{exc} = 313$ nm and $\lambda_{lum} = 338$ nm yielded the ratio of the areas under the spectra, which were converted into the edge luminescence efficiency allowing

---

*The scatter of the experimental points exceeded somewhat the experimental error ($\sim 5\%$).

† The fine structure was observed by us in the absorption spectra of the same samples, deduced from the reflection spectra, which agreed closely with the results reported in [134] for hexagonal ZnS phosphors.

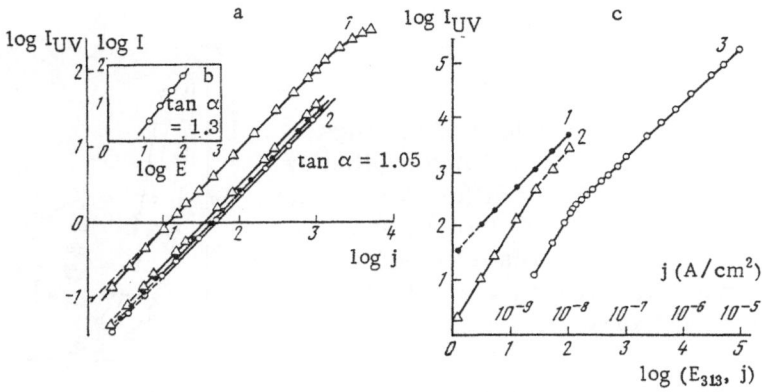

Fig. 9. Dependences of the intensity of the luminescence
of zinc sulfide on the rates of electron-beam and optical
excitation at various temperatures: a) dependence on the
electron-beam current density and T = 77°K plotted for
the whole spectrum (curve 1) and for the individual lines
(curves denoted by 2); b) dependence on the photoexcitation intensity; c) dependence on the photoexcitation intensity at T = 17°K (curve 1) and at T = 77°K (curve 2) and
on the electron-beam current density at T = 77°K (curve 3).

for the photomultiplier sensitivity and for the partial ($\sim$15%) reflection of the exciting light
from ZnS.

According to our estimates, the absolute energy efficiency of the edge photoluminescence
of unactivated ZnS is $\eta_{ph} \sim 6.5 \pm 0.5\%$ at T = 77% (excitation power densities $\sim$5 $\times 10^{-5}$ W/cm^2).
An estimate of the energy efficiency of the edge cathodoluminescence of ZnS, obtained by com-
paring with the luminescence of CaWO$_4$ (the cathodoluminescence efficiency of CaWO$_4$ is given,
for example, in [135]) at the same excitation power density of $\sim$1 $\times 10^{-2}$ W/cm^2 (j = 5 $\times 10^{-7}$
A/cm^2, V = 20 keV), shows that $\eta_{cl} \sim 3.5 \pm 0.5\%$.

The quantum efficiency $\eta_q = \eta_e (\bar{\varepsilon}_{eh}/h\nu_{lum})$, where η_e is the energy efficiency and $\bar{\varepsilon}_{eh}$
is the energy of formation of an electron–hole pair, obtained when a ZnS phosphor is excited
with the 313 nm mercury line is practically identical with the energy efficiency because the
Stokes losses in the photoexcitation are small ($\varepsilon_{eh} \sim h\nu_{313}$). Popov et al. [112] calculated the
specific thermalization losses in the cathodoluminescence during the initial stage of the sharing
of energy between carriers and found that in order to form one electron-hole pair one needs
$\varepsilon_{eh} \simeq 3E_g$. Thus, the efficiency of the recombination stage alone is higher under the electron-
beam excitation conditions, $\sim$10-15%, than under the photoexcitation conditions, $\sim$7%.

This difference is explained by the differences between the surface (and, consequently,
volume) power densities of the excitations employed[*]) and by the dependence of the luminescence
efficiency on the excitation power density. In fact, measurements indicate that the intensity of
the ultraviolet luminescence is a nonlinear function of the excitation power density when the
value of the latter is low. This is demonstrated by the curves plotted in Fig. 9a on a double
logarithmic scale.

It is clear from this figure that in the range of the photoexcitation power densities em-
ployed we have I $\propto$ E$^{1.3}$, whereas in a wide range of the electron-beam current density (1 $\times 10^{-7}$-5 $\times$

---

[*]The depth of penetration of a beam of V $\sim$15-20 keV electrons is $\sim$ 1 μ and it corresponds to
the depth of penetration of the exciting light with λ_{exc} = 313 nm (k $\geq$ 1 $\times 10^{-4}$ cm^{-1}).

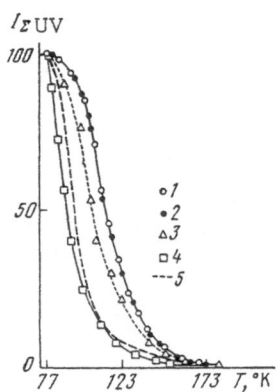

Fig. 10. Thermal quenching of the edge luminescence of unactivated ZnS excited by 15 keV electron beams of different densities $j = 4 \times 10^{-6}$, 4×10^{-7}, 4×10^{-8} and 1×10^{-8} A/cm^2 (points 1-4 respectively) and by photoexcitation with λ_{exc} = 313 nm, E ~ 50 μW/cm^2 (curve 5).

10^{-5} A/cm^2) the luminescence intensity is practically a linear function of the excitation power density $I \propto E^{1.05}$.

Additional investigations have shown that this dependence is satisfied by the integrated luminescence intensity and by individual lines, i.e., in this range of the excitation power density the luminescence spectrum does not change. At low electron-beam current densities there is some deviation from the linear dependence.

An extension of the range of the photoexcitation power densities and a reduction of the electron-beam current density made it possible to cover the range where the dependence on the excitation power density changed from nonlinear to linear for both types of excitation (Fig. 9c). At T = 77°K the transition from the dependence $I \propto E^{1.8}$ to a linear one occurred at α_{exc} ~ 10^{19} pairs · cm^{-3} · sec^{-1}. At lower temperatures (T = 17°K) the linear dependence was observed at much lower excitation rates. The ratio of the absolute efficiencies* in the linear region was ~ 1.5-2. Thus, in the linear range the optimal efficiency of the 77°K edge luminescence of ZnS could reach ~ 15%.

The fall of the luminescence efficiency at high electron-beam current densities may be explained either by the thermal action of the beam or by the saturation of the luminescence centers. As shown in [88], under steady-state excitation conditions corresponding to ~1-2 W/cm^2 the heating of thin screens does not exceed several degrees, which has little effect on the luminescence of activated crystal phosphors. However, in the case of the ultraviolet luminescence of unactivated ZnS such an increase in the temperature of the sample may reduce considerably the luminescence efficiency.

Thermal quenching of the luminescence begins in the T = 77°K range. The quenching depends strongly on the excitation power density in the nonlinear region but is independent of this density in the range $j \geq 10^{-7}$ A/cm^2 (Fig. 10). When the volume power densities of the excitation in the cathodoluminescence and photoluminescence case are equal, the thermal quenching behaves similarly, which is further evidence of the identity of the quenching kinetics for very different types of excitation.

The curves obtained for the photoluminescence and cathodoluminescence are very similar and they resemble also the curves obtained in the case of external quenching at different rates of excitation of activated crystal phosphors [108]. The activation energy of the quenching process, deduced from the behavior of the thermal quenching in the linear region of the dependence $I_{UF} = f(E)$, is $\Delta E_t \approx 0.11 \pm 0.02$ eV, i.e., it is considerably less than the difference between the

_____
*The lower value (~1.5), was obtained after allowing for the reduction in the width of the edge luminescence spectrum at liquid helium temperature.

TABLE 6. Dependence of the Duration
of Initial Decay Regions (down to 10%)
on Electron-Beam Current Density and
Temperature

j, A/cm^2	τ, sec	
	$T = 77°$ K	$T = 100°$ K
$5 \cdot 10^{-7}$	$4.7 \cdot 10^{-4}$	—
$5 \cdot 10^{-6}$	$4.5 \cdot 10^{-4}$	$2.6 \cdot 10^{-4}$
$5 \cdot 10^{-5}$	$3.7 \cdot 10^{-4}$	—

energies corresponding to the forbidden band width and to the first head line of the edge luminescence spectrum ($E_g - h\nu_{334.5} \sim 0.20 \pm 0.01$ eV). This difference should be attributed to the difference between the depths of a donor level and an electron trap (~ 0.1 eV).[*]

The intensity of the edge luminescence of ZnS decreases by about two orders of magnitude by the time the temperature T $\sim$ 150°K is reached, but we extended our study of the quenching to room temperature ($\eta_{293}/\eta_{77} = 10^{-3}-10^{-4}$). It should be noted that the thermal quenching of the ultraviolet edge luminescence is accompanied by the growth of the visible bands. The intensity of the red luminescence increases between 77 and 110°K (i.e., in the region where the ultraviolet luminescence is strongly quenched) and the activation energy of this luminescence is ~ 0.03 eV, indicating the liberation of electrons from the excited states of the ultraviolet luminescence centers. This energy transfer, like the results reported above, is evidence of the recombination mechanism of the edge luminescence of ZnS under strong external quenching conditions.

Thus, the experimental data on the nonlinear dependence of the intensity of the edge luminescence of unactivated ZnS and on the dependence of the thermal quenching on the excitation power density indicate a qualitative agreement between the kinetics of the edge luminescence and the kinetics of the activator luminescence emitted under strong external quenching conditions [107, 108], i.e., it is evidence of the thermal liberation of nonequilibrium carriers from local luminescence centers followed by nonradiative recombination at quenching centers.

The recombination mechanism of the edge luminescence of ZnS follows also from the dependence of the late stages of the decay curves on the excitation density [2]. The initial stages of the decay curves (up to 10%) obtained at T = 77°K are described satisfactorily by the exponential formula $I = I_0 \exp(-t/\tau)$ with $\tau \sim 5 \times 10^{-4}$ sec.

In order to determine the processes which govern the duration of the initial stage of the decay (afterglow), we carried out a study of the dependence of the decay on T and j. When the current density was increased from $\sim 5 \times 10^{-7}$ A/cm^2, the duration of the initial decay region changed by not more than 10%. A more noticeable fall of τ in the range j $\sim 5 \times 10^{-5}$ A/cm^2 (Table 6) could be due to the de-exciting action of the incident electrons [125], i.e., it could be due to a reduction in the luminescence efficiency at the moment of excitation.

---

[*]In the general case of an external thermal quenching the behavior of the quenching may depend strongly on the power density E of exciting light and may differ from the internal quenching described by the Mott formula. However, in certain ranges of E characterized by the constancy of the luminescence efficiency, i.e., by the absence or weak influence of the de-exciting action of the incident light, the quenching is described by an expression similar to the Mott formula [108] in which the activation energy has a different meaning (this energy now represents the difference between the depth of the principal luminescence level and of an electron trap participating in the luminescence).

A fairly strong temperature dependence (and an increase of the duration of afterglow) makes it possible to attribute this value of τ to the thermal liberation of holes from ultraviolet edge luminescence centers.

§2. Spectral Composition of Ultraviolet Luminescence

We have considered so far the general form of the luminescence spectrum of unactivated ZnS, which shows that the edge luminescence is the main channel of the radiative recombination in the investigated samples. In this section we shall consider in detail the structure of the edge luminescence spectrum and the origin of weaker bands observed in the ultraviolet range.

A fine structure in the edge luminescence of ZnS phosphors was observed by Kröger [1] and it has been the subject of a theoretical analysis [67]. However, this structure has been investigated less thoroughly than in the case of other compounds such as CdS.

A comparison of the edge luminescence spectra obtained at T = 77 and 17°K is made in Fig. 11a. Cooling not only reduces the widths of the maxima but results in a gradual disappearance of the line with the shortest wavelength ($\lambda_1 = 334.5$ nm). The separation (~ 0.035 eV) of this line from the next maximum ($\lambda_2 = 338$ nm) is less than the energy of a longitudinal optical phonon in ZnS and less than the separation between subsequent lines. Initially, this shortwavelength line was attributed by Kröger to a weak mercury line present in the spectrum of the lamp used as the excitation source. Our investigations indicated that it was also observed under electron-bombardment conditions and that the cathodoluminescence and photoluminescence spectra obtained at 77°K were identical. This line can be regarded as the head (zero-phonon) line of the second edge luminescence series, analogous to a similar second series in the edge luminescence of CdS [72, 75].

Separate studies of the temperature dependences of the intensities of the $\lambda_1 = 334.5$ nm and $\lambda_2 = 338$ nm lines between 17 and 77°K confirmed that these lines were of different origin (Fig. 11b). The intensity of the 338 nm line (2) was found to decrease monotonically and that of the 334.5 nm line (1) increased nearly exponentially when the temperature was raised from 17 to 50°K.

The thermal activation energy of the enhancement of the short-wavelength (334.5 nm) line, deduced from the temperature dependence of the ratio of the intensities of this and the 338 nm lines, was found to be ~ 0.03 eV, i.e., close to the difference between the photon energies corresponding to these lines. The disappearance of the first short-wavelength edge luminescence series of ZnS at liquid helium temperature could be explained not only by the absence of the ~ 334.5 nm maximum but also by a considerable improvement of the resolution of the spectrum ($I_{min}/I_{max} \sim 0.4$). The difference between the frequencies of the maxima in the edge lumines-

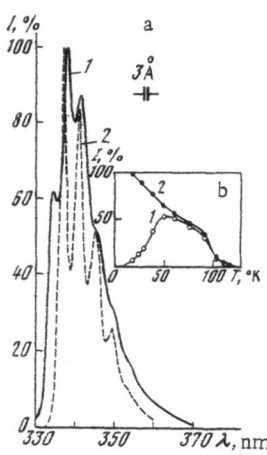

Fig. 11. a) Edge luminescence spectrum of unactivated zinc sulfide: 1) electron-beam excitation at T = 77°K; 2) photoexcitation at T = 17°K. b) Temperature dependences of the intensities of the photoluminescence lines at λ = 334.5 nm (1) and 338 nm (2).

TABLE 7. Positions of Maxima in Edge Luminescence Spectra of
Unactivated Zinc Sulfide

T, °K	λ_{max}, nm						Reference
77	334.5	338.00	341.5	345.0	349.0	353.5	Our results
17	—	337.00	341.0	345.5	349.5	353.5	Ditto
77	334.0	337.60	341.5	345.4	348.8	353.2	[85]
4	—	337.00	341.1	345.4	349.5	353.9	[136]
4	—	337.04	341.2	345.0	—	—	[76]
4; 77	334.5	337.04	341.2	345.0	—	—	[76]
77	334.8	338.00	342.0	347.0	—	—	[76]

cence spectrum at T = 17°K, equal to 346 cm^{-1}, was very close to the frequency of longitudinal optical phonons (ν_l = 349 cm^{-1}) in the hexagonal lattice of ZnS.

Thus, the position of the maxima in the spectrum with two series are in good agreement with the phonon replica model, i.e., they may be attributed to transitions accompanied by the emission of from zero to five longitudinal optical phonons, in agreement with the experimental results obtained by other workers (Table 7).

However, according to our results, the distribution of the intensities of the individual maxima differs considerably from the Poisson distribution, which is predicted by the theory and is in good agreement with the experimental results for CdS [67].

Investigations of the edge luminescence spectra of ZnS are complicated not only by the presence of two series but also by the coexistence of two crystalline phases (hexagonal and cubic). The edge luminescence spectra of the wurtzite and sphalerite modifications are slightly shifted relative to one another but they still overlap. This can be seen in Fig. 12, which shows the luminescence spectra of ZnS after heating at 1200 and 900°C, i.e., above and below the phase transition temperature, respectively. The shift of the maxima is ~0.07 eV and the 341 nm line (cubic phase) is an analog of the 334.5 nm line (hexagonal phase).

It should pointed out that in most cases, even well above and below the phase transition temperature, there is always a considerable admixture of the other phase [7]. This may explain the deviation from the Poisson law in the distribution of the intensities of the line maxima which — in the simplest case, i.e., in the presence of one series — should represent transitions accompanied by the emission of different numbers of phonons. If the intensities of the maxima corresponding to the Poisson distribution are subtracted from the experimentally obtained spectrum, we obtain an additional spectrum which is in good agreement with the luminescence emitted by the cubic phase (Fig. 13). The head line (λ = 341 nm) of the low-temperature series of the cubic modification is missing and the ratio of the intensities of the maxima in the second series of the cubic phase is described satisfactorily by the Poisson distribution (Table 8). Naturally, this separation of the spectra is not very accurate but it confirms qualitatively the presence

Fig. 12. Edge luminescence spectra recorded at T = 77°K for different phases of ZnS: 1) hexagonal phase (sample heated to a temperature of 12000°C); 2) cubic phase (sample heated to 900°C).

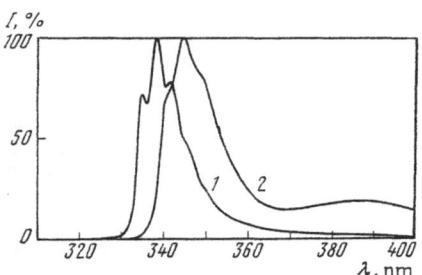

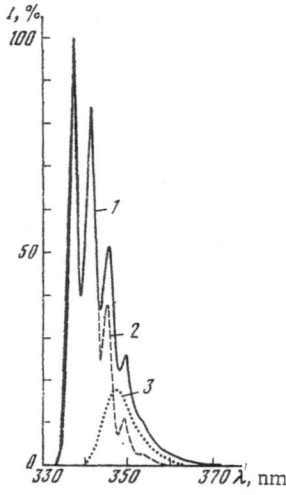

Fig. 13. Edge photoluminescence spectra: 1) hexagonal zinc sulfide at T = 17°K; 2) spectrum corrected for the Poisson distribution with $\bar{N}$ = 0.87; 3) admixture of the second (cubic) phase of ZnS.

of a considerable amount of the second (cubic) phase in the investigated hexagonal samples. The coexistence of the two phases is a very important point in the studies of sublimated ZnS screens.

Thus, the existence of two edge luminescence series and of two crystal phases may be one of the main reasons for the broadening of the edge luminescence spectra of ZnS and of the departure from the predicted statistical distribution of the intensities of the maxima; it may even account for the slight shift (within several angstroms) which is observed from sample to sample. The changes in the edge luminescence spectra observed for the same sample when the excitation conditions are altered may partly be due to the same factors.

The edge luminescence bands of our samples of unactivated zinc sulfide are much stronger than the visible and other ultraviolet bands. However, other types of luminescence can be seen in the ultraviolet range. In particular, lines in the region of ~ 322 and 328 nm, whose intensities amount to a few tenths of a percent of the main luminescence band, can be seen superimposed on the short-wavelength wing of the edge luminescence band of hexagonal and cubic zinc sulfide. Similar short-wavelength lines are observed for ZnS sublimates and single crystals, whose spectra exhibit these lines more clearly (see §3).

A structure-free band extending from 360 to 390 nm is located much further from the fundamental absorption region. There have been many suggestions (Zn, ZnO, Cl) as to the origin of this band (see, for example [98]); the maximum of this band is not constant in all the samples and this suggests the existence of different centers due to the postulated defects. This

TABLE 8. Relationships between Intensities in the Second Series of Bands of the Edge Luminescence of Zinc Sulfide for Various Numbers of Emitted Phonons (T = 17°K)

Phonons	Theory	Hexagonal phase		Cubic phase	
		λ_{max}, nm	Line intensity	λ_{max}, nm	Line intensity
0	1,000	337.0	1.00	346.0	1.00
1	0.870	341.0	0.85	349.5	0.80
2	0.380	345.5	0.52	353.0	0.40
3	0.110	349.5	0.26	356.5	0.15
4	0.024	353.5	0.10	360.0	0.05
5	0.004	357.5	0.03	363.5	—

band is probably not associated with the lattice of zinc sulfide and we did not investigate its properties in any detail.

§3. Comparison of the Cathodoluminescence of Powders, Single Crystals, and Sublimates of Unactivated ZnS

In order to determine whether the ultraviolet luminescence emitted by different forms of ZnS is identical, we compared the luminescence spectra of ZnS recorded under identical excitation conditions at T = 77°K (Figs. 14a and 14b).

We also compared our results with the results of a study of the spectral and thermal properties of ultraviolet luminescence of sublimated phosphors, which was carried out by Gutan et al. [137]. According to earlier investigations [2, 84], powders of hexagonal zinc sulfide exhibit clearly overlapping bands at 334.5 nm (3.707 eV), 338.5 nm (3.663 eV), and 342.5 nm eV) with separations close to the energy of longitudinal optical phonons in the ZnS lattice. The luminescence spectrum of ZnS single crystals has a similar but a poorly resolved structure shifted in the direction of longer wavelenths and exhibiting maxima at ~341.0 nm (3.636 eV), 345.0 nm (3.594 eV), and 349 nm (3.553 eV), in good agreement with the maxima of the bands exhibited by cubic ZnS single crystals [77]. It is natural to assume that the ~0.07 eV shift between the spectra of powders and single crystals is due to the difference between the same edge luminescence spectra of the hexagonal and cubic modifications.

Growth of cubic crystals at ~1380°C is somewhat unexpected. However, it is known [138] that the cubic lattice may form above the phase transition temperature (~1050°C) in the presence of excess sulfur. Moreover, the slow cooling of a single crystal may also exert a considerable influence on the lattice structure. The more blurred structure of the spectrum of single crystals may be due to a considerable admixture of the second (hexagonal) phase.

In the case of sublimated screens the edge luminescence spectrum is usually somewhat wider than the spectra of the hexagonal or cubic samples and the maximum occupies an intermediate position at ~341.5 nm, which suggests that the amounts of the two phases are comparable. An edge luminescence spectrum, close to the spectrum of a hexagonal powder, is obtained when the conditions of preparation of the screens favor the formation of the hexagonal phase (substrate temperature ~800°C, excess Zn, heating of the original charge to ~1200°C). A spectrum resembling the edge luminescence of cubic powders and single crystals (see Fig. 14) is obtained under different conditions (substrate temperature 400°C, excess S, 900°C).

Thus, the nature of the low-temperature 330-360 nm luminescence emitted by different forms of ZnS can be assumed to be identical and the differences may be attributed to the different lattice structures.

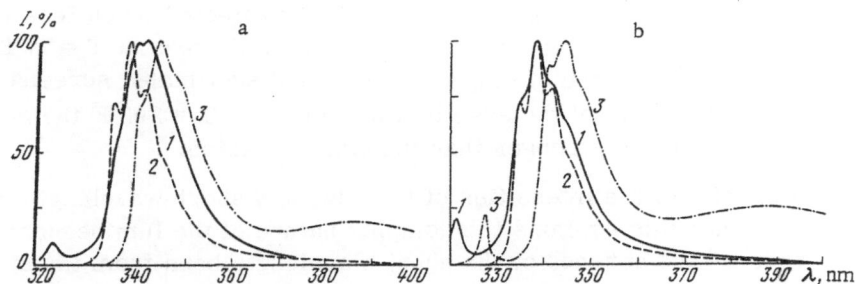

Fig. 14. Normalized ultraviolet luminescence spectra of various forms of unactivated zinc sulfide recorded at 77°K. a: 1) Sublimate No. 1; 2) powder (heated to 1200°C); 3) powder (heated to 900°C); b: 1) Sublimate No. 2; 2) powder (heated to 1200°C); 3) single crystal.

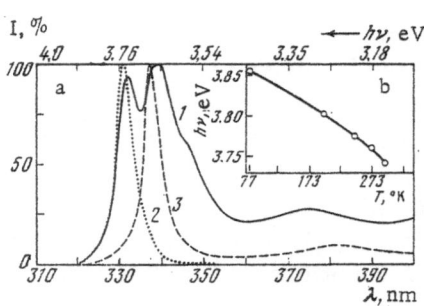

Fig. 15. a) Normalized ultraviolet luminescence spectra of various samples of unactivated zinc sulfide at 293°K: 1) powder (heated to 1200°C); 2) sublimate No. 2; 3) single crystal. b) Temperature shift of the short-wavelength luminescence.

Single crystals, like sublimates, emit also luminescence of shorter wavelengths than the edge band. The positions of the short-wavelength bands at 77°K are 321.5 or 327.5 nm, for the hexagonal and cubic ZnS samples. These bands are observed in the reflection spectra of ZnS powders composed of different modifications [8]. A shift of both types (edge and short-wavelength) of the ultraviolet luminescence indicates that the difference between the energies of the photons emitted by the hexagonal and cubic phases is 0.068 ± 0.002 eV, which is close to the difference between the postulated optical widths of the forbidden band of the hexagonal ($E_g = 3.917$ eV [4]) and cubic ($E_g = 3.84$ eV [77]) forms of zinc sulfide.

We must stress particularly the considerable difference between the influence of temperature on the edge and short-wavelength luminescence bands (shift, quenching, and half-width) and the relative constancy of the thermal properties of each type of luminescence irrespective of the nature of the object being investigated. The ultraviolet luminescence spectra of powders, single crystals, and sublimates recorded at T = 293°K (Fig. 15) are located in the same region as the edge luminescence at T = 77°K. However, the luminescence bands of the single-crystal ($\lambda_{max} = 337.5$ nm) and sublimate ($\lambda_{max} = 331.5$ nm) samples are considerably narrower than the edge luminescence recorded at 77°K. The half-widths of these bands are ~ 0.05 eV, i.e., about 2kT. We shall show later that these bands correspond not to the edge luminescence but to the short-wavelength bands at 327.5 and 321.5 nm, if the thermal shift is allowed for.

The measured temperature coefficient of the shift $dE/dT \approx -5.2 \times 10^{-4}$ eV/deg is in agreement with the temperature coefficient of the optical width of the forbidden band [139].

At T $\approx$ 300°K the edge luminescence of single cyrstals and sublimated screens is too weak to be recorded. Therefore, the electroluminescence of zinc sulfide single crystals observed at T = 293°K [5] is due to radiative transitions different from those responsible for the edge luminescence of ZnS. The intensity of the edge luminescence of powders decreases by a factor of several thousand when the temperature is raised from 77 to 293°K. This should be constrated with the short-wavelength luminescence which is affected much less (its intensity decreases only by a factor of 5-6 in the same range). Thus, whereas at T = 77°K the intensity of the short-wavelength luminescence of single crystals and sublimated screens is an order of magnitude lower than the intensity of the edge luminescence, at T = 293°K the shifted short-wavelength luminescence is much stronger than the edge emission.

At T = 293°K the ratio of the intensities of the edge and short-wavelength luminescence bands of powders is ~1 and this explains the complex nature of the luminescence spectrum of this form of zinc sulfide. The absence of the short-wavelength band from the spectrum in Fig. 14a is due to the fact that at T = 77°K the edge luminescence of powders is over two orders of magnitude stronger than the intensity of the short-wavelength bands. This difference between the intensities of the edge and short-wavelength luminescence observed at low temperatures is approximately compensated by the different thermal quenching of the two types of band between 77 and 293°K.

TABLE 9. Comparison of the Efficiencies of the
Short-Wavelength and Edge Luminescence of Different
Forms of Zinc Sulfide at Various Temperatures

ZnS	$T = 77°$ K		$T = 293°$ K	
	Edge	Short-wavelength	Edge	Short-wavelength
Powder	1	$1 \cdot 10^{-3}$	$2.5 \cdot 10^{-4}$	$2 \cdot 10^{-4}$
Sublimate	$2 \cdot 10^{-3}$	$4 \cdot 10^{-4}$	—	$6 \cdot 10^{-5}$
Single crystal	$1 \cdot 10^{-3}$	$2 \cdot 10^{-4}$	—	$3 \cdot 10^{-5}$

The absolute intensities of the edge luminescence of powders, single crystals, and sublimates excited to the same extent by electron bombardment at T = 77°K are quite different. The luminescence efficiency of ZnS powders is about 10% [84]; the efficiency of ZnS sublimates and single crystals is much lower (< 0.1%, Table 9).

Since the chemical purity of all the samples is approximately the same, the reduction in the efficiency of the luminescence emitted by single cyrstals and sublimates is evidently due to a smaller number of the radiative recombination centers formed by associated lattice defects (V_{Zn}, V_S) [2].

Since the difference between the energies of the edge luminescence photons of the cubic and hexagonal forms is close to the difference between the optical widths of the forbidden bands of these two modifications, it follows that the corresponding recombination centers are located at regular lattice sites and the carriers captured by these centers are weakly localized. The spectrum of the luminescence emitted by samples with mixed lattices is not the result of the superposition of the spectra of the different phases but of the spectra of the centers with average lattice parameters because the radii of the orbits of the trapped carriers amount to several lattice constants.

The difference between the nature of the edge and short-wavelength ultraviolet luminescence bands follows primarily from the different behavior under thermal quenching conditions. The strong thermal quenching of the edge luminescence of ZnS is explained by the thermal liberation of carriers from shallow levels (V_{Zn}, V_S). The much weaker quenching of the short-wavelength luminescence means that we cannot use a similar model of shallow levels; for example, we cannot use the model of isolated lattice defects. The short-wavelength luminescence of ZnS, differing by ≤ 0.1 eV from the energy corresponding to the forbidden band width of ZnS, is best explained by the exciton model [89, 140].

If the exciton model is used, the weak thermal quenching is explained because the exciton lifetime is governed primarily by the nonradiative annihilation at lattice defects and not by the thermal annihilation.

§ 4. Influence of Temperature on the Ultraviolet Luminescence of Unactivated Zinc Sulfide

We have reported above that the influence of temperature on two types of the ultraviolet cathodoluminescence bands of unactivated zinc sulfide is quite different in the range 77°K ≤ T ≤ 293°K.

We have mentioned that when the temperature is increased the short-wavelength luminescence shifts toward longer wavelengths at a rate $\overline{dE}/dT \approx -5.2 \times 10^{-4}$ eV/deg, in close agreement with the reduction in the forbidden band width of ZnS [139].

The edge luminescence losses its fine structure but it shifts much less rapidly (the rate of shift of the edge maximum is $\leq 1.3 \times 10^{-4}$ eV/deg).

The intensity of the edge luminescence decreases by a factor of more than 10^3 between 77°K and 293°K, whereas the short-wavelength luminescence decreases only by a factor of 5–7. This strong difference between the influence of temperature on the two types of luminescence indicates that the mechanisms of the corresponding radiative recombination processes are different. Bearing in mind the results of our previous investigations of the edge luminescence and the published information on the reflection spectra, we may conclude that the edge luminescence of ZnS is associated with local states formed by intrinsic lattice defects, whereas the short-wavelength luminescence si due to excitons. This interpretation is in basic agreement with the results obtained for other II–VI compounds [75, 87]. Further studies have raised new questions that had to be resolved. Since $(dE/dT)_{sw} \gg (dE/dT)_{edge}$ in the 77°K < T < 293°K range, it follows that when the temperature is raised the short-wavelength ("sw") and edge luminescence bands may either merge, or the edge luminescence may disappear, or — beginning from some temperature — the two types of band may move in parallel.

It would be interesting to analyze in detail the positions of the edge luminescence maxima at T > 77°K because in the range 17°K < T < 77°K there are two series of bands which behave quite differently when the temperature is varied. Moreover, it would be interesting to study the cause of the differences between the positions of the reflection and absorption maxima and the changes in the positions of the maxima and in the intensity of the luminescence bands at T > 293°K.

In view of this we investigated the changes in the ultraviolet luminescence spectra of un-activated ZnS (hexagonal powders, cubic powders, and single crystals [141]) in the range 77 and ~600°K. The experimental results are plotted in Figs. 16–19; the investigation was carried out using steady electron-beam excitation (V = 15 keV, $j \approx 10^{-5}$ A/cm²). The spectral width of the slit of a ZMR-3 monochromator with quartz optics was ~0.3 nm at 77°K; at high temperatures this width was ~1–1.5 nm. The wavelength was determined to within ≈0.3–0.5 nm; the temperature was determined to within ±3 deg K.

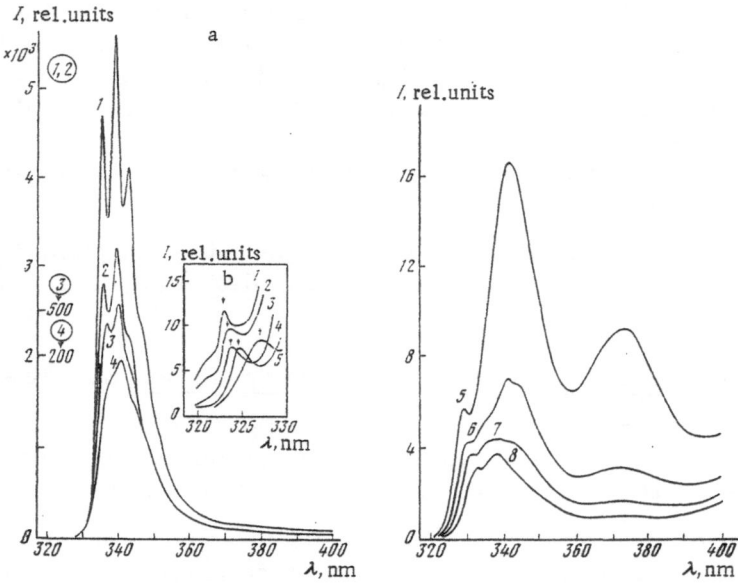

Fig. 16. Ultraviolet luminescence spectra of powders of hexagonal ZnS recorded at different temperatures (°K): 1) 77; 2) 98; 3) 120; 4) 150; 5) 220; 6) 245; 7) 265; 8) 293.

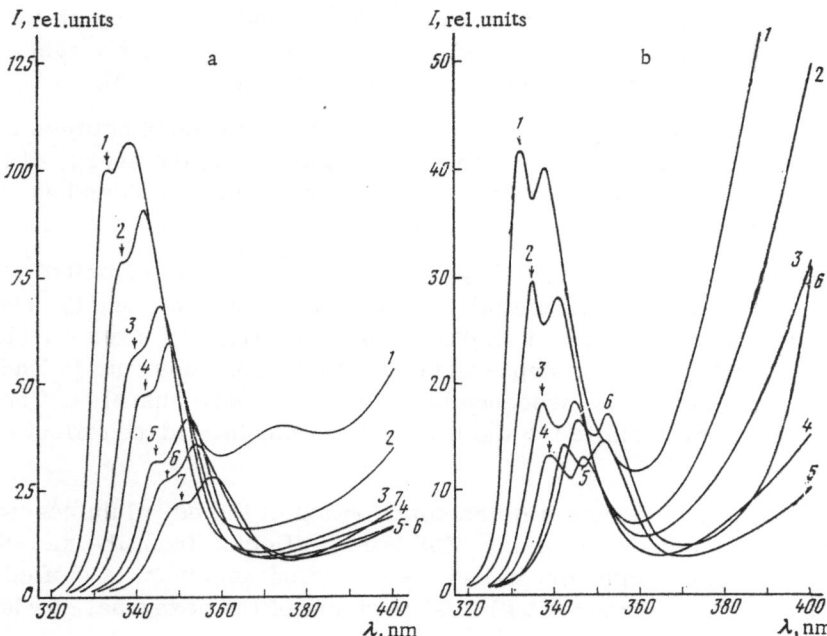

Fig. 17.   Ultraviolet luminescence spectra of a powder of hexagonal zinc sulfide deposited on reflecting (a) and absorbing (b)
substrates, recorded at different temperatures (°K): 1) 293;
2) 350; 3) 400; 4) 445; 5) 510; 6) 550; 7) 625.

Figure 16a shows the changes in the edge luminescence spectrum of hexagonal ZnS powders. As the temperature is increased, the fine structure is gradually lost, the short-wavelength series ($\lambda_{max} \approx 335$ nm) becomes more important, and both series exhibit a small thermal shift. At 150°K the luminescence spectrum loses its structure, the band profile becomes
asymmetric, and the half-width reaches ~14 nm. At room temperature the band with a maximum at $\lambda_{max} \approx 335$ nm occupies the position of the band with $\lambda_{max} \sim 338$ nm.

At low temperatures the edge luminescence of polycrystalline samples predominates
strongly over the short-wavelength luminescence; the latter appears as a step in the short-
wavelength wing of the spectrum (Fig. 16b). When the temperature is increased, the ratio of
the intensities changes in favor of the short-wavelength luminescence; the two types of luminescence can now be shown easily on the same scale. Between 77 and 293°K there is a definite
gradual thermal shift of the short-wavelength luminescence band to ~332 nm at room temperature. The differences between the positions of the short-wavelength luminescence and absorption maxima is due to the reabsorption of this luminescence. In fact, when the luminescence spectra are measured at 293°K using different voltages to accelerate the incident electrons, it is found that an increase in the energy of these electrons alters the ratios of the band
intensities in favor of the edge luminescence. The short-wavelength luminescence maximum
shifts from 330.5 to 332 nm. A shift by ~1.5 ± 0.5 nm appears because of the reabsorption in
this region (the absorption coefficient is k > 10^3 cm^{-1}). The edge luminescence is characterized
by k ≤ 10^2 cm^{-1} and, therefore, the reabsorption is insignificant at the depth of penetration of
electrons (~10^{-4} cm).

At higher temperatures (T > 293°K) both types of luminescence shift gradually toward
longer wavelengths at the same rate: dE/dT ≈ (5-6) × 10^{-4} eV/deg (see Figs. 17a and 17b).

Thus, in spite of the large differences between the thermal shifts of the edge and short-
wavelength luminescence bands observed in the range T < 293°K, the two bands do not merge
at higher temperatures.

The difference between the forbidden band width E_g and the energy of the edge luminescence photons at room temperature ($\sim 0.12-0.14$ eV) is close to the activation energy of the thermal and optical quenching of the edge luminescence of ZnS [84, 142].

When the temperature is increased still further, the difference between the optical and thermal depths of the edge luminescence centers disappears and the energy of the edge luminescence photons decreases in agreement with the reduction in the forbidden band width, as observed earlier for some shallow local levels [45].

The maxima of both types of ultraviolet luminescence are not exhibited clearly in the spectra of ZnS powders deposited on metallic (reflecting) substrates and the short-wavelength luminescence maximum is even shifted slightly in the direction of longer wavelengths (Fig. 17a). When a powder layer $\sim 10\mu$ thick is deposited on an absorbing substrate (a ZnS single crystal), the exciton (short-wavelength) luminescence band becomes more sharply defined and increases in intensity; this maximum is shifted to the left by ~ 1.5 nm toward the reflection maximum (Fig. 17b).

This is explained by a reduction in the contribution of the edge luminescence because in this case the luminescence penetrating into ZnS is not reflected from the metallic substrate but is absorbed in the cubic single crystal whose reflection spectrum is shifted by ~ 0.07 eV toward longer wavelengths, compared with the spectrum of the hexagonal powder of ZnS.

The total half-width of the luminescence spectrum is now 2kT and it increases with rising kT but more slowly than kT itself.

The intensity falls by more than an order of magnitude in the $293°K \leq T \leq 600°K$ range and the intensity of the short-wavelength luminescence band decreases more slowly than the intensity of the edge luminescence.

The short-wavelength luminescence is visible clearly in the spectrum of a cubic ZnS single crystal at $T \geq 293°K$ (Fig. 18a). The position of its maximum is in good agreement with

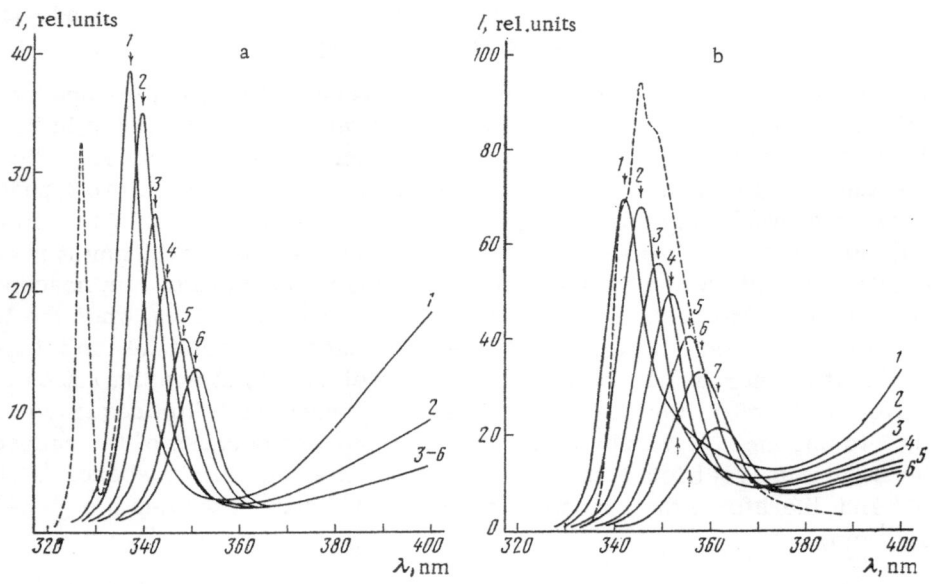

Fig. 18. Ultraviolet luminescence spectra of a single crystal (a) and a powder heated to 900°C (b) of cubic zinc sulfide, recorded at different temperatures (°K): 1) 293; 2) 350; 3) 400; 4) 445; 5) 510; 6) 550; 7) 625. The dashed curves show the spectra recorded at 77° K.

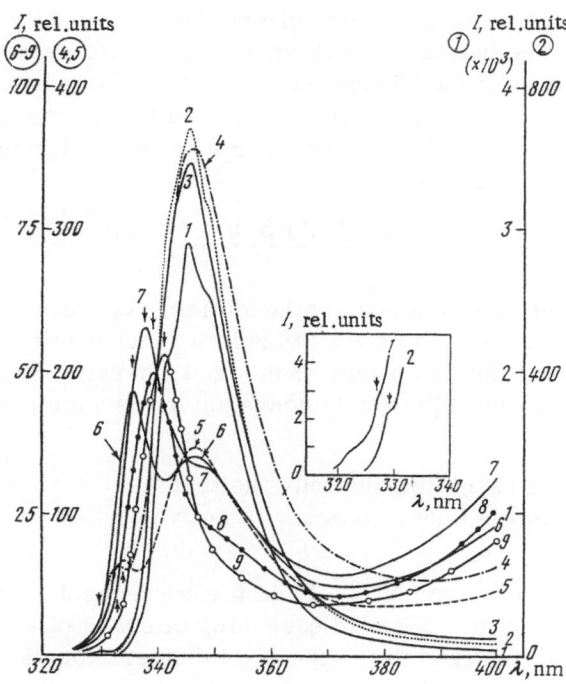

Fig. 19. Edge luminescence spectra of a powder of cubic zinc sulfide, recorded at different temperatures (°K): 1) 77; 2) 98; 3) 110; 4) 160; 5) 190; 6) 220; 7) 245; 8) 265; 9) 293.

the reflection peak of ZnS (336.5–337.0 nm) of the cubic modification [6]. The short-wavelength luminescence spectrum is almost elementary. The half-width, which is (2.4 ± 0.2) kT throughout the range 293°K ≤ T ≤ 600°K, is evidence of the elementary nature of this band. At 77°K the half-width of this band ~4kT, which may be explained by the presence of bound and free excitons. In fact, at T ≤ 77°K the structure of the short-wavelength band is quite complex [89, 140].

The thermal shift of the ultraviolet luminescence spectrum of ZnS is observed also (Fig. 18b) for cubic powders of ZnS (however, the distinction between the two types of the luminescence is less clear). The spectrum obtained at T ≥ 293°K to liquid nitrogen temperature (Fig. 19). Such cooling results in the gradual separation of the exciton (short-wavelength) luminescence band and a shift toward shorter wavelengths. The intensity of the edge luminescence rises when the temperature is lowered.

Thus, our investigation of the ultraviolet luminescence of various forms of unactivated ZnS carried out in the range 77°K ≤ T ≤ 600°K has revealed a gradual but parallel shift, amounting to ~0.3 eV (i.e., ~30 nm), and changes in the structure of the short-wavelength and edge bands. The short-wavelength band represents the properties of the lattice of ZnS itself, in the form of free excitons with a binding energy $\Delta E_{ex} \sim 0.04 \pm 0.01$ eV. When the temperature is increased (T > 293°K) the short-wavelength band shifts toward longer wavelengths ahead of the position of the edge band at liquid nitrogen temperature. The position of the edge band remains apparently unaffected in the range 77°K ≤ T ≤ 293°K. Actually, a short-wavelength series, due to the recombination of free electrons with holes trapped by acceptor levels (V_{Zn}), splits off from the edge luminescence spectrum and is slightly shifted. At higher temperatures T > 293°K) both the short-wavelength and edge luminescence bands experience a similar shift and the quenching slows down due to the empltying of deep nonradiative recombination levels.

The similar thermal shifts of both bands are explained by the shallowness of the levels and the consequent strong correlation of the shift with the reduction in the forbidden band width.

Our high-temperature investigation revealed deep traps in zinc sulfide phosphors. At T ~ 500°K the quenching of the ultraviolet luminescence bands slows down and the green luminescence grows strongly in intensity. This is evidence of the liberation of carriers from deep nonradiative recombination levels. The activation energy of the rise of the green luminescence is ΔE ~1.1 eV, which is in agreement with the depth of the electron traps in ZnS deduced from the conductivity [143], ESR spectra [144], and by other methods [145].

§ 5. Quenching of the Edge Cathodoluminescence of ZnS by Infrared Radiation

Earlier investigations have shown that the kinetics of the edge cathodoluminescence and photoluminescence of ZnS is governed by the radiative recombination at shallow local levels under strong external quenching conditions, i.e., under the conditions of thermal liberation of nonequilibrium carriers from local luminescence centers followed by nonradiative recombination at quenching centers.

This quenching mechanism was confirmed by investigating not only the thermal but also the optical quenching by infrared radiation [142], which is known to occur in activated ZnS phosphors.

In view of the shallowness of the levels of the active centers [84] and the consequent strong thermal liberation of carriers from these levels, the infrared quenching effect is observed only if the infrared radiation intensity is sufficiently high for the optical liberation of carriers to compete successfully with the thermal process. This experimental difficulty obviously explains the lack of published information on the infrared quenching of the edge luminescence of ZnS.

Using the known values of the cross sections for the capture of holes in CdS (σ_h ~10^{-15}–10^{-16} cm^2) [75], we can estimate the power of an optical source necessary for the probabilities of the optical ($w_{\nu \, IR}$) and thermal (w_t) liberation of carriers to be comparable ($w_{\nu \, IR} \approx w_t$).

The probability of the thermal liberation of carriers at 77°K deduced from the formula[*] $w_t = \sigma_h N^*_{eff} v_t \exp(-\Delta E_t/kT)$ is ~10^2–10^3 sec^{-1}.

The probability of the optical liberation $w_{\nu IR} = I_\nu \sigma_{\nu \, IR}$, governed by the cross section $\sigma_{\nu IR}$ for the absorption of infrared radiation[†] ($\sigma_{\nu IR}$~10^{-15} cm^2 if the oscillator strength is f = 1), is comparable with ~10^2–10^3 sec^{-1} if the infrared radiation power is ~0.01–0.1 W/cm^2.

The use of a ribbon lamp (SI-8-200) with filters selecting the range 2–2.5μ, i.e., 0.5 eV < hν_{IR} < 0.6 eV, made it possible to quench the edge luminescence of ZnS (during excitation with an electron beam) by ~30% at T = 77°K and the measured infrared radiation power density[‡] was ~0.018 ± 0.002 W/cm^2.

However, it was necessary to ensure that this infrared radiation power density did not heat the investigated luminescent screen and did not cause thermal quenching because the change in the ultraviolet luminescence brightness at low excitation intensities was known to be dI/dT ≤ 10% per 1 deg K (i.e., a temperature rise by ~3–4 deg K would be sufficient to explain the observed quenching).

We assumed that the thermal conductivity of zinc sulfide powder was K ~ (2 ± 1) × 10^{-4} cal · cm^{-1} · deg^{-1}, sec^{-1} and the thickness of the luminescent screen was d ~20μ (~8 mg/cm^2)

---

[*]N^*_{eff} = $2(2\pi m^* kT)^{3/2}/\hbar^3 = 4.8 \cdot 10^{15} [(m^*/m)T]^{3/2}$ is the density of states in the relevant allowed band; $v_t = \sqrt{3kT/m^*}$ is the thermal velocity of carriers; m^* is the effective mass of carriers, assumed to be m on the basis of [29].

[†]The data on the absorption in Ge and Si give 0.5×10^{-16} cm^2 ≤ σ_{IR} ≤ 4×10^{-15} cm^2.

[‡]This power density was determined with an IMO-1 power meter.

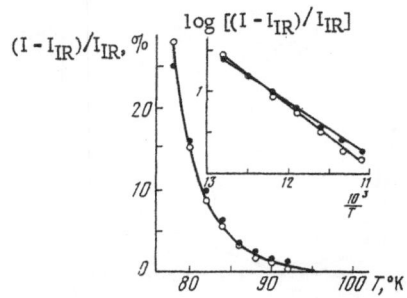

Fig. 20. Temperature dependences of the infrared quenching (by radiation of ~ 0.02 W/cm²
power density) of the ultraviolet edge luminescence
(λ_{max} = 338 nm) of unactivated zinc sulfide excited by bombardment with 15 keV electrons
(j = 10^{-8} A/cm²). Here, I is the intensity of the
ultraviolet luminescence and I_{IR} is the same intensity in the presence of infrared radiation
(points represent experimental values).

and we found that $\Delta T = 0.24 W_{IR}d/K \leq 0.2$ deg K, i.e., the thermal effect could not explain
quenching in excess of 1-2% and if allowance was made for the incomplete absorption of infrared radiation this value should be even smaller. Therefore, the thermal effect was at least an
order of magnitude smaller than the optical quenching.

The temperature dependence of the infrared quenching was found to be extremely strong
(the quenching effect fell by an order of magnitude in a range of ~ 15 deg K). The results of
several measurements of the temperature dependence of the infrared quenching effect are
plotted in Fig. 20. In spite of the fact that the investigated temperature range was narrow, the
experimental points fitted well a common curve. Applying the Fok method [107] and plotting
the dependence $\log[(I-I_{IR})/I_{IR}] = f(10^3/T)$, we determined the activation energy of the optical
quenching as a function of temperature and found that ΔE_{IR} ~ 0.14 ± 0.02 eV.

This activation energy was only slightly greater than ΔE_t representing the thermal quenching effect, which suggested that the optical and thermal quenching of the edge luminescence was
due to the same mechanism in which localized carriers were liberated from the ultraviolet luminescence centers and recombined nonradiatively at the quenching centers.

The edge luminescence spectra of ZnS were not greatly affected by infrared radiation,
which indicated the absence of any significant redistribution of the two series of bands composing the spectrum. This was to be expected because the probability of the thermal liberation of electrons from shallow levels at ~ 0.03 eV (V_S) was several orders of magnitude greater than the probability of the optical liberation and, therefore, the action of infrared radiation
reduced mainly to the emptying of the acceptor levels.

Infrared radiation not only quenched the edge luminescence but also stimulated a blue
luminescence band. The temperature dependence of the intensity of this band (Fig. 21) made
it possible to determine its activation energy, which was ~ 0.065 ± 0.01 eV. It was probable,
as suggested in [146], that the stimulation was due to the liberation of electrons and not due to
the transfer of holes from the edge luminescence centers. The absolute reduction in the number of the ultraviolet edge luminescence photons was of the order of magnitude of the increase
in the number of the blue luminescence photons. However, this important point would need a
further study.

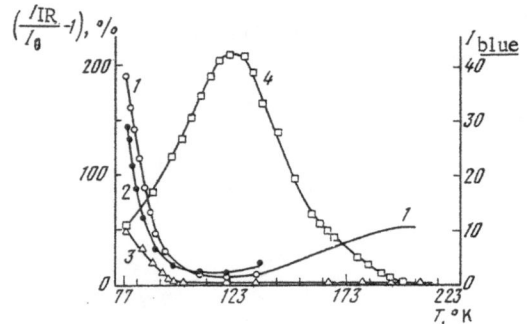

Fig. 21. Temperature dependences of the intensity of the blue (λ_{max} = 465 nm) luminescence of unactivated zinc sulfide: 1) flash produced by infrared radiation of λ = 1.0-1.5μ wavelength; 2) flash produced by infrared radiation of λ = 2.0-3.0 μ wavelength; 3) steady-state enhancement of the intensity of the blue band by infrared radiation; 4) normal blue luminescence (right-hand scale).

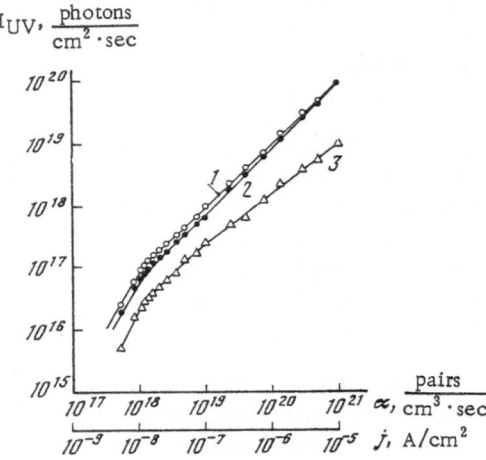

Fig. 22. Dependences of the ultraviolet edge luminescence intensity on the electron excitation rate or electron-beam current density: 1) in the absence of infrared radiation; 2) in the presence of infrared radiation; 3) number of quenched quanta (difference between curves 1 and 2 for the same excitation rate).

We also investigated the dependence of the quenching of the edge luminescence of ZnS on the intensity of the electron excitation (Fig. 22). A double logarithmic scale is used in this figure to show the dependence of the intensity of recombination via the edge luminescence centers on the excitation rate,[*] allowing for the penetration of electrons into ZnS [112]. The abscissa in Fig. 22 gives the number of carriers excited by an electron beam in 1 cm³ per 1 sec, whereas the ordinate gives the number of recombination events allowing for the efficiency [84] (in the linear region, $\eta_{cl} \sim 10\%$).

It is clear from Fig. 22 that when the electron-beam current density is j $\approx 10^{-5}$ A/cm², the number of the quenched photons reaches $\approx 10^{19}$ photons·cm⁻¹. This value represents the

---

[*]The excitation rate was determined from the formula

$$\bar{\alpha} = \{[(\overline{dE}/dx)\, i_0]\, /\bar{\varepsilon}_{eh}e\} \simeq 1.2 \cdot 10^{26} \text{ (pairs·cm}^{-3}\cdot\text{sec}^{-1}\text{)} \; i_0\,(a).$$

minimum number of the excited centers n^* in which infrared photons are absorbed giving rise to quenching:

$$n^* \, (\text{cm}^{-3}) \geqslant \frac{N \ (\text{quenching photons} \cdot \text{cm}^{-3} \cdot \text{sec}^{-1})}{I_{\nu \text{IR}} \ (\text{photons} \cdot \text{cm}^{-2} \cdot \text{sec}^{-1}) \ \sigma_\nu \, (\text{cm}^2)} \gtrsim 10^{16} \ \text{cm}^{-3} \ .$$

This number is very close to the thermodynamic estimates of the concentration of associated defects ($\sim 10^{15}$ cm^{-3}) but it is less than the number of simple defects ($\sim 10^{17}$ cm^{-3}).

Thus, the depth of the acceptor level and the concentration of the excited edge luminescence centers in ZnS deduced from the infrared quenching effect are in agreement with the donor–acceptor model of the centers [83, 86] formed from intrinsic lattice defects.

The relatively high concentration of the edge luminescence centers in unactivated zinc sulfide is evidence that the centers responsible for the ultraviolet luminescence are composed of defects separated by fairly large (~ 25 Å) distances from one another.

CHAPTER IV

INFLUENCE OF CONDITIONS DURING SYNTHESIS ON THE EFFICIENCY AND PROPERTIES OF EDGE LUMINESCENCE OF ZnS

§ 1. Influence of Stoichiometry on the Edge Luminescence of ZnS

Since one of the basic assumptions about the origin of the edge luminescence is the radiative recombination at intrinsic lattice defects, it would be interesting to consider the influence of stoichiometry on the properties of such luminescence. The results of the initial investigation of the influence of excess Zn and S on the edge luminescence intensity are given in [2]. An analysis of these results shows that single defects cannot act as the edge luminescence centers because the intensity of the edge luminescence decreases on introduction of excess zinc or excess sulfur. It has been suggested that the edge luminescence is due to sulfur and zinc vacancies forming a dipole pair which can act as an electron trap, i.e., the Williams model has been assumed.

Fuller investigations of the luminescence of phosphors with excess zinc and sulfur were carried out later [80].

The ultraviolet cathodoluminescence spectra obtained using 15 keV electrons (j $\approx 10^{-6}$ A/cm^2) at 77°K are plotted in Fig. 23, where the scale for the original ZnS sample is reduced by a factor of 2 in order to show clearly the structure of the spectra of the nonstoichiometric samples.

An examination of these spectra demonstrates that the main influence of the excess Zn and S is a reduction in the intensity of the edge luminescence. In spite of the broadening of the individual edge luminescence bands, a slight shift of the maxima (by not more than 1 nm), and a change in the ratio of the peak intensities, the luminescence spectrum retains its general form, i.e., it is not greatly affected by the presence of excess zinc or sulfur. Thus, the model postulating isolated stoichiometric defects (Zn_i, S_i, V_{Zn}, V_S) cannot explain the origin of the edge luminescence because the addition of Zn or S should have increased the number of the edge luminescence centers and it should have raised the luminescence intensity, contrary to the experimental observations. Therefore, the results in Fig. 23 support the associative nature of the centers (either V_{Zn}–V_S or Zn_i–S_i).

The hypothesis that the concentration quenching appears because of an increase in the number of the luminescence centers is unacceptable because thermodynamic calculations of

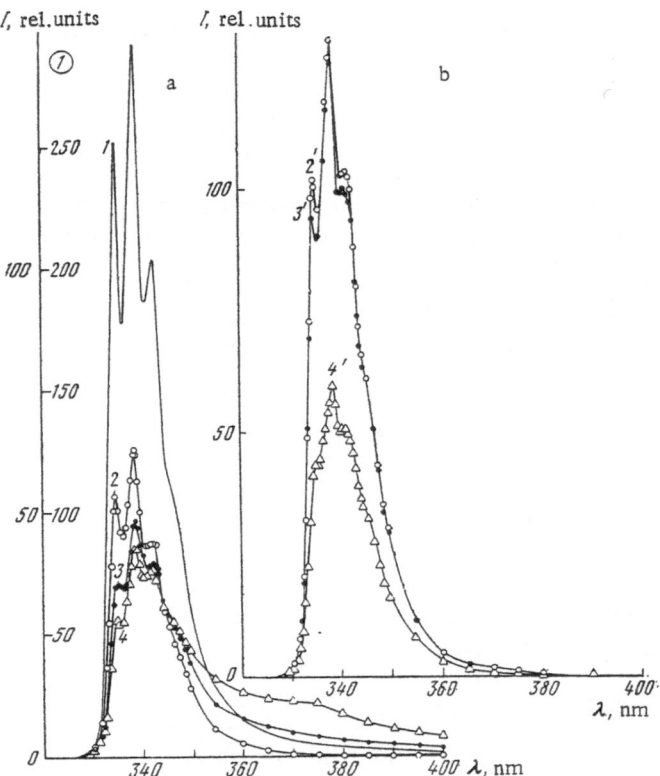

Fig. 23. Ultraviolet edge luminescence spectra recorded at 77°K for ZnS phosphors prepared at different zinc (p_{Zn}) and sulfur (p_S) vapor pressures. a: 1) Original ZnS; 2) p_{Zn} = 10 atm; 3) 20 atm; 4) 60 atm. b: 2') p_S = 10 atm; 3') 20 atm; 4') 60 atm.

the concentration of the centers give values not exceeding 10^{17} cm^{-3}, which are outside the range of the usual concentration quenching and which support the associative nature of the edge luminescence centers in ZnS.

We shall consider the question of the changes in the edge luminescence due to deviations from stoichiometry employing the associative model of the V_{Zn}–V_S centers. The number of such centers is governed by the product of the concentrations of the donor and acceptor defects and, in the first approximation, it is independent of the stoichiometry of a sample:

$$\sqrt{N_c N_a} = N_0 \exp\left[-\left(\Delta_c + \Delta_a\right)/2kT\right],$$

where N_c and N_a are the numbers of the cation and anion defects; Δ_c and Δ_a are the energies of formation of these defects which are not known exactly but range from 1 to 2 eV; N_0 is the total number of atoms ($\sim 3 \times 10^{22}$ cm^{-3}); T is the temperature at which the thermodynamic equilibrium is "frozen."

It follows that the reduction in the efficiency of the ultraviolet luminescence of zinc sulfide on introduction of Zn and S results from an increase in the concentration of the nonradiative recombination centers and of the trapping levels [147], which compete with the edge luminescence centers. Our experiments show that the deviation from the stoichiometric amounts of Zn and S increases the overall efficiency of the visible luminescence of unactivated ZnS phosphors.

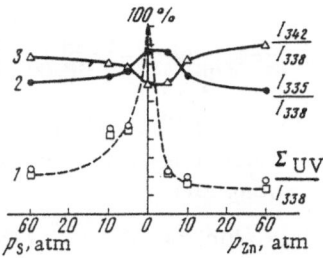

Fig. 24. Changes in the intensity of the edge luminescence of zinc sulfide on deviation from stoichiometry: 1) integrated luminescence (open circles) and intensity of the λ_{max} = 338 nm band (squares); 2) ratio of the intensities of the ~ 335 and ~ 338 nm bands; 3) ratio of the intensities of the 342 and 338 nm bands.

However, the absolute reduction in the efficiency of the edge luminescence is considerably greater than the increase in the efficiency of the visible bands i.e., the major reason for the fall of the energy of the edge luminescence is its redistribution between the nonradiative channels.

Changes in the edge luminescence spectra of ZnS are shown schematically in Fig. 24. The total intensity (Σ_{UV}) of the edge luminescence of ZnS represented by the areas under the bands decreases by a factor of 5 when Zn and S are introduced; the first portions of Zn and S have the greatest effect.

Introduction of the excess Zn has a somewhat different effect than the introduction of sulfur; however, since the exact amounts of sulfur and zinc are not known, it is not possible to draw a definite conclusion that the quenching centers resulting from the introduction of Zn have a stronger effect.

The hypothesis [85] on the need for the excess sulfur in the formation of the edge luminescence centers in unactivated ZnS and on the absence of structured edge luminescence in the spectra of phosphors with excess Zn is in conflict with our results.

The changes in the edge luminescence spectra which occur on introduction of excess Zn or S can be explained by the existence of two overlapping series of edge luminescence bands [76, 84].

The intensity of the $\lambda_{max} \sim 338$ nm line, which is strongest at 77°K, decreases for all the investigated ZnS samples in close agreement with the general reduction in the edge luminescence efficiency. The absolute intensities of the other lines also decrease in the presence of excess Zn or S but the relative intensities of the ~ 335 and 342 nm lines and their ratio to the strongest line (~ 338 nm) change on addition of Zn or S.

Since the edge luminescence of ZnS recorded at T ≤ 77°K [76, 84] consists of two superimposed phonon replica series separated by the phonon energy $h\nu_{phon}$ ~ 0.04 eV (the head lines of the short- and long-wavelength series are located at $I_0^{sw} \approx 334.5$ nm and $I_0^{lw} \approx 337$ nm), it follows that the change in the luminescence intensity on deviation from stoichiometry corresponds to the changes in the ratio between these two series.

If we assume that the spectral profiles are the same in the first and second edge luminescence series of ZnS, as proved for other II–VI compounds [148], and calculate the ratios of the zeroth and first phonon maxima in both series from the experimental results, we find that [84]

$$\frac{I_0^{sw}}{I_0^{lw}} = \left(\frac{I_{335}}{I_{338}}\right)_{exp}\left[1 - \alpha\left(\frac{I_{335}}{I_{338}}\right)_{exp}\right]^{-1},$$

where I_0^{sw} and I_0^{lw} are the intensities of the head lines of the short- and long-wavelength series; the factor α is equal to the ratio of the intensities of the first phonon replica and the zero-phonon head line, which is 0.87 for ZnS phosphors.

According to these calculations, introduction of excess Zn or S should reduce the intensity of the short-wavelength series more rapidly than that of the long-wavelength series. It should be noted that an increase in the relative importance of the long-wavelength series may impair the resolution of the spectrum and cause a slight shift of the maxima in the edge luminescence.

These changes are almost the same for excess Zn and excess S; moreover, an analysis of these changes suggests that even the original ZnS samples contain excess sulfur. This agrees with the conditions during preparation of zinc sulfide (heating in an H_2S atmosphere).

When the temperature is raised, the ultraviolet radiation is quenched very rapidly and it loses its fine structure. However, if electron bombardment is used, the ultraviolet luminescence spectrum could be studied even at 293°K (Fig. 25).

When the temperature is raised, the visible luminescence band also weakens strongly, not by three orders of magnitude as in the case of the edge luminescence, but by just one order. Consequently, at room temperature the visible luminescence is the main type of emission from ZnS phosphors.

In the wavelength range 330-360 nm (i.e., in the range of the edge luminescence of ZnS at 77°K) the spectra recorded at 293°K for all the phosphors consist of two structure-free bands

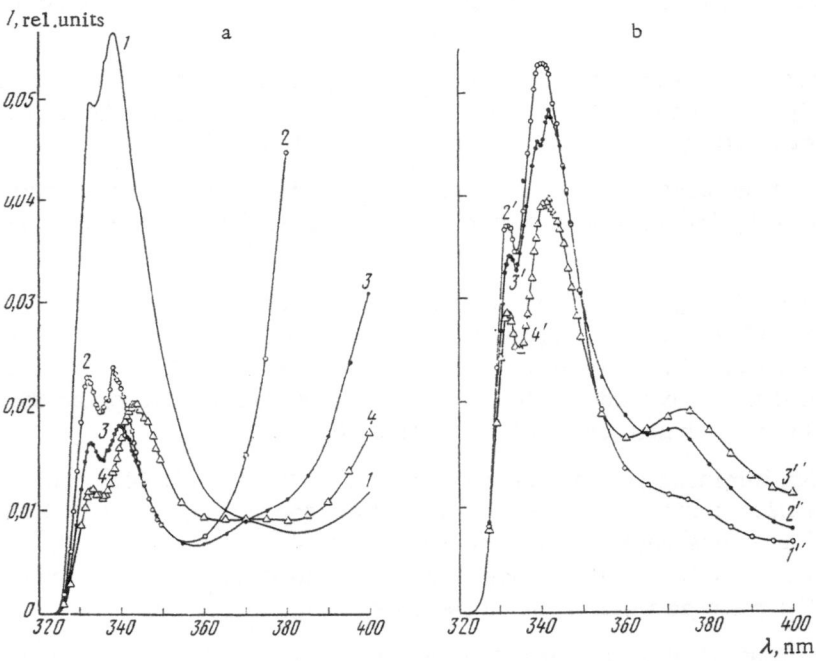

Fig. 25. Luminescence spectra of ZnS at 293°K obtained for samples with different amounts of excess zinc and sulfur. The designations are the same as in Fig. 23.

whose maxima (~ 332 and ~ 340 nm) shift by a small (2-3 nm) but regular amount on introduction of excess zinc and sulfur.

An investigation of the temperature dependence of the luminescence of various forms of zinc sulfide (powders, single crystals, and sublimates) [137] shows that the short-wavelength luminescence (~ 332 nm) is due to the exciton radiative recombination with a band at about 321.5 nm at 77°K (this band is not visible on the scale used in Fig. 23). The presence of excess Zn or S in ZnS phosphors reduces also the intensity of this luminescence but much less strongly than the intensity of the edge luminescence. Therefore, the intensity of the ~ 332 nm band becomes comparable with the intensity of the ~ 340 nm band, where the latter represents a somewhat modified edge luminescence. However, whereas in the case of the original ZnS the edge luminescence maximum is located at 338 nm, a deviation from stoichiometry in the direction of Zn or S not only reduces the density of this band (which is in qualitative agreement with the reduction at 77°K) but also shifts gradually the maximum by 1-4 nm in the direction of longer wavelengths. This also agrees with changes in the spectra at liquid nitrogen temperature and with the reduction in the importance of the short-wavelength series. The position of the exciton maximum at ~ 332 nm changes less (within 1.5-2 nm) and the change is different for excess Zn or S; introduction of S shifts this maximum from 332.5 nm for the original ZnS to 331.5 nm; introduction of Zn shifts the maximum to 333.5 nm. This may be due to changes in the reabsorption of the luminescence or changes in the energy band structure of the host lattice due to considerable deviations from stoichiometry.

The asymmetric effects of small amounts of Zn and S on the edge luminescence, which cannot be attributed to the different nature of the quenching (because of actual number of impurities introduced is not known) or to changes in the spectral composition (because this composition changes basically in the same way), is manifested clearly in the dependences of the luminescence intensity on the excitation power density. As mentioned earlier, the intensity of the edge luminescence generated employing conventional excitation sources ($E_{exc} \leq 10^{-4}$ W/cm^2) varies nonlinearly with the power density of the optical excitation. Within two orders of magnitude of E_{exc} the intensity of the ultraviolet luminescence emitted by the original samples can be approximated satisfactorily by the formula $I_{UV} = C E_{exc}^n$, where n = 1.8.

Measurements on a batch of samples with added Zn and S demonstrated that in this range of the excitation power densities the exponent n in the above formula depended on the stoichiometry (Fig. 26a). The first portions of S increased the nonlinearity of the above dependence (n reached 2), whereas the first portions of Zn reduced the nonlinearity. Thus, the luminescence quenching processes resulting from the introduction of Zn and S are different, which is in

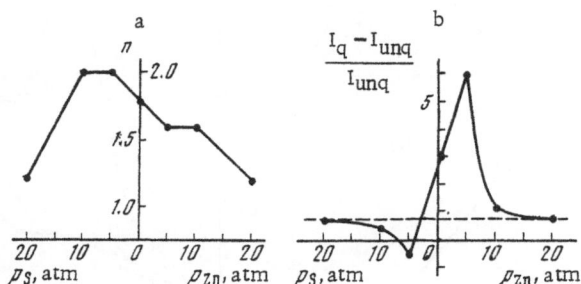

Fig. 26. Changes in the nonlinearity parameter n in the formula $I_{UV} = CE^n$ (a) and the influence of quenching (rapid cooling) on the intensity of the edge luminescence of ZnS phosphors (recorded at 77°K) containing different amounts of excess zinc and sulfur (b).

agreement with the asymmetric structure of the edge luminescence centers and with the consequent differences between the influence of additional donor and acceptor levels on the ultraviolet luminescence. The asymmetric influence of small amounts of Zn and S is also reflected to some extent in the ratio of the intensities of the luminescence emitted from quenched (rapidly cooled) and unquenched samples of different stoichiometry (Fig. 26b). The quenching of samples has a favorable influence on the edge luminescence intensity (particularly when the cooling time $t_{cool} \sim 12$ h is reduced to $t_{cool} \sim 15$ min): the quenching effect appears more clearly in the case of samples with small amounts of Zn (the intensity increases by a factor of 7), whereas large amounts of Zn or S increase the intensity only by a factor of 1.5. In the case of samples with small amounts of sulfur the edge luminescence intensity may even decrease as a result of rapid cooling and this may be attributed to a considerable increase in the number of single defects formed by excess sulfur and frozen-in by quenching. The reported results indicate that single defects cannot be responsible for the edge luminescence.

Moreover, a deviation from stoichiometry produces centers which can quench the ultraviolet luminescence, and the properties of the edge and short-wavelength luminescence of the samples with small amounts of Zn and S differ considerably. We did not investigate particularly the origin of the visible luminescence whose intensity at 77°K was always at least 5% of the edge luminescence (Fig. 27). The experiments carried out established the presence of the following structure-free bands: ~ 360, 460, 510, and ~ 600 nm (sometimes the last band splits into two). It should be noted that in the presence of small amounts of Zn and S, which reduce strongly (by a factor of up to 3-5) the intensity of the ultraviolet luminescence, the intensity

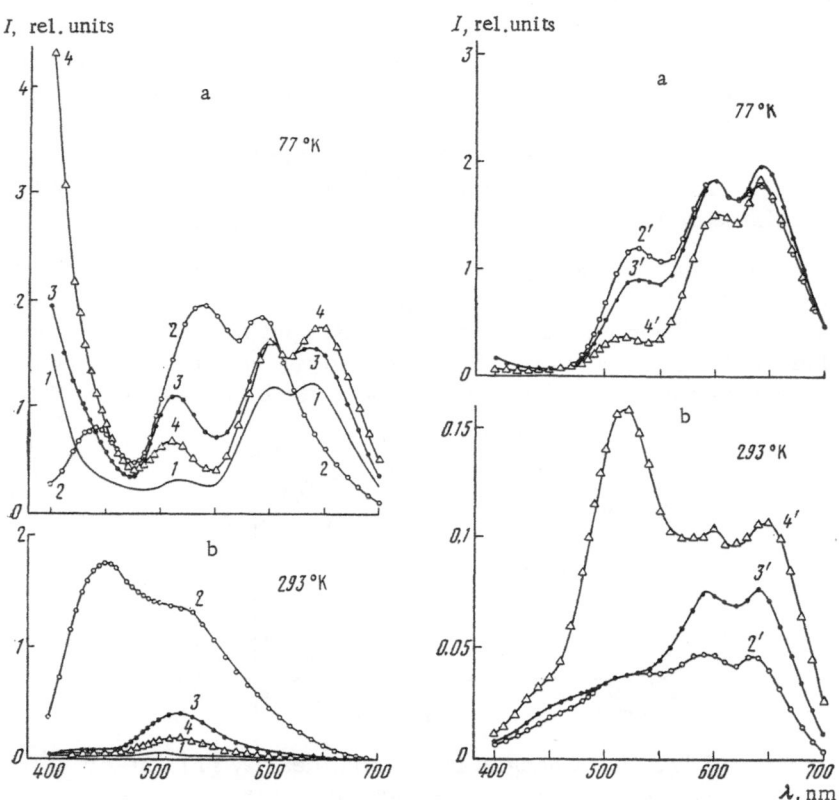

Fig. 27. Visible luminescence spectra of unactivated ZnS samples with different amounts of excess zinc and sulfur, recorded at different temperatures: a) 77°K; b) 293°K. The designations are the same as in Fig. 23.

of the visible bands taken as a whole increased; further portions of Zn and S quenched all these bands but their intensities remained at least as high as the intensity of the visible luminescence of the original sample.

The spectrum of the original samples and those with excess sulfur was dominated by the red group of bands. In the samples with excess Zn this group retained approximately its original intensity but the long-wavelength ultraviolet band and the blue luminescence developed more rapidly. The ratios of the intensities of the visible bands obtained at liquid nitrogen temperature did not agree with the corresponding ratios at room temperature. The luminescence of the samples with excess zinc decreased by a factor of 2-3, whereas that of the samples with excess sulfur decreased by a factor of 5-10.

These visible bands are not simply due to lattice defects but are due to various complexes formed by these defects with residual impurities. The luminescence of unactivated ZnS phosphors of different compositions may vary from red to violet with all the intermediate shades. The origin of the visible bands could not be determined without a high-precision chemical analysis but this was unnecessary because the total contribution of these bands did not exceed $\leq 0.1\%$ and the competition between the visible- and edge-luminescence centers could not be responsible for the quenching of the visible luminescence.

It should be pointed out that these experiments established definitely the quenching action of the excess zinc and sulfur on the edge luminescence of ZnS and that the excess zinc and sulfur atoms cannot act as the ultraviolet luminescence centers.

The quenching of the edge luminescence of ZnS by the excess Zn is not equivalent to the quenching by the excess S, as established directly from the dependence of the luminescence intensity on the excitation power density and from the influence of rapid cooling on the luminescence intensity.

The changes observed in the edge luminescence spectra of ZnS due to excess zinc and sulfur, including broadening and a redistribution of the intensities, are due to an increase in the importance of the long-wavelength edge luminescence series.

§ 2. Thermoluminescence of Unactivated Zinc Sulfide Excited by Electron Bombardment

In recent years much attention has been paid to studies of the luminescence of unactivated ZnS of different compositions [2, 80, 85] in order to determine the nature of the visible [98] and ultraviolet [1] luminescence discovered earlier and to determine the relationship between such luminescence and intrinsic lattice defects. The present section reports the experimental results obtained in an investigation of the thermoluminescence of hexagonal ZnS powders with excess Zn or S, which was carried out in the temperature range 80-400°K. The powders were excited first by bombardment with electrons of initial energy $E_0 = 6\text{-}20$ keV using a current of density from 10^{-8} to 10^{-6} A/cm^2 [147]. The main experimental results are plotted in Fig. 28, where the ordinate gives the luminescence intensity in units which are arbitrary but comparable for all curves. In all cases the main storage of the energy occurred in the levels that were de-excited at $T_{max} \sim 125$, 155, and 250°K ($\beta \approx 10$ deg/min), which has been observed earlier [77] for ZnS:Cl crystals.

The results indicated that the thermoluminescence curves, i.e., the total light sum and its distribution, depended on the stoichiometry of the samples and on the energy of the electrons used in the preliminary excitation.

The main thermoluminescence band is green ($\lambda_{max} \sim 510$ nm) with a weaker blue band superimposed on its short-wavelength wing. The edge ultraviolet band is not observed in the thermoluminescence spectra but at the moment of excitation its intensity is about 100 times higher than the intensity of the visible bands. This is due to the fact that the energy depth of

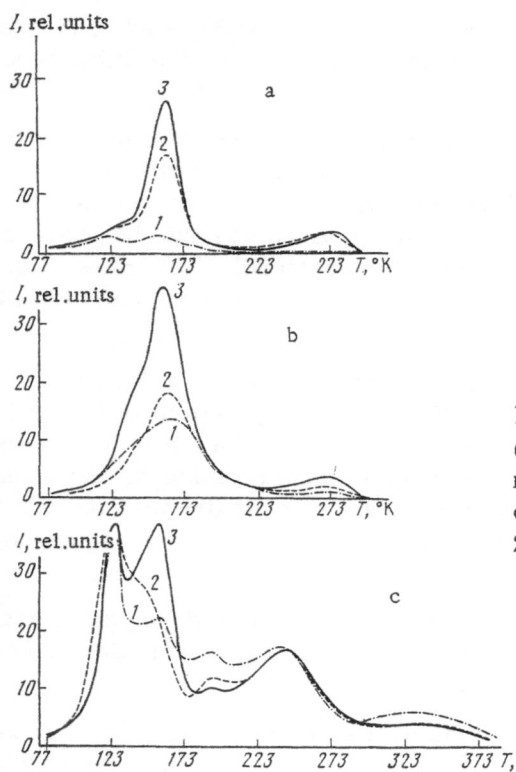

Fig. 28. Thermoluminescence spectra of ZnS (a), ZnS:S (b), and ZnS:Zn (c) after a preliminary excitation by bombardment with electrons of different energies E_0 (keV): 1) 6; 2) 12; 3) 20.

the trapping levels responsible for the observed thermoluminescence peaks exceeds the depth of the ultraviolet luminescence centers (≤ 0.2 eV) [84]. Temperatures above 100°K cause thermal quenching of the edge luminescence even at the moment of excitation, i.e., even when the probability of radiative recombination is considerably enhanced. The energy depth of the level responsible for a clear-low-temperature peak (Fig. 28c) was determined by various methods and it was $\Delta E \sim 0.23 \pm 0.02$ eV. In obtaining estimates of this kind we used the approximate formula E (eV) = $T_{max}/540$ (T_{max} is in °K), whose validity was recently confirmed in the 100–200°K range by the fractional thermal de-excitation method [149]. Estimates were also obtained by applying the Antonov-Romanovskii formula [108] to the rising branch of this thermoluminescence peak and using the Lushchik formula [150], which allowed for the peak position and half-width.

Since the light sum stored at this level increases by a factor of ~ 10 on introduction of excess Zn, it is natural to assume that excess Zn, whose presence is indicated by the ESR measurements, participates in the centers responsible for this level. An investigation of cubic ZnS single crystals with different departures from stoichiometry [104] has also led to the conclusion that a low-temperature peak resulting from photoexcitation is due to excess Zn. However, according to the estimates given in [104], the depth of the level ($\Delta E = 0.11$ eV) is considerably less than that obtained in the present investigation or in earlier studies [77]. It is possible that the level observed in different investigations is due to different charge states of the excess Zn. The ESR signal of our samples was observed without a preliminary excitation, whereas in the case of ZnS crystals studied in [104] it was observed only in the presence of a background illumination with $\lambda_{b.i.} = 365$ nm.

There is a relationship between the increase in the stored light sum on departure of ZnS from stoichiometry, observed in the present investigation, and the reduction in the intensity of the edge ultraviolet luminescence observed earlier [2] for the same series of samples on in-

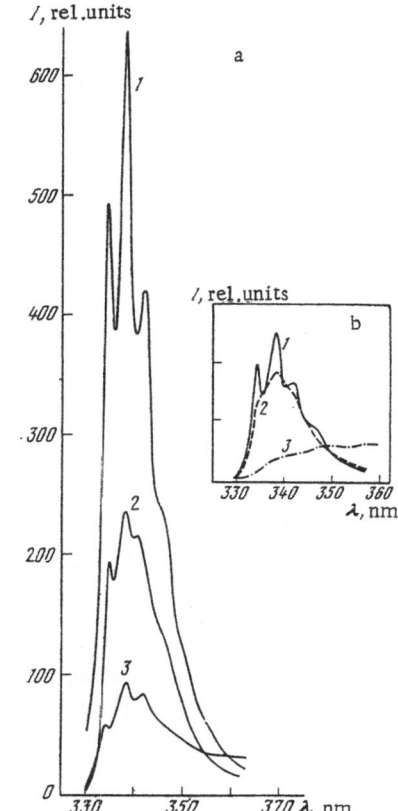

Fig. 29. Edge luminescence spectra of ZnS (1), ZnS:S (2), and ZnS:Zn (3) obtained using exciting electrons (j = 10^{-7} A/cm^2) of different energies: a) 20 keV; b) 6 keV.

troduction of additional Zn or S. This relationship can be explained on the basis of a model in which the thermoluminescence levels are due to single defects and the ultraviolet luminescence centers are due to associated lattice defects of opposite sign [84, 85]. The ratio of the number of the associated defects should be highest also in sotichiometric samples. An increase in the measured light sum on deviation from stoichiometry indicates that introduction of excess Zn or S does not raise significantly the concentration of nonradiative recombination centers.

We shall now consider the dependence of the thermoluminescence curves on the voltage used to accelerate the electrons employed in the excitation stage. Since an increase in the initial energy E_0 of the exciting electrons increases the depth of penetration in accordance with the formula CE_0^n, where $1 < n < 2$, it follows that if the spatial distribution of the trapping levels is homogeneous and there are no secondary effects, the light sum should also increase. This dependence is indeed observed for the peak at $T_{max} \sim 155°K$, which is the main level for the storage of the light sum in the original ZnS phosphor. However, the peak at $T_{max} \sim 125°K$, attributed to the excess Zn, and some of the other thermoluminescence peaks exhibit an anomalous dependence on the initial electron energy. The light sum stored at the level corresponding to the $\sim 125°K$ peak in ZnS:Zn hardly changes when the exciting electron energy E_0 is increased from 6 to 20 keV and in the original sample the $\sim 125°K$ peak is clearly visible only at low energies E_0 when the second peak at $\sim 155°K$ is weak. Since the energy absorbed in the surface layers does not increase when E_0 is made larger (in fact, it decreases somewhat [112]), these results can be explained by a surface distribution of the excess Zn. Then, an increase of the initial electron energy from 6 to 20 keV corresponds to an increase in the depth of penetration of electrons by about an order of magnitude.[*] Using the appropriate data on the penetration of electrons into ZnS, we find that the depth of diffusion in our experiments is $\sim 0.1-0.2 \mu$.

---

[*]The maximum depth of penetration of $E_0 \sim 20$ keV electrons is $\sim 1-1.5 \mu$, which corresponds approximately to the depth of the optical excitation with $\lambda_{exc} = 313$ nm and is considerably less than the dimensions of the individual crystallites in ZnS ($\sim 20 \mu$).

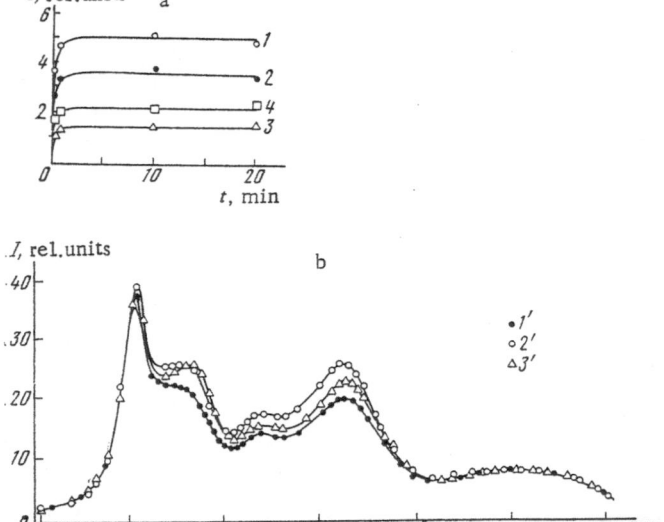

Fig. 30. Dependences of the light sums stored in different (1-4) thermoluminescence peaks: a) on the duration of excitation, $V = 10$ keV, $j = 2 \times 10^{-7}$ A/cm^2; b) on the current density, $V = 10$ keV, $t_{exc} = 20$ min (1': 2×10^{-3} A/cm^2; 2': 2×10^{-7} A/cm^2; 3': 2×10^{-6} A/cm^2).

The inhomogeneity of the energy structure of the investigated crystal phosphors with depth is supported also by the dependence of the ultraviolet luminescence spectra on the accelerating voltage applied to the exciting electrons (Fig. 29). This inhomogeneity is manifested most clearly by the ZnS:Zn phosphor whose cathodoluminescence spectrum excited with 6 keV electrons differs considerably from the usual edge luminescence spectrum.

Thus, investigations of the thermoluminescence spectra obtained after excitation with electrons of different initial energies can be used in qualitative and even quantitative (calculations of the dependence of the energy absorption on the depth) analyses of the diffusion of various impurities and intrinsic lattice defects in ZnS phosphors. In such studies allowance must be made for the possibility of secondary effects that can alter the shape of thermoluminescence curves when the accelerating voltage of the incident electrons is varied. The most important among them are related, in the final analysis, to the change in the average volume excitation density with the initial electron energy E_0: a) the absence of saturation of the maximum light sum during the excitation period; b) the dependence of the distribution of localized carriers on the excitation density. However, we can see from Fig. 30a that when the excitation period is $t_{exc} \approx 20$ min, the stored light sum reaches its saturation limit (to within an error not exceeding 5%); the changes (increase or decrease) in the relative light sums stored in various thermoluminescence peaks with changes in the electron-beam current density by two orders of magnitude do not exceed 10% (Fig. 30b). Therefore, these factors do not upset our conclusion that the trapping levels attributed to the excess Zn in the investigated ZnS phosphors are located near the surface.

§3. Influence of Quenching on the Efficiency of Edge Luminescence of ZnS Single Crystals

In comparing the edge luminescence of various samples of ZnS it is found (§3, Chap. III) that in the case of single crystals which — in principle — should have fewer defects the efficiency of the edge luminescence is usually less than that of polycrystalline samples (powders) and the difference is over two orders of magnitude. Therefore, it became necessary to carry out a detailed comparison of the luminescence emitted by the original charge and that emitted by single crystals prepared from it. This was needed in order to determine the cause of such very large difference between the intensities of the luminescence emitted by these forms of ZnS and to find the conditions for synthesis of ZnS single crystals which would favor a higher efficiency of the ultraviolet luminescence [151].

The difference between the efficiencies of the edge luminescence of these very different samples could have been due to differences between the conditions in the excitation and extraction of luminescence or due to differences in the structure of the samples, i.e., due to different ratios of the radiative and nonradiative recombination centers. In order to determine the possibility of the electron-beam heating of a ZnS single crystal, which was considerably thicker ($\sim$10-20μ), a layer of the original powder charge was deposited on the surface of a single crystal. However, even under these conditions the edge luminescence spectrum of the original polycrystalline charge had a structure typical of the hexagonal powder of ZnS and a good resolution (Fig. 31), which was only slightly inferior to the resolution obtained at 77°K for screens deposited directly on metal substrates and much better than the resolution obtained for such screens at 100°K. Consequently, the temperature drop across the powder layer was less than 20 deg K. Therefore, the lower efficiency of the luminescence emitted by single crystals could not be due to the electron-beam heating.

The absolute intensities of the edge luminescence emitted by a powder deposited directly on a single crystal or a metal substrate differed by a factor not exceeding 2-2.5. This difference could be due to the different conditions for the emergence of the luminescence from these objects (this could account for the factor of 1.5-2) and due to the possible charging of single crystals under electron excitation conditions. It had been reported in [121] that the charging of ruby single crystals may reduce the luminescence efficiency by a factor of up to 10; however, when the angle of incidence of electrons relative to the normal was increased, the charging became

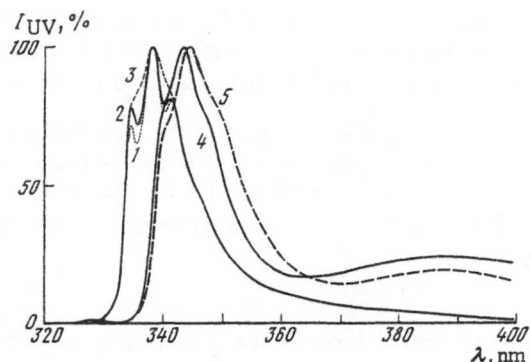

Fig. 31. Edge luminescence spectra of unactivated zinc sulfide recorded at 77°K: 1) polycrystalline charge (heated at T = 1200°C) deposited on a metal substrate; 2) the same charge deposited on a single crystal; 3) same charge deposited on a metal substrate studied at 100°K; 4) rapidly cooled single crystal prepared from a charge heated at T = 1200°C; 5) polycrystalline charge heated at T = 900°C on a metal substrate.

much weaker because the secondary emission became much stronger. The use of an appropriate excitation method confirmed the possibility of some charging of the surface of ZnS crystals by an electron beam but such charging reduced the luminescence efficiency by a factor not exceeding ~1.5. Thus, the results of our experiments indicated that the difference by a factor of ≥ 100 between the efficiencies of the ultraviolet luminescence emitted by polycrystalline and single-crystal samples could not be explained by the difference between the forms of the objects or the difference between the excitation conditions. The change occurred during the actual recombination stage of energy transfer and it could be due to a lower probability of radiative transitions in single-crystal samples.

According to our investigations [2, 80, 84] and the information on the edge luminescence of CdS [73, 75] and of GaP [87], the most likely origin of the edge luminescence is the radiative recombination at intrinsic lattice defects so that in order to increase the efficiency of ultraviolet luminescence we must raise the number of centers which produce this luminescence.

The number of lattice defects rises exponentially with the thermodynamic equilibrium temperature (§1, Chap. IV). Consequently, we shall attribute the difference between the efficiencies of the ultraviolet edge luminescence of powders (~10% [84]) and single crystals ($\leq 0.1\%$ [137]) primarily to the lower number of the luminescence centers in ZnS single crystals, which is due to their slower cooling and the correspondingly lower temperature of thermodynamic equilibrium. Therefore, we may expect a considerable increase in the edge luminescence efficiency as a result of quenching (rapid cooling) of single crystals. Conversely, annealing should reduce strongly the edge luminescence [152]. Since ZnS crystals prepared by the usual method (Chap. II) were cooled in a period of ~12 h and the original charge was cooled in ~30 min, we could assume that the thermodynamic equilibrium temperature of single crystals was considerably higher (cubic modification of ZnS) than in the case of powders (hexagonal ZnS). In order to freeze-in the high-temperature chemical equilibrium in grown ZnS crystals, we subjected them to rapid cooling (~15 min).

The results of measurements of the intensity of the ultraviolet and visible luminescence bands of different types of sample are presented in Table 10. The 100% intensity was the maximum intensity of the ultraviolet luminescence emitted by the original charge (heated to a temperature T = 1200°C). The proportion of the recombination events occurring at the edge luminescence centers was calculated bearing in mind the differences between the photon energies of the ultraviolet luminescence (~345 nm) and visible luminescence (~460 nm).

According to the investigations reported in [137], the intensity of the ultraviolet luminescence of single crystals prepared from a particular charge (heated to T = 1100°C) was only 0.2-0.3% of the intensity of the luminescence emitted by the original charge. The difference between the intensities of the visible bands was much less and, moreover, the intensities were higher in the case of single crystals.

TABLE 10. Comparison of the Intensities of the Edge and Visible Luminescence Emitted by Different Forms of ZnS

Sample	I_{UV}, %	I_{vis}, %	Proportion of edge-luminescence recombination, %
Charge heated at T = 1100°C	80	3.8	94
Single crystal	0.22	9.2	~2
Quenched single crystal	2.2	3.8	31
Charge heated at T = 1200°C	100	4.2	95
Single crystal	3.4	21	11
Quenched single crystal	9.8	3.6	68

Thus, the proportion of the recombination events occurring at the ultraviolet luminescence centers, relative to the total number of radiative recombination events, was 95% in the case of the original charge and 2% in the case of single crystals.

As expected, a rapid cooling of single crystals increased the ultraviolet luminescence intensity by about an order of magnitude and at the same time weakened the visible bands by a factor of about 3. Thus, it was possible to increase considerably the proportion of the recombination events occurring at the edge luminescence centers, which then reached ~30% of the total number of radiative recombination events. Similar results were also obtained for single crystals prepared from a charge which was heated to T = 1200°C. Once again, rapid cooling increased the ultraviolet luminescence intensity and reduced the intensity of the visible bands. However, in the former case the change was dominated by the increase in the ultraviolet luminescence intensity, whereas in the latter case the reduction in the intensity of the visible bands was an important effect.

In this way it was possible to increase considerably the intensity of the ultraviolet luminescence and the corresponding proportion of the recombination events occurring at the centers which, according to our hypothesis, were complexes or associates formed from lattice defects (V_{Zn}, V_S).

Rapid cooling produced not only quantitative but also qualitative changes in the nature of the radiative transitions in single crystals. In the case of the slowly cooled single crystals the recombination occurred mainly at the visible luminescence centers, whereas in the rapidly cooled ZnS single crystals the radiative recombination events occurred mainly (to the extent of ~70%) at the edge luminescence centers.

The relative changes in the intensities of the visible and ultraviolet luminescence bands due to rapid cooling are illustrated clearly in Fig. 32.

It should be pointed out that the ultraviolet part of the spectrum includes not only the edge luminescence and the much weaker short-wavelength luminescence, but also a wide band in the region of 360–400 nm, which in some cases is comparable with the intensity of the edge luminescence. Rapid cooling had little effect on the absolute intensity of this band but when the

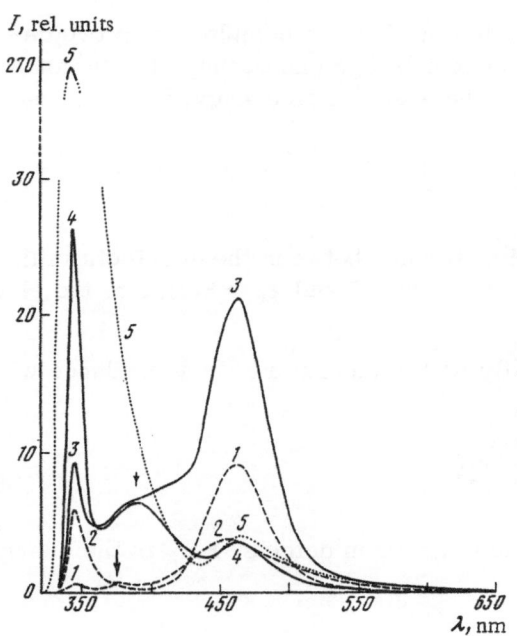

Fig. 32. Changes in the ratio of the intensities of the visible and ultraviolet luminescence bands (recorded at 77°K) of unactivated ZnS due to quenching of single crystals and other samples: 1) single crystal prepared from a charge heated at T = 1100°C; 2) rapidly cooled single crystal prepared from the same charge; 3) single crystal prepared from a charge heated at T = 1200°C; 4) rapidly cooled single crystal prepared from the latter charge; 5) polycrystalline charge heated at T = 1200°C.

firing temperature was increased from 1100 to 1200°C, the intensity and position of this band were affected (the maximum shifted from 365 to 390 nm). This band has been observed earlier and is known as the SAL band [41]; it is very likely to be a complex band with several components.

The edge luminescence spectrum of the investigated ZnS crystals corresponds to the cubic phase, in contrast to the original charge which has the hexagonal structure. Some broadening of the short-wavelength part of the spectrum of single crystals as a result of quenching may be due to an increase in the amount of the second (i.e., hexagonal) phase.

An increase in the intensity of the edge luminescence due to the quenching of the samples can be explained by an increase in the number of the recombination centers associated with the intrinsic lattice defects.

It should be noted that if we adopt the opposite interpretation of the edge luminescence as the direct result of interband transitions [5, 143], an increase in the number of defects in a crystal should reduce the edge luminescence intensity.

Naturally, a question arises of the absolute numbers of such defects in zinc sulfide. An estimate of the number of intrinsic defects in ZnS can be obtained by thermodynamic calculations. Such a calculation is given in [153] for ZnS:Cl and CdS:Cl phosphors and the results are used to deduce the degree of association of point defects. Such calculations make it possible to analyze the formation of an interaction between lattice defects due to the dissolution of impurities (for example, $ZnCl_2$ in ZnS) and other processes, to explain the influence of several preparative factors on the luminescence and photoconductivity of crystals, etc. Our calculations of the possible concentration of defects in the crystal lattice of zinc sulfide have made it possible to estimate the number of associated centers (made up from intrinsic oppositely charged defects V_{Zn} and V_S) in ZnS crystals prepared at different temperatures.

We shall carry out calculations in the same way as in [153] using the equation deduced by Lidiard [154, 155]:

$$K_{eq} = g \exp\left(\frac{E_{ass}}{KT}\right). \tag{1}$$

Here, K_{eq} is the equilibrium constant of the reaction which produces a complex from two defects; g is the number of equivalent states which a defect of one sign can occupy at a distance R_0 from a defect of opposite sign (g = 4 for ZnS); E_{ass} is the association energy, i.e., the energy of the Coulomb interaction between two defects:

$$E_{ass} = \frac{Z_1 Z_2 e^2}{\varepsilon R_0}, \tag{2}$$

where Z_1 and Z_2 are the charges of the defects; R_0 is the distance between these defects in the complex; ε is the permittivity (in the case of ZnS we have $\varepsilon \sim 5.07$ and $\varepsilon_0 \sim 8.3$); e is the electronic charge.

The number of lattice defects C rises exponentially with temperature, in accordance with the equation

$$C = \sqrt{N_c N_a} \propto \exp\left(-\frac{E_{def}}{kT}\right), \tag{3}$$

where N_c and N_a are, respectively, the numbers of cation and anion defects; E_{def} is the energy of formation of a defect pair; $E_{def} = (E_a + E_c)/2$.

According to [153], the degree of association α can be determined if we know E_{eq} and the total defect concentration C:

$$K_{eq} = \frac{\alpha C}{[(1-\alpha)C]^2} = \frac{\alpha}{(1-\alpha)^2 C}.$$ (4)

The degree of association is defined by

$$\alpha = \frac{N_{ass}}{C},$$ (5)

where N_{ass} is the concentration of defects associated into complexes.

Then, if we assume that $\alpha \ll 1$, we obtain

$$N_{ass} = \alpha C = K_{eq} C^2.$$ (6)

Substituting Eqs. (1) and (3) into Eq. (6), we obtain the following expression for the concentration of defects associated into complexes:

$$N_{ass} \propto \exp \frac{E_{ass}}{kT} \exp\left(-\frac{2E_{def}}{kT}\right) \propto \exp\left[-\left(\frac{2E_{def} - E_{ass}}{kT}\right)\right].$$

We can see that the concentration of associated defects, i.e., the concentration of complexes, depends strongly on the energy of formation of defects E_{def} and on their association energy E_{def}.

In fact, the number of associated defects rises with temperature if $2E_{def} - E_{ass} > 0$ and it is practically independent of temperature if $2E_{def} - E_{ass} = 0$. If $2E_{def} - E_{ass} < 0$, the concentration of associated defects falls with rising thermodynamic equilibrium temperature.

The energy of formation of defects in zinc sulfide can be estimated from the heat of formation of the ZnS lattice [156, 157]: Q = 48.6 kcal/mole at T $\approx$ 300°K. At T ~ 1500°K allowance for the temperature dependence of the energy gives Q = 40 kcal/mole, i.e., the energy of formation of a pair of defects in zinc sulfide is E_{def} ~ 1.5 eV. This estimate of E_{def} is quite admissible because the heat of formation of the lattice and the forbidden band width are often correlated [158].

An estimate of the association energy E_{ass}, i.e., of the energy of the Coulomb interaction between two oppositely charged lattice defects (V_{Zn} and V_S), in zinc sulfide gives 0.75 eV $\leq$ E_{ass} $\leq$ 1 eV if allowance is made mainly for the interaction with the nearest defects (R_0 of the order of one lattice constant).

The results of calculations are given in Table 11, which shows that the degree of association rises with temperature and with increasing concentration of lattice defects.

TABLE 11. Dependences of the Lattice Defect Concentration, Degree of Association of Defects, and Number of Associated Defects in Zinc Sulfide on Thermodynamic Equilibrium Temperature

T, °K	kT, eV	K_{eq}	C	α	N_{ass}
1200	0.1	10^3–10^4	10^{-6}	10^{-3}	10^{-9}–10^{-8}
1500	0.125	10^3	10^{-5}	10^{-2}	10^{-7}
1800	0.150	10^2–10^3	10^{-4}	10^{-2}–10^{-1}	10^{-6}

The concentration of associated defects at $T \sim 1500°K$ is comparable with the concentration of residual impurities in zinc sulfide (10^{14}-10^{15} cm^{-3}) [159].

An increase in the defect concentration as a result of rapid cooling of ZnS single crystals is manifested by an increase in the light sums stored during the excitation, which may be attributed to an increase in the density of the localization levels formed by the intrinsic lattice defects.

Since the lattice defects may act as the trapping levels, their number in the ZnS lattice can be deduced from the thermoluminescence data, i.e., from the number of photons emitted as a result of thermal de-excitation, if allowance is made for the relationship between the thermoluminescence and steady-state luminescence intensities of a single crystal. The maximum number of photons emitted in the visible part of the spectrum ($\lambda_{max} \sim 460$ nm) in the course of thermal de-excitation shows that the concentration of the intrinsic defects in quenched ZnS is at least 10^{16}-10^{17} cm^{-3}.

An estimate of the minimum density of the localization levels deduced from the thermoluminescence curves is in qualitative agreement with the thermodynamic estimates of the numbers of defects in ZnS and exceeds considerably the concentration of accidental impurities in ZnS.

The light sums are stored throughout the thickness of a crystal (~ 1 mm) and not only in a layer equal to the depth of penetration ($\sim 1\mu$) of the exciting 20 keV electrons. Consequently, the light sums de-excited from ZnS single crystals are greater than those de-excited from polycrystalline screens by a factor of about 150-200.

Special experiments[*] indicated that the excitation of the whole thickness of a crystal was due to the reabsorption of the intrinsic ultraviolet luminescence, which should be taken into account in studies of the thermal de-excitation after electron-beam or optical excitation of the edge luminescence centers.

Thus, studies of the thermal de-excitation (thermoluminescence) indicated that the edge luminescence increases in strength with the number of defects, which confirms its relationship to the intrinsic lattice defects.

The rapid cooling experiments indicated that the difference between the efficiencies of the edge luminescence of ZnS single crystals and powders can be reduced to a factor of 10 (and if the conditions for the emergence of the ultraviolet luminescence are allowed for, this factor can be lowered to ~ 2-3).

A further increase in the efficiency requires the preparation of ZnS single crystals with the correct proportions of zinc and sulfur, because a deviation from the stoichiometry can reduce the efficiency of the ultraviolet edge luminescence by an order of magnitude [80].

§4. Quenching of the Edge Luminescence of ZnS as a Result of Introduction of Luminescence and Quenching Centers

As shown in the preceding sections, the ultraviolet edge luminescence due to the cation-anion complexes of lattice defects is the main radiative recombination channel in unactivated zinc sulfide and the efficiency of this luminescence is comparable with the efficiency of the impurity luminescence centers.

---

[*]A single crystal of ZnS was covered by a thick dielectric (sapphire) film, which was transparent in the ultraviolet region, and this film was coated with a thin layer of the ZnS powder. The powder was known to exhibit a strong edge luminescence and its visible thermoluminescence was negligible compared with the thermoluminescence of the single crystal itself.

Therefore, it would be interesting to compare the development of the luminescence of various activators with the quenching of the edge luminescence in a series of phosphors with different activator concentrations. Such an investigation was carried out at 77°K using 15 keV electrons ($j = 5 \times 10^{-7}$ A/cm^2), i.e., the excitation was applied in the region where the ultraviolet luminescence efficiency was independent of the excitation density.

We started with hexagonal zinc sulfide powders prepared using a flux (MgCl$_2$), as usually employed in the preparation of high-efficiency activated ZnS phosphors. The activators were typical representatives of the heavy metals (Ag) and of the rare-earth elements (Sm).

The investigation carried out demonstrated that the intensities of the edge luminescence of unactivated ZnS prepared without and with the MgCl$_2$ flux were practically identical (to within 10-20%). This was evidence of a considerable difference between the influence of the MgCl$_2$ and HCl fluxes, i.e., the results were different when Cl$^-$ was introduced simultaneously with the Mg$^+$ cation and when Cl$^-$ was introduced as a result of heating in the HCl atmosphere, because the latter treatment quenched strongly the edge luminescence of ZnS. The difference between the actions of HCl and MgCl$_2$ on the development of the blue luminescence and the activator luminescence was known; the different action of these fluxes on the edge luminescence of ZnS was discovered for the first time. However, whereas small amounts of Cl$^-$ impaired only slightly the fine structure of the edge luminescence [160], in the case of samples containing MgCl$_2$ the structure disappeared practically completely but the position of the maximum and the general shape of the luminescence band were retained (the same nature of centers). Thus, the influence of the simultaneous introduction of anions and cations on the edge luminescence was quite different from the influence of anions alone. These experimental results could be explained using the donor-acceptor pair model and assuming that in the case of introduction of the anion alone the number of the edge luminescence centers ($V_{Zn}-V_S$ complexes) decreased, whereas in the case of simultaneous introduction of the cation and anion the number of these centers remained constant but they were disturbed by the flux (in other words, the lattice was deformed and, consequently, the frequency of longitudinal optical phonons was no longer equal to the frequency which governed the structure of the edge luminescence spectra in the unperturbed lattice). It should be noted that a similar noncorrespondence between the changes in the luminescence intensity and the structure of the spectra was observed when ZnS departed from stoichiometry (see §1, Chap. IV) as a result of introduction of different amounts of zinc and sulfur.

Activation with Ag and Sm quenched the edge luminescence but the luminescence was still observed right up to the optimal amounts of the activator corresponding to the maximum activator luminescence intensity. The nature of the influence of the ions was quite different, as indicated, for example, by the different nature of the concentration curves which were much more complex in the case of samarium (Fig. 33).

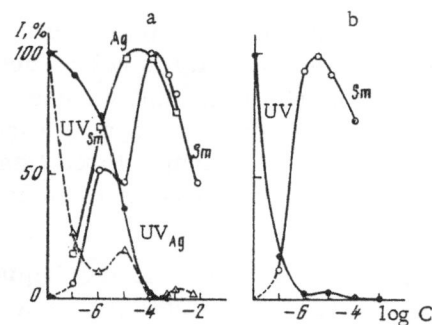

Fig. 33. Intensities of the ultraviolet edge luminescence and of the luminescence of Ag and Sm in ZnS:Ag:MgCl$_2$ and ZnS:Sm:MgCl$_2$ (a), and in ZnS:Sm without a flux (b) plotted as a function of the concentration C of the Ag and Sm activators.

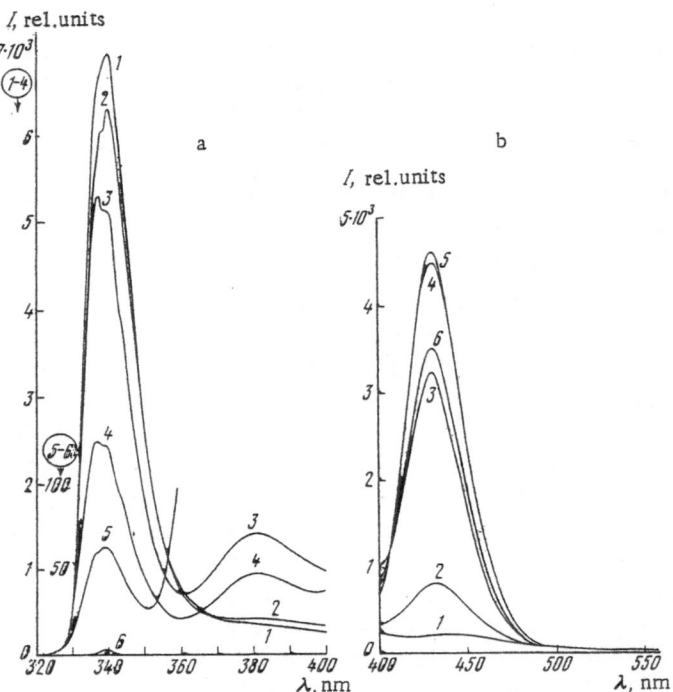

Fig. 34. Edge (a) and silver (λ_{max} = 430 nm) (b) luminescence spectra of ZnS:Ag:MgCl$_2$ phosphors with different activator concentrations, recorded at 77°K using 15 keV electron excitation (j = 5 × 10^{-7} A/cm^2): 1) Ag-free; 2) 10^{-7} g/g; 3) 10^{-6} g/g; 4) 10^{-5} g/g; 6) 5 × 10^{-4} g/g.

We shall first consider the simpler results obtained for the ZnS : MgCl$_2$ phosphors* in a wide range of concentrations of silver (from 10^{-7} to 5 × 10^{-4} g/g), plotted in Fig. 34. An examination of the spectral and concentration curves (Figs. 33 and 34a) shows that introduction of silver reduces monotonically the efficiency of the ultraviolet edge luminescence and increases the intensity of the blue Ag luminescence (λ_{max} = 430 nm) (Fig. 34b) and of the long-wavelength ultraviolet luminescence ($\lambda_{max} \sim 380$ nm), attributed to different types of silver centers (Ag-Cl, Ag-V [161-163]. The absolute efficiencies of such phosphors are listed in Table 12, which shows that the latter two types of luminescence are of comparable efficiency and, as the concentration of Ag increases, the intrinsic luminescence (338 nm) is gradually replaced with the impurity luminescence bands.

The first portions of silver (10^{-7}-10^{-6} g/g) have a weak quenching influence on the edge luminescence (the intensity is not reduced by more than 25%) and result in a strong development of the impurity (Ag) luminescence ($\sim$ 75% of the eventual maximum). The intensity of the total luminescence increases by a factor of 1.5 and it is optimal for C$_{Ag}$ = 10^{-6} g/g, which corresponds to the maximum stored light sum (represented by the thermoluminescence peak at T$_{max} \sim$ 160°K, corresponding to an energy level at a depth of ΔE $\sim$ 0.33 eV) [164]. The

*The concentration of silver in the original phosphor was not determined but the curves obtained indicated that it was $\leq$ 10^{-8} g/g, in agreement with the analyses of the purity of the ZnS charge [159].

TABLE 12. Dependences of the Absolute Efficiencies of Various Luminescence Bands of ZnS:Ag:MgCl$_2$ Phosphors on the Activator Concentration

Sample No.	C_{Ag}, g/g	n, cm^{-3}	Quantum efficiency, %				I_{UV}/I_Σ
			λ_{max}			Intensity of total luminescence I_Σ, %	
			338 nm	380 nm	430 nm		
1	10^{-8}	10^{14}	11	0.7	1.2	13	0.9
2	10^{-7}	10^{15}	10	1	2.6	14	0.7
3	10^{-6}	10^{16}	8.2	3	8.5	20	0.4
4	10^{-5}	10^{17}	4	2	12	18	0.22
5	10^{-4}	10^{18}	2	1.8	12	16	0.15
6	$5 \cdot 10^{-4}$	$5 \cdot 10^{18}$	0.1	1.5	9	11	0.01

highest rate of quenching of the edge luminescence is observed for $C_{Ag} \sim 10^{-5}$ g/g, i.e., when the impurity (Ag) luminescence is strongest. In this range of silver concentrations the reduction in the absolute quantum efficiency of the edge luminescence is approximately equal to the absolute increase in the quantum efficiency of the impurity luminescence.

A further increase in the amount of silver results in a strong quenching of the ultraviolet luminescence and a fall of the intensity of the total luminescence; in fact, the efficiency of the visible luminescence decreases at $C_{Ag} \approx 5 \times 10^{-4}$ g/g, i.e., quenching centers are formed and these may be large complexes of silver ions (or even precipitates of AgCl).

Quenching of the edge luminescence by the silver centers can be used in estimating the number of the edge luminescence centers. Since it is usually assumed that silver centers act as donor–acceptor pairs [98], i.e., they are analogous to the edge luminescence centers, the cross sections of the silver centers should not be very different from the cross sections of the edge luminescence centers. Hence, we may assume that the approximate equality of the efficiencies of the two types of luminescence observed at $C_{Ag} \sim 10^{-6}$ g/g is an indication that the numbers of these two types of center are comparable and we can thus estimate roughly the concentration of donor–acceptor pairs (C $\sim 10^{16}$ pairs/cm^3).

However, it follows from the calculations of the luminescence kinetics of two types of center, carried out by Fok [107], that under conditions of equilibrium occupancies of these centers the ratio of their efficiencies is given by

$$\frac{\nu_{Ag}\beta_{Ag}}{\nu_{UV}\beta_{UV}} \exp\left(\frac{\Delta E_{Ag} - \Delta E_{UV}}{kT}\right),$$

where ν_{Ag} and ν_{UV} are the concentrations of the relevant centers; β_{Ag} and β_{UV} are the respective radiative recombination probabilities; ΔE_{Ag} and ΔE_{UV} are the depths of the levels of the silver and ultraviolet luminescence centers.

The exponential factor in the above expression is very large ($\sim 10^3$-10^4) at 77°K because the difference between the depths of the levels is ~ 0.1 eV ($\Delta E_{Ag} \approx 0.2$-$0.25$ eV, $\Delta E_{UV} \approx 0.12 \pm 0.02$ eV).

It follows from these considerations that the number of the edge luminescence centers is $\geq 10^{19}$ cm^{-3}, which is greater than the concentration of defects in stoichiometric phosphors.

This contradiction can be removed if we consider an energy (and not a geometric) model of a phosphor, i.e., if we use the diffusion theory of luminescence put forward by Antonov-

Romanovskii [108]. In the case of a random (independent) distribution of two types of center, their effective interaction begins only at concentrations such that the distance between the individual centers becomes comparable with the diffusion length of free carriers. At lower concentrations the luminescence emitted by the two types of center develops almost independently. It should be noted that the mean free path of carriers in ZnS at 293°K is ~120 Å, whereas at 77°K it is slightly longer [30].

This explanation is in good agreement with the strong initial growth of the luminescence efficiency of the silver centers in the range $C_{Ag} \leq 10^{-5}$ g/g, which affects only slightly the ultraviolet luminescence efficiency (in this range the conversion coefficient is $\eta_{Ag}/\eta_{UF} > 2$).

The order of magnitude of C_{Ag} at which the efficiencies of the two types of luminescence are close ($\eta_{Ag}/\eta_{UF} \sim 1$) is in agreement with the value deduced from the diffusion length of carriers. However, the absence of any interaction in the Ag and edge luminescence centers (or its extreme weakness), requires that the distance between the edge luminescence centers be greater than the mean free path of free carriers.

On the other hand, since the efficiency of the ultraviolet luminescence is fairly high ($\geq 10\%$), the distance between the edge luminescence centers cannot be much greater than the mean free path. Thus, rough estimates of the numbers of the edge luminescence centers in ZnS give 10^{14} cm^{-3} $\leq C_{UV} \leq 10^{16}$ cm^{-3}.

The quenching of the edge luminescence by the introduction of Sm^{3+} differs strongly from the quenching influence of silver, and the concentration quenching curves with and without a flux are also different (Fig. 33).

A specific difference between silver and samarium is the strong quenching of the edge luminescence (by a factor of 3-4) by small amounts of Sm ($\sim 10^{-7}$ g/g), when the luminescence of Sm^{3+} is still very weak (its efficiency at these concentrations is ~ 0.1 of the optimal value). Another difference is observed in the changes of the half-widths of the edge luminescence spectra resulting from the introduction of these ions: the half-width remains almost constant (it changes from 13 to 14 nm) on introduction of silver, but it increases considerably on introduction of Sm^{3+} (from 10.5 to 17 nm). The half-width varies in a nonmonotonic manner in the same way as the concentration dependences of the efficiencies of the edge and samarium luminescence [164].

These observations can be explained bearing in mind the correlation between the distribution of the Sm^{3+} and the edge luminescence centers, and the formation of several different samarium centers.

The assumption of a close correlation is fully justified because the incorporation of trivalent rare-earth ions in a divalent lattice must involve charge compensation. In particular, at low Sm^{3+} concentrations (10^{-7}-10^{-6} g/g), the charge may be compensated by intrinsic lattice defects which are components of the edge luminescence centers. Thus, introduction of Sm^{3+} may not only generate centers which quench the edge luminescence but also reduce the number of the edge luminescence centers. A quenching effect by a factor of about 3 is observed for $C_{Sm} \sim 10^{-7}$ g/g. This shows that at such concentrations, samarium reduces the number of the edge luminescence centers. The estimate is in qualitative agreement with the estimate given below for ZnS:Ag phosphors.

It is interesting to note that the flux reduces somewhat the Sm^{3+} luminescence efficiency but maintains a fairly high edge luminescence efficiency, i.e., it reduces the interaction between the samarium and edge luminescence centers. The protective role of the flux can be explained by the fact that the compensation of the Sm^{3+} charge can be performed not only by the intrinsic lattice defects but also by the flux ions (obviously, the Mg^+ ions and not Cl^-, by analogy with the compensation in divalent laser crystals of the $CaWO_4$ type, which results from the introduc-

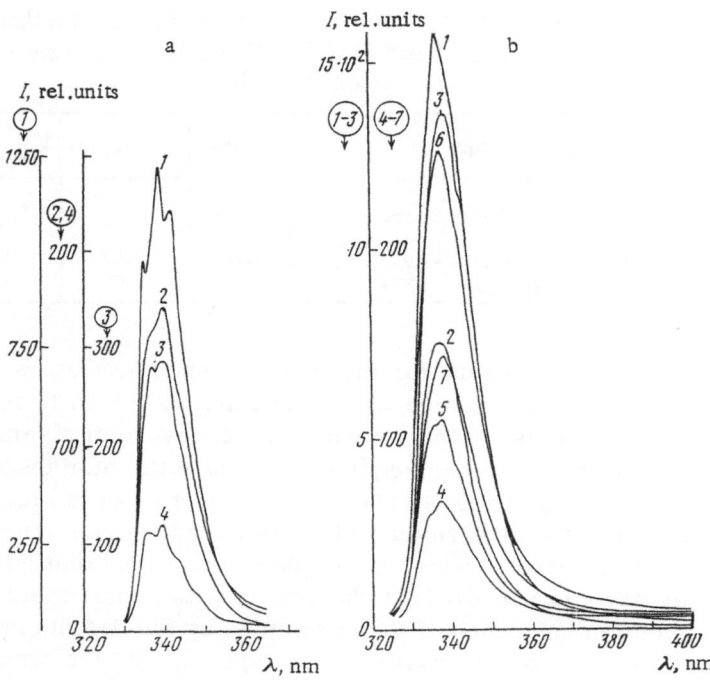

Fig. 35. Edge luminescence spectra of ZnS:Sm phosphors with different activator concentrations, recorded at 77°K using 15 keV electron excitation ($j = 5 \times 10^{-7}$ A/cm^2). a) Samples prepared without a flux: 1) 10^{-7} g/g; 2) 10^{-6} g/g; 3) 10^{-5} g/g; 4) 10^{-4} g/g. b) Samples prepared with MgCl$_2$ flux: 1) 10^{-7} g/g; 2) 10^{-6} g/g; 3) 10^{-5} g/g; 4) 10^{-4} g/g; 5) 5×10^{-4} g/g; 6) 10^{-3} g/g; 7) 5×10^{-3} g/g.

tion of Na$^+$). A somewhat different fine structure of the Sm^{3+} luminescence bands is observed in the presence of the flux (Fig. 35b). It is important to note also another feature of the quenching of the edge luminescence by the ions of silver and samarium present in very high concentrations (C > 10^{-4} g/g): the concentration quenching of the Ag luminescence is observed and the quenching of the ultraviolet edge luminescence becomes even greater; the concentration quenching by samarium with the flux may even enhance the ultraviolet luminescence. The phosphors with $C_{Sm} > 10^{-4}$ g/g without the flux exhibit a reduction in the half-width of the edge luminescence spectrum and a clearly resolved fine structure of the spectrum (Fig. 35). At these concentrations the perturbing effect of Sm^{3+} on the edge luminescence becomes somewhat weaker because the Sm^{3+} ions can now form complex centers (two Sm^{3+} ions with three Zn^{2+} ions) or they can form a complex ZnS · SmS lattice, which is reflected by a shift of the edge luminescence spectrum toward longer wavelengths.

Thus, our investigations have confirmed that the efficiency of the low-temperature edge luminescence of ZnS is comparable with the efficiency of the impurity luminescence of such typical activators as the heavy metals (Ag) and rare-earth elements (Sm), and this should be taken into account in an analysis of the total luminescence efficiency of a phosphor.

On the other hand, the interaction between the edge luminescence centers and impurities depends strongly on the nature of the impurities.

It should also be pointed out that the edge luminescence of ZnS exhibits an irreversible quenching after prolonged electron bombardment, which indicates formation of the quenching centers.

TABLE 13. Reduction in the Intensity of the Edge and Visible Luminescence of Unactivated Zinc Sulfide after Electron Irradiation
$(Q = 0.05 \ C/cm^2)$

Sample No.	Form of ZnS	I_{UV}	$I_{UV.irr}/I_{UV}$	$I_{vis.irr}/I_{vis}$
1	Single crystal prepared from charge heated at T = 1100°C	0.002	0.17	0.31
2	Quenched single crystal (T = 1200°C)	0.10	0.40	0.35
3	Polycrystalline charge (T = 1200°C)	1.00	0.65	0.34

The reduction in the efficiency of the luminescence of such activators as Ag or Cu is
usually attributed to the damage of the lattice itself (knocking out of Zn from the regular sites),
which produces nonradiative recombination centers. The exposure of a sample, i.e., the
charge which crosses 1 cm^2 of the surface, required to reduce the luminescence brightness
by a factor of 2 should be 10-20 C/cm^2 [165, 166]. The reduction of the luminescence efficiency of unactivated zinc sulfide in the ultraviolet and visible bands begins much earlier. The
hypothesis of a large number of defects which give rise to the ultraviolet edge luminescence
in quenched ZnS single crystals (§2 in present chapter) is — to some extent — supported by
the weaker quenching of the luminescence of these samples by the defects generated in the lattice as a result of prolonged electron bombardment, compared with the similarly induced quenching in slowly cooled samples. In our experiments (Table 13) an exposure of ∼0.05 C/cm^2 (V =
15 keV, j = 10^{-5} A/cm^2, t = 60 min) reduced the intensity of the visible luminescence bands by
a factor of 3, irrespective of the nature of the sample. The intensity of the ultraviolet edge
luminescence was reduced most (by a factor of 6) in the case of ZnS crystals prepared from
a charge heated at T = 1100°C; the least effect (by a factor of 2.5) was observed for quenched
ZnS crystals [151] and the reduction was only by a factor of 1.5 in the case of a strongly luminescing
polycrystalline charge. The simplest explanation of these results is the hypothesis of the more
successful competition between the ultraviolet luminescence centers and the newly formed
quenching centers in those cases when the number of the ultraviolet luminescence centers is
large and the intensity of the luminescence emitted by these centers is high: this is true of
the quenched crystals and polycrystalline charge (heated at T = 1200°C) and less true of the
slowly cooled ZnS single crystals.

CHAPTER V

NATURE OF ULTRAVIOLET LUMINESCENCE OF UNACTIVATED ZINC SULFIDE

It is difficult to interpret the nature of the ultraviolet luminescence of unactivated ZnS
observed near the fundamental absorption edge and to compare the experimental results with
the various theoretical models.

The experimental results reported above and those which appeared in the last few years
indicate that the various bands observed near the fundamental absorption edge are typical not
only of the lattice of unactivated zinc sulfide but also of all the II-VI compounds and of the lattices of various other binary compounds (GaP, SrS, etc.).

Extensive but very varied and difficult to compare experimental material has been accumulated because many investigations have not been sufficiently extensive in respect of the
characteristics that were determined (the luminescence spectra were usually obtained at temperatures T ≤ 77°K) and in respect of the nature of the objects (polycrystalline samples in
some cases, single crystals in others), and particularly in respect of the excitation conditions

(photoluminescence or cathodoluminescence or electroluminescence have been usually determined separately).

The investigations reported above were carried out under a great variety of the excitation conditions (E_{exc} was varied by six orders of magnitude; $17°K \leq T \leq 600°K$) and of the synthesis or growth conditions (powders and single crystals of zinc sulfide; influence of rapid cooling, deviations from stoichiometry, and various activators). This produced a wide range of luminescence efficiencies (from 10 to $\geq 0.01\%$) and luminescence spectra (including luminescences with different properties in the same spectral range but at different temperatures) and provided the basis for a considerable progress in the understanding of the ultraviolet luminescence of zinc sulfide.

Since these changes enabled us to alter strongly the ratio of the probabilities of the various transitions and to reproduce the spectra observed by other workers, we were thus able to utilize more fully the published data.

Moreover, the kinetic properties of ZnS were found to be very close to the kinetics of the corresponding luminescence of other compounds and particularly of CdS, which is the closest analog of zinc sulfide in the general family of compounds with similar types of edge luminescence.

§1.   Various Types of Radiative Transitions

It is no longer possible to regard the ultraviolet luminescence of ZnS as a band with components of the same origin, accompanied by a series of different but similar radiative transitions [5, 143]. Moreover, the same radiative transition mechanisms can give rise to photons of different energies in the case of the cubic and hexagonal lattices, and the photon energy may decrease considerably when the temperature is raised.

Our studies of the cathodoluminescence of unactivated zinc sulfide carried out under the same excitation conditions, on powders cooled rapidly to temperatures below and above the phase transition and on ZnS single crystals, have shown that the luminescence of the hexagonal and cubic forms of ZnS is separated by $\sim 0.07\,eV$. This shift is in close agreement with the data on the absorption in the hexagonal and cubic phases of ZnS [167-169], as well as with the theoretical estimates and experimental data on the forbidden band widths of these two phases of ZnS [4, 23, 24, 170].

It should be stressed that the identity of the ultraviolet luminescence spectra of the cubic powders and single crystals has shown that unactivated ZnS crystals prepared at the Lebedev Physics Institute as well as crystals employed in earlier investigations [5, 143] are not hexagonal — as suggested in [6, 143] on the basis of their preparation conditions (firing at temperatures $T \geq 1400°K$) — but cubic. This has been confirmed later [171] on the basis of an investigation of the reflection spectra and qualitatively on the basis of birefringence studies (these have indicated that the admixture of the hexagonal phase is $<10\%$). In the case of comparable ratios of the cubic and hexagonal phases one may observe intermediate edge luminescence spectra (§3 in Chap. III).

All the samples investigated in the present study exhibit (allowing for their phase composition and measurement temperature) three main ultraviolet luminescence bands (Fig. 36).

In the case of hexagonal ZnS we find that at $T = 77°K$ the spectrum has the components listed below.

1. A narrow luminescence band at the shortest wavelengths with λ_{max} is located at 320-322 nm, i.e., it adjoins the reflection maximum (under the same conditions this maximum is located at ~ 320.8 nm). At $T < 77°K$ this band sometimes splits into several narrow sub-bands [89, 140].

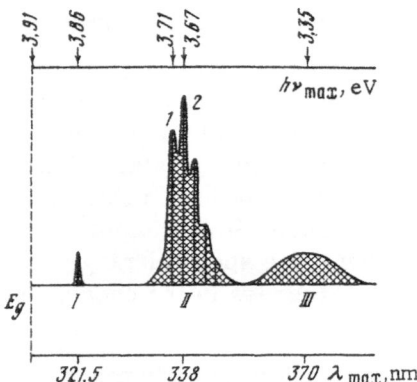

Fig. 36. Schematic representation of the three
ultraviolet luminescence bands of unactivated
hexagonal zinc sulfide observed at 77°K: I) short-
wavelength luminescence, II) edge luminescence
(330-360 nm), consisting of two series of bands
(1 and 2 denote the head lines of the short- and
long-wavelength series); III) wide structure-
free luminescence band.

2. A structured luminescence band in the region of 330–360 nm is called, after Kröger
[1], the edge luminescence band because at the time of its discovery bands with shorter wavelengths were not known and accurate information on the optical width of the forbidden band of
ZnS was not available (the data on the absorption edge suggested that the forbidden band width
was ~ 3.7 eV).

This structured band consists of two superimposed series with head lines at λ_{max} = 334.5
nm and λ_{max} = 337 nm and of their phonon replicas. It is best to call this band the pseudo-edge
luminescence. However, in order to avoid confusion, we shall follow the old terminology adopted by other workers.

3. A wider structure-free luminescence band with a flat maximum is observed in the
range 360–390 nm. This shows that the band is not simple. This band is far removed from
the fundamental absorption edge of ZnS and we shall not consider it in detail.

Numerous studies have shown that the selection of the excitation conditions (particularly
of the temperature and volume excitation rate) makes it possible to enhance selectively one of
the three types of luminescence described above (bearing in mind the temperature and phase
shifts) for any samples of zinc sulfide, even those activated with the heavy metals (Ag) or rare-
earth elements (Sm).

In particular, measurements of the luminescence spectra of different types of sample
at temperatures gradually increasing from 77 to 293°K have demonstrated a considerable difference between the effects of temperature on the edge and short-wavelength luminescence as
well as the possibility of transition from one type of predominant edge luminescence to another.
The shift of the short-wavelength luminescence maximum is ~ 0.11 eV, i.e., it is close to the
temperature shift of the absorption and reflection spectra (dE/dT $\approx -5 \times 10^{-4}$ eV/deg). The
shift of the edge luminescence is much smaller ($\leq$ 0.03 eV). In the 77-293°K range the short-
wavelength luminescence is quenched much less strongly (by less than an order of magnitude)
than the edge luminescence (by more than three orders of magnitude). At 77°K the intensity
of the short-wavelength luminescence is only slightly greater than 1% of the intensity of the
edge luminescence, whereas at room temperature the former is stronger than the edge lumi-

nescence. Moreover, at room temperature the short-wavelength luminescence is located closer to the edge luminescence because of the temperature shift. If we ignore this important point (in addition to the errors in the determination of the phase composition), we may compare incorrectly [5, 143] the edge luminescence bands of hexagonal powders at 77°K and the short-wavelength luminescence of cubic single crystals of ZnS at 293°K, because of the identical positions of the maxima ($\sim$ 337.5-338 nm).

These points should be remembered in any comparison of the luminescence emitted by different objects (powders, single crystals, and sublimates). At 77°K the edge luminescence usually predominates in the case of polycrystalline samples and the short-wavelength luminescence appears as a step on the high-energy side of the edge luminescence band. However, it has been reported [136] that the edge luminescence is emitted only by polycrystalline objects and not by ZnS single crystals. Our investigation of the influence of temperature (quenching, thermal shift, half-width of the spectrum and its temperature dependence) has shown convincingly that it is incorrect to oppose the properties of powders and single crystals, i.e., it is necessary to indicate the limits of the ratio of the band intensities and the conditions for their excitation because of the completely different influence of such conditions on these bands. This is a direct evidence of the different origin of the edge and short-wavelength luminescence. Our results on the effect of temperature on the ratio of the intensities of the short-wavelength and edge luminescence bands of ZnS are in good agreement with the experimental data [6] on the change in the ratio of the intensities of the two types of luminescence obtained at T $\leq$ 77°K at very high electron-beam excitation densities. In this case the edge luminescence becomes saturated and the short-wavelength luminescence rises superlinearly.

We shall now consider in greater detail the edge and short-wavelength luminescence of ZnS.

§2. Detailed Consideration of the Model of Donor—Acceptor Pairs and Discussion of the Polaron Edge Luminescence Model

The edge luminescence observed under optimal conditions (unactivated ZnS powder with equivalent amounts of zinc and sulfur at T $\leq$ 77°K) is characterized by an efficiency up to 10-15% (deduced from the number of recombination events per number of ionization events). This value is less than an order of magnitude below the efficiency of the best activated ZnS crystal phosphors and it follows that studies of the edge luminescence of ZnS may be of practical importance.

The edge luminescence is the main radiative recombination channel but even so about 80-90% of the energy is lost in nonradiative channels. This is evidence of the presence of nonradiative recombination centers which may be quenching centers formed by residual impurities (Co, Ni, Fe, etc.) or intrinsic defects of the lattice structure (dislocations, single defects, grain boundaries, etc.).

However, ZnS single crystals prepared by sublimation under ordinary conditions in which the concentration of the quenching centers cannot be greater than in the original charge are characterized by a lower luminescence efficiency. There have been reports that the crystals with very low numbers of imperfections do not emit the edge luminescence at all [136]. Moreover, it is known that the slow cooling method [152] which reduces the number of defects also reduces the edge luminescence efficiency. Hence, the lower efficiency of single crystals compared with that of powders can be explained by the smaller number of defects.

The number of lattice defects rises exponentially with increasing temperature in accordance with the formula

$$\sqrt{N_c N_a} = N_0 \exp\left(-\frac{E_{def}}{kT}\right),$$

where T is the temperature at which the thermodynamic equilibrium is frozen (depending on the rate of cooling, this temperature may vary from the heating temperature to the temperature at which defects cease to move).[*]

In fact, an increase in the rate of cooling of ZnS crystals increases the edge luminescence efficiency by one or two orders of magnitude (§3 in Chap. IV). Allowance for the difference between the efficiencies of the ultraviolet luminescence of objects as different as powders and single crystals shows that in the case of edge luminescence this difference does not exceed a factor of 2-3.[†]

Thus, an increase in the concentration of defects resulting from rapid cooling of ZnS crystals and manifested directly by an increase in the stored light sums (i.e., as an increase in the concentration of single defects which act as localization levels) may raise the luminescence efficiency of single crystals to the values obtainable for powders. This is an important confirmation of the association of the edge luminescence with intrinsic lattice defects and is in conflict with the attribution of the edge luminescence of ZnS to the direct interband recombination [5, 143].

Moreover, we can say that these defects are not boundaries separating the cubic and hexagonal phases or grain boundaries, because in these cases the maximum efficiency of the edge luminescence of ZnS would have been obtained for samples with a mixed structure (heated at ~1100°C) and the efficiencies of powders and single crystals should differ considerably.

Moreover, dislocations cannot act as the luminescence centers because the efficiency of the edge luminescence is high and the differences between the maxima of the phonon spectra agree well with the frequency of optical phonons in the unperturbed ZnS lattice, which is unlikely to be true of large-scale distortions of the lattice.

An important class of intrinsic defects, which are discussed in the modern literature on the physical chemistry of solids, are the Schottky defects (empty lattice sites) and the Frenkel defects (excess interstitial ions and atoms). According to [85], the edge luminescence centers in ZnS are interstitial zinc atoms whose existence in ZnS phosphors has been demonstrated on many occasions [168].

However, our study of zinc sulfide samples with excess zinc and sulfur, which increase the number of lattice defects, have indicated that the presence of excess atoms of either of the components reduces the edge luminescence efficiency (§1 in Chap. IV).

An investigation of the spectral composition of the ultraviolet luminescence shows that in spite of the broadening of the individual maxima, their small shift (~1 nm), and changes in the relative intensities of the maxima, the spectrum retains its original form in samples with excess zinc or sulfur, i.e., the luminescence is of the same type as before. These results indicate that the Frenkel and Schottky defects cannot be responsible for the edge luminescence and that, conversely, they form trapping levels and nonradiative recombination centers.

The results of the experiments can be explained using the Williams hypothesis on the associative nature of the edge luminescence centers formed from intrinsic lattice defects of opposite sign (V_{Zn}, V_S).

---

[*]Studies of the preparation of ZnS phosphors at low temperatures suggest that this temperature is 600-700°C.

[†]Since samples with plane-parallel faces emit only a small proportion of the luminescence ($1/\varepsilon^2$), the experimentally observed difference between the ultraviolet luminescence efficiencies of ZnS powders and single crystals, which amounts to a factor of 10, should be reduced approximately by a factor of 5.

The concentration of such centers is governed by the product of the concentrations of single cation and anion defects, which is independent of stoichiometry and — as demonstrated above — is related to the temperature at which thermodynamic equilibrium is frozen in. Therefore, a strong dependence of the edge luminescence efficiency on the excess of zinc cations or sulfur anions is explained not by a reduction in the concentration of the edge luminescence centers but by an increase in the concentration of the nonradiative recombination centers formed from single defects which compete with the ultraviolet luminescence centers. This increase in the quenching centers is also manifested by the quenching of the short-wavelength luminescence. Such quenching is accompanied by some increase in the intensity of the visible bands (which is much weaker than the quenching of the edge luminescence) but this can be explained by an increase in the concentration of the centers in the form of various complex composed of lattice defects and residual impurities in ZnS (the existence of such centers is postulated, for example, in [98]).

The introduction of excess zinc is not equivalent to the introduction of excess sulfur in respect of the quenching of the edge luminescence, as deduced directly from changes in the quenching kinetics, decay, and influence of rapid cooling on the efficiency of the luminescence emitted by ZnS. It should be pointed out that a change in the thermodynamic equilibrium temperature of crystals may result in an increase or a reduction in the concentration of intrinsic defects (§ 3 in Chap. IV). Heating increases considerably the defect concentration but the equilibrium constant K_{eq} falls. A general reduction or increase in the concentration of associated defects (complexes) N_{ass} depends on the ratio of the energy of formation of defects to their binding energy. Since the energy of formation of a pair of defects in ZnS is in any case > 1 eV ($E_{def} \sim 1.5$ eV) and the binding energy of defects does not exceed ~ 0.6-0.8 eV (even for complexes with the shortest internal separations), the degree of association rises with increasing thermodynamic equilibrium temperature and N_{ass} should increase, in agreement with our experiments.

Thus, the experimental results obtained in an investigation of the influence of the preparation technology (deviation from stoichiometry and rate of cooling) on the edge luminescence efficiency and the conclusions that follow are an important argument in support of the Williams associative model of the edge luminescence centers, which has been applied by many workers to ZnS and other binary compounds (CdS, CdTe, GaP) but only on the basis of physical investigations. Therefore, although the associative nature of the edge luminescence centers is accepted by many workers, the actual composition of these centers has not been discussed much [136].

Among the published results the most important evidence in support of the Williams model is the shift of the luminescence spectra in the direction of longer wavelengths during afterglow [76], which is explained — in accordance with the Williams theory [83, 86] — as a manifestation of the variation of the binding energy of donor–acceptor pairs with the internal separation between the donor and acceptor in a pair, because this separation governs the value of λ_{max} and the radiative transition probability.

The most complex and important question in the theory of donor–acceptor pairs is the determination of the absolute number of such pairs and their relative number compared with the total number of defects and residual impurities, as well as the nature of association of such pairs ("close" or "distant" pairs).

Direct estimates of the number of the Schottky defects in ZnS crystal phosphors can, in principle, be deduced from statistical thermodynamics [172]. However, such estimates are not easy to obtain because the temperature of freezing of the thermodynamic equilibrium of a phosphor is difficult to determine and it may be considerably lower than the temperature at which phosphors are prepared (~ 1400-$1500°$K). If we use the experimental observation that the hexagonal phase of ZnS is formed only at T > 1450°K and if we postulate that defects are frozen in at this temperature, we find that in stoichiometry ZnS phosphors the concentration of single defects with the formation energy $E_{def} \sim 1.5$ eV is $C_{def} \sim 10^{16}$-10^{17} cm^{-3}.

Similar estimates have been obtained by other workers [172] from calculations of the type described above and from similar physicochemical calculations. In particular, such defect concentrations are in order-of-magnitude agreement with the results of our study of the thermoluminescence spectra (§3 in Chap. IV) and with estimates based on the optical absorption of unactivated zinc sulfide [173].

There are no published direct estimates of the defect concentration based on absorption but the available information on the absorption coefficient in the near ultraviolet region (k = 10-100 cm^{-1}, depending on the preparation technology) can be used to obtain the lower limit of the concentration of defects in ZnS: since the absorption cross section is $\sigma_{abs} \sim 10^{-15}$cm^2 ($f_{osc} = 1$), it follows that $N_{def} = k/\sigma_{abs} \geq 10^{16}$-$10^{17}$ cm^{-3}. The concentration of associated centers N_{ass} can only be lower than this concentration.

Direct estimates based on the infrared quenching of the edge luminescence of ZnS give $N_{ass} \sim 10^{16}$ cm^{-3}, which is in agreement with the lower limit of the defect concentration (§5 in Chap. III).

The maximum number of the edge luminescence centers can also be estimated from the saturation of the edge luminescence of ZnS under strong electron-beam excitation conditions, employing the approximate Brill formula [126]:

$$\frac{N}{\tau} \simeq \alpha_p \eta,$$

where N is the concentration of the luminescence centers; τ is the excited-state lifetime; α_p is the rate of generation of electron–hole pairs (per 1 cm^3 per 1 sec); η is the luminescence efficiency.

According to our results,* the saturation of the edge luminescence under steady-state electron excitation conditions is not observed to least j $\sim 5 \times 10^{-5}$A/cm^2.

Thus, we find that N $\approx 5 \times 10^{16}$ cm^{-3} if we adopt the values $\tau \sim 10^{-4}$ sec and $\eta \sim 0.1$. According to Nolle et al. [6] who used electron excitation pulses ($\tau_p \sim 3 \times 10^{-7}$ sec), the edge luminescence of ZnS becomes saturated at j $\approx 10^{-1}$ A/cm^2 (this value is obtained for 150 keV electrons which can penetrate to a depth of $\sim 50 \mu$ and for $\eta \sim 0.1$). In this case the concentration of the centers under consideration is N $\sim 10^{17}$ cm^{-3}. A similar or lower (down to $\sim 10^{15}$ cm^{-3}) concentration of the luminescence centers is obtained by considering the quenching of the edge luminescence of ZnS at high Ag and Sm impurity concentrations (§4 in Chap. IV).

Consequently, taking all these approximate estimates together, we can say with assurance that the range of possible values of the concentration of the edge luminescence centers in ZnS is 10^{15}-10^{17} cm^{-3} and the most probable value is $N_{ass} \sim 10^{16}$ cm^{-3}.

Therefore, bearing in mind that the residual impurity concentration is $\sim 10^{14}$-10^{15} cm^{-3} [159], we may assume that the concentration of the edge luminescence centers is relatively high because it is one or two orders of magnitude greater than the concentration of the residual impurities in ZnS. Further reduction in the concentration of residual impurities in ZnS is very desirable because it should make it possible to increase the edge luminescence efficiency.

Thus, a comparison of the concentrations of intrinsic lattice defects and of the residual impurities with the experimental estimates of the edge luminescence centers shows that these centers are associated defects (V_{Zn}–V_S) with a degree of association $\alpha = N_{ass}/C \sim 0.1$.

---

*The deviation of the intensity of the edge luminescence of ZnS from the linear dependence on the excitation rate, observed at j $\approx 10^{-4}$ A/cm^2 for 20 keV electrons, is due to the heating of a sample.

This degree of association is much higher than that deduced from calculations ($\alpha \sim 10^{-2}$-10^{-3}, as given in Table 12 in Chap. IV) based on the hypothesis of the participation of only the closest associates in the edge luminescence in which the components of a center are separated by just a few lattice constants (up to 10 Å).

The rms distance between oppositely charged defects (V_{Zn} and V_S) present in a concentration of $\sim 10^{17}$ cm^{-3} is $\overline{R} = 1/\sqrt[3]{(\pi/2) \times 10^{17}} \sim 200$ Å and the total number of such distant pairs is in agreement with the total concentration of lattice defects.

The luminescence of similar distant "unassociated" pairs in activated Zn phosphors was observed by Riehl [61, 62] at very low temperatures (T $\sim 4°$K).

In this case the luminescence events would be separated by hour-long intervals because of the small overlap of the wave functions of a donor and an acceptor separated by these distances and because of the consequent low radiative tunneling probability. Therefore, we cannot attritute the edge luminescence to such distant donor–acceptor pairs because the probability of the thermal liberation to an allowed band at 77°K is considerably greater than the probability of a radiative transition in such pairs; moreover, the lifetime of the centers deduced in the experiments reported in [76] does not exceed $\sim 10^{-3}$ sec at 4°K.

Thus, in the edge luminescence we are dealing with an intermediate case such as suggested already for CdS, GaP, etc., when the average distance between donor and acceptor defects R that dominate the edge luminescence spectrum should satisfy the condition $d_{latt} \ll \overline{R} \ll R_{rms}$, where d_{latt} is the lattice constant.

This average distance can be made more precise by comparing the Williams theory [83, 86] with the experimentally obtained edge luminescence spectra of ZnS and with the duration of afterglow.

According to the Williams theory, the difference between the values of the association energy and the corresponding difference between the energies of photons emitted by two pairs separated by n and n + 1 lattice sites decreases with increasing values of n. In the case of close associates (n = 1 or 2), the difference between the photon energies is ~ 0.1 eV, whereas for pairs with n > 10 it is only ~ 0.001 eV, i.e., the depths of a donor and an acceptor are almost equal to the depths of unassociated defects.

Since the spectra shift by ~ 0.01 eV during afterglow [76] and this shift corresponds to the width of the phonon replica peaks (~ 0.015 eV at $\sim 17°$K), it follows that the difference between the energies of photons emitted from the active donor–acceptor pairs should lie within these limits.

Applying the Williams theory, we have to assume that the average distance between the pairs which can act as the edge luminescence centers is ~ 6 lattice constants, i.e., this distance is ≥ 20-30 Å. In this case the association energy is ~ 0.1 eV, i.e., it is comparable with the value of kT corresponding to the phosphor preparation temperature (T $\sim 1200°$C), and the Coulomb interaction is important in the formation of donor–acceptor pairs. This distance is not in conflict with the duration of intracenter transitions ($\sim 10^{-13}$ sec), because we can expect a tunneling probability of this order of magnitude (in allowed transitions) if we bear in mind that the internal separation in a pair is somewhat larger than the sum of the Bohr orbit radii.

Calculations of the Bohr orbit radii of donors ($R_e \sim 15$ Å) and acceptors ($R_h \sim 9$ Å) carried out for ZnS using the effective mass theory (§ 1 in Chap. I) and calculations of the Bohr orbit radius of an acceptor ($R_h \sim 11$ Å), based on the value of $\overline{N} = 0.87$ for ZnS [67] and the formula $\overline{N} = (e^2/R_h)(1/h\omega_0)(1/\sqrt{2\pi})(1/n^2)(1/\varepsilon_0)$, show that the maximum separation between a donor and acceptor, defined approximately as the sum of the orbit radii (overlap of the wave functions), is ~ 25 Å. A more accurate estimate of this important parameter would require a detailed analysis of the edge luminescence line profiles and of the excitation conditions.

We shall now consider in greater detail the experimental data on the energy structure of the edge luminescence centers in ZnS and on the kinetics of the action of these centers.

It is clear from the excitation spectrum (§ 1 in Chap. III) that the difference between the maximum in this spectrum and the short-wavelength luminescence maximum (zero-phonon line of the first series) is $\sim 0.20 \pm 0.01$ eV and the position of the maximum in the excitation spectrum is in good agreement with the interband transition edge, i.e., $h\nu_{exc} \approx E_g$.

Thus, under optical and electron excitation conditions the radiative recombination in donor-acceptor pairs occurs after the ionization of the host substance and capture of free carriers by the vacancies V_{Zn} and V_S, which are both in singly charged states and which become neutral as a result of such excitation. The neutrality of the acceptor and donor in the excited state follows from the small difference between the width of the forbidden band and the edge luminescence energy (~ 0.2 eV), because the binding energy of charged defects should be considerably higher. A theoretical analysis of the edge luminescence spectra of zinc sulfide is made with this point in mind in [67-69].

The luminescence spectrum of ZnS at 77°K is a narrow band with a series of equidistant maxima. It should be stressed that the separation (~ 0.035 eV) of the shortest-wavelength (λ_{max} = 334.5 nm) from the next maximum (λ_{max} = 338 nm) is less than the energy of longitudinal optical phonons in the unperturbed ZnS lattice ($h\nu_l$ = 0.043 eV) and the distance between the subsequent maxima.

Investigations of the luminescence spectra at lower temperatures have shown that the head line (334.5 nm) is absent at liquid helium temperature although the resolution of the edge luminescence spectrum improves at this temperature.

A separate study of the temperature dependences of the intensities of the lines with maxima at 334.5 and 338 nm carried out in the 17-77°K range has confirmed that these transitions are of different origin. The intensity of the 338 nm line decreases monotonically whereas that of the 334.5 nm line rises nearly exponentially between 17 and 50°K.

An analysis of the ratio of the intensities of these two lines as a function of temperature shows that this ratio varies with an activation energy of ~ 0.03 eV, which is close to the corresponding difference between the photon energies of these lines. The intensities of the two series associated with these lines become equal at T $\sim 40°$K, i.e., when kT $\sim 0.1E_d$. At 77°K the intensity of the short-wavelength maximum is less than the intensity of the 338 nm maximum.

However, according to the theory given in [68, 69], the first maximum (zero-phonon head line) should be stronger than the other maxima (§ 2 in Chap. I). This suggests the presence of two overlapping series with zero-phonon head lines 1 (334.5 nm) and 2 (337 nm). The partial coincidence of the first phonon replica of the first series with the head line of the second series (this also applies to the higher phonon replicas) is responsible for the broadening of the edge luminescence spectrum at liquid nitrogen temperature.

It is possible that apart from the second edge luminescence series observed at low temperatures (T $\sim$ 17°K), only one line is observed at higher temperatures rather than the first equidistant series with the head line at λ_{max} = 334.5 nm. It is reported in [77] that this line represents the luminescence associated with the absorption rather than emission of a phonon; however, this hypothesis is in conflict with the theoretical calculations of Hopfield [67] (the probability of such a process is far too low).

Irrespective of the nature of this line, it follows from the experimental results that an equidistant series and not a line is observed at higher temperatures because the half-width of the spectrum in which the luminescence of the first series is concentrated does not decrease but remains constant. Deterioration of the fine structure at 77°K in the short- and long-wave-

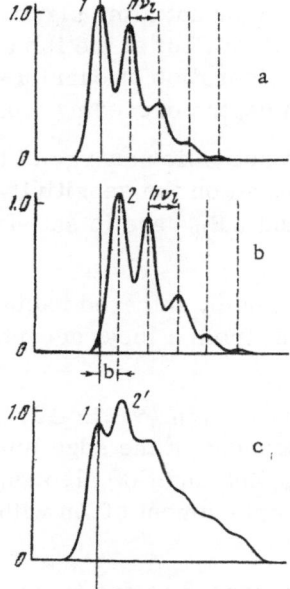

Fig. 37. Fine structure of the phonon replicas of the two edge luminescence series of unactivated zinc sulfide: a) first series (λ_{max} = 334.5 nm); b) second series (λ_{max} = 337 nm) at 17°K; c) first and second series at 77°K.

length parts of the edge luminescence spectra also supports this conclusion. The two luminescence series can be observed separately in the spectra of CdS [75] and CdTe [148], provided the technology used in the preparation of samples is appropriate.

The two edge luminescence series of zinc sulfide and their positions at 77°K are shown schematically in Fig. 37. Apart from the existence of the two series, the edge luminescence spectrum can be complicated further by the coexistence of the two crystal phases of ZnS.

The two edge luminescence series and the redistribution of intensities in these series when the temperature is varied as well as the thermal and infrared quenching of the luminescence are explained by a very simple model suggested by the present author for ZnS [84] and confirmed subsequently for CdS and CdTe (without consideration of the details of the nature of donors and acceptors).

Possible explanations of the origin of the two series based on the donor–acceptor model of the luminescence centers formed from intrinsic lattice defects in ZnS (V_{Zn} and V_S) are given in Fig. 38. The difference between the schemes in Figs. 38a and 38b is important in the case of the short-wavelength series and it results from the difference of the relative depths of the donor and acceptor levels. The evidence supports the scheme in Fig. 38a, i.e., it supports a deeper acceptor level and a shallower donor level. According to this model, we have $E_a > E_d$ (and not the converse inequality as suggested in the Williams theory), which follows from the difference between the effective masses of carriers in ZnS ($m_h^* \sim 3m_e^*$) [31]. This is in agreement with the usual n–type conduction of ZnS.

The deeper level (V_{Zn}^-) is common to the two series because of the approximate identity of the thermal quenching of both series at T > 77°K and because infrared radiation does not al-

Fig. 38. Possible models of a center composed of two intrinsic lattice defects in zinc sulfide, used to explain the existence of two edge luminescence series.

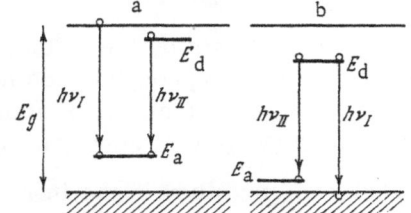

ter the edge luminescence spectrum. At 77°K infrared radiation cannot alter significantly the rate of thermal liberation of carriers from levels of depth ~ 0.3 eV, but at the infrared radiation intensities used in our study the probability of the optical liberation of carriers is fully comparable with the probability of the thermal liberation from depths exceeding ~ 0.1 eV.

Estimates of the depth of the acceptor level based on the activation energy of the thermal quenching of ZnS give $\Delta E_t \sim 0.11 \pm 0.02$ eV. Estimates based on the sensitivity to infrared quenching give $\Delta E_{IR} \sim 0.14 \pm 0.02$ eV, i.e., the values of ΔE_t and ΔE_{IR} are in satisfactory agreement.

Thus, the quenching of the edge luminescence of zinc sulfide by infrared radiation is due to, like the thermal quenching, the liberation of carriers (holes) from a local acceptor level located at a depth $\Delta E_a \sim 0.12$-0.14 eV.

It should be noted that a similar value of the acceptor level depth (~ 0.12 eV) is obtained in [75] for CdS from an investigation of the temperature dependences of the edge luminescence and photoconductivity. The similarity of the values of ΔE_t for ZnS and CdS also supports the selected model because the replacement of a cation (i.e., the replacement of Zn with Cd) should affect primarily the depth of the donor levels.

As mentioned earlier, the hypothesis that the edge luminescence is due to recombination at single defects (or impurities) of different sign is in clear conflict with the general form of the ultraviolet luminescence intensity and with the same nature of the changes in the edge luminescence spectra when the composition of ZnS is altered in the direction of excess zinc or excess sulfur (§ 1 in Chap. IV).

According to the donor–acceptor model, the difference between the energy positions of the two edge luminescence series of ZnS may be due to the difference between the gaps separating the donor from the conduction band and the acceptor from the valence band and due to the consequent difference between the Coulomb interaction energies in the formation of the donor–acceptor pairs.

The experimentally obtained energy of the donor level $E_d \sim 0.03$ eV and the difference between the positions of the head lines $\Delta h\nu_{edge} = h\nu_{334.5} - h\nu_{338} \sim 0.03$ eV suggest that the position of one level varies.

This is possible if the donor and acceptor depths differ considerably and if the changes in these depths due to association are relatively small. Then, it is more likely that the deep level changes but the infrared quenching of the edge luminescence suggests that the deep level should be common to both series. Hence, it follows that the acceptor level should be deep, i.e., it should be separated by a greater gap from the nearest allowed band. This is supported by the absence of the third, fourth, and other series of lines.

According to this model a $V_{Zn}^{=}$ acceptor located at a depth $E_a \sim 0.12 \pm 0.02$ eV first captures a hole from the valence band (this hole is generated by the direct excitation of the lattice).

The radiative transitions responsible for the first series occurs as a result of a direct recombination of free electrons with localized holes (free–bound transitions), whereas the second series is due to the recombination of electrons captures by the shallow (~ 0.03 eV) donor level V_S^+ with holes localized at the deeper acceptor level (bound–bound transitions).

The experimental data on the influence of the deviation from stoichiometry on the ultraviolet luminescence of ZnS phosphors do not allow us to attribute not only the second but also the first series to single defects, i.e., both the edge luminescence series are due to transitions in associated center ($V_{Zn}-V_S$), common to both series.

The results of our study of the infrared quenching of ZnS (constancy of the edge luminescence spectrum) and the published information on the edge luminescence of CdTe [148] ($N \approx 0.4$

for both series, each of which is observed separately depending on the preparation technology) indicate the presence of a common bound state (V_{Zn}^-) for both series of the edge luminescence of ZnS.

The presence of the second defect (V_S^+) favors radiative transitions of free electrons to the acceptor level (first series) by raising the density of free electrons near the donor–acceptor centers.

A local increase in the density of electrons may be due to their attraction by the donor level followed by their thermal liberation to the conduction band and recombination in the free state.

The origin of the two series cannot be explained by an inverted model of the center [173], allowing for the splitting of the valence band of the hexagonal phase of ZnS, which is close to $\Delta h\nu_{edge} = h\nu_{334.5} - h\nu_{338} \sim 0.03$ eV, because there are two analogous edge luminescence series in the spectrum of the cubic zinc sulfide. The origin of the two edge series may be considerably more complex, such as the radiative recombination of free and bound carriers of opposite sign (or of the same sign but at different levels) or internal bound–bound transitions between different donors and acceptors [73]. However, these interpretations are in conflict with the model of associated centers formed from intrinsic lattice defects, whose validity follows from the dependences of the luminescence efficiency on the conditions during synthesis and growth of zinc sulfide (these conditions include the rate of cooling and deviations from stoichiometry).

The polaron theory cannot be used to explain the edge luminescence of ZnS for a number of reasons: $E_g - h\nu_{334.5} \gg E_{polaron}$ in the case of ZnS; emission of the edge luminescence by such compounds as CdTe; small shift of the edge luminescence (~ 0.03 eV) in the temperature range 77-293°K; influence of quenching. Although the assumption of the importance of the lattice ionicity is fully justified not only in the case of ZnS but also in the case of other II–VI compounds, this point has not been considered in the case of other compounds such as CdS. As pointed out in reviews of the literature, the presence of three structure-free bands in the ultraviolet electroluminescence spectrum recorded at 293°K for unactivated ZnS single crystals [5, 143] can be attributed to the recombination of free carriers before establishment of a thermal equilibrium with the lattice (first band), transitions to the polaron state of one carrier (second band with $\lambda_{max} \sim 335$ nm, which coincides with the edge luminescence of unactivated ZnS), and luminescence of electron–hole polarons (third band). This interpretation is based on the results given in [29] where it is reported that the energy of the optical annihilation of polarons in ZnS is ~ 0.27 eV. However, according to the calculations reported in [68, 174],[*] the eigenvalue of the ground-state energy of a polaron E_0, equal to the energy of its optical polarization, is

$$E_0 = -0.164 \frac{m^* e^4}{\hbar^2 \varepsilon^{*2}} = -0.328 \left(0.5 \frac{m e^4}{\hbar^2}\right) \frac{m^*}{m \varepsilon^{*2}},$$

where $-0.5 \, (m e^4/\hbar^2) = \varepsilon_H = -13.5$ eV, which is the ionization energy of the hydrogen atom; m is the mass of a free carrier; m^* is the effective mass; $1/\varepsilon^* = (1/n^2) - (1/\varepsilon_0)$ is the Pekar parameter which is $\sim 0.0767 \approx 0.08$ for ZnS [81]. Using the values of the effective masses of carriers in ZnS (see Table 4), we find that $E_{0_e} \sim 0.01$ eV, $E_{0_{h_1}} \sim 0.014$ eV, and $E_{0_{h_2}} \sim 0.042$ eV.

Thus, the energies of electron (e) and hole (h_1 and h_2) polarons deduced from the Pekar theory are not equal and lie within the range ~ 0.01-0.05 eV for ZnS.

---

[*]We carried out these calculations for the case of tight-binding polaron, i.e., for $E_0 > h\nu_1$ (the latter quantity is the energy of a lattice phonon), but the values of E_0 obtained were ≤ 0.01-0.05 eV. Thus, weak-binding polarons appear in ZnS and their energy is much less than the energy of a lattice phonon (~ 0.043 eV).

Therefore, the equidistant nature of the three ultraviolet luminescence bands observed at 293°K for unactivated ZnS single crystals [143] not only does not support the polaron model but is even in direct conflict with this model because the polaron energy is proportional to the effective masses of the carriers.

On the polaron model we may expect the thermal shift of these three bands to be in agreement with the thermal shift of the forbidden band of ZnS. However, according to our results, such a shift is not exhibited by the edge luminescence of zinc sulfide in the range $T \leq 293°K$ (this is also true of CdS).

The polaron model is not needed to explain why the separation between the equidistant bands is equal to the frequency of a longitudinal optical phonon in the unperturbed lattice: this can be explained by the weak localization of carriers (due to the large radius of the Bohr orbits) and the regularity of the lattice.

The influence of the preparation conditions and particularly of the cooling rate is also a direct evidence of the participation of lattice defects in the edge luminescence and it would be more difficult to explain by the polaron model. Since $E_g - h\nu_{edge} > \Delta E_t$, carriers may be located at the donor–acceptor pair levels in the polaron state and the energy of this state may be $\Delta E = \Delta E_{h\nu} - \Delta E_t \leq 0.1$ eV.

The luminescence band in the $\lambda_{max} \sim 335$ nm range observed in the electroluminescence spectrum of ZnS at 293°K is located close to the edge luminescence of zinc sulfide and, as demonstrated in our study, represents the short-wavelength luminescence of the cubic ZnS, which shifts by ≈ 0.11 eV when the temperature is raised from 77 to 293°K.

The edge luminescence of ZnS, which we have explained by the Williams model of donor–acceptor pairs [86], could be attributed also to the recombination of bound excitons. However, this would give meaningless results because $E_d + E_a \gg E_{ex}$, where E_{ex} is the exciton binding energy. No information is available on the migration of the excitation between the edge luminescence centers. However, there are no maxima in the excitation spectrum of the edge luminescence of ZnS which would correspond to the exciton absorption peaks and this is additional evidence of the absence of a direct relationship between the edge luminescence and the excitation of excitons.

§ 3. Exciton Mechanism of the Short-Wavelength Bands and the Influence of Temperature on Various Types of Luminescence of ZnS

The short-wavelength luminescence band of zinc sulfide also does not fit well the polaron model because its maximum almost coincides (to within ≤ 1 nm) with the position of the maxima in the absorption and reflection spectra (if the comparison is made for the same phase of ZnS and at the same temperature). The difference between the photon energy corresponding to the short-wavelength luminescence band and the forbidden band width of ZnS is ~ 0.05 eV. This difference may be identified with the depth of an electron trap deduced from the thermoluminescence spectra of unactivated ZnS at $T \sim 50°K$ [60, 61]. This trap may be a sulfur vacancy, which becomes associated with V_{Zn} (in accordance with the Williams theory) and approaches an allowed band to within $\sim 0.03-0.05$ eV. However, the relatively weak thermal quenching of the short-wavelength luminescence is difficult to explain by a model of a center with such a shallow level.

It is more likely that the short-wavelength luminescence band is due to the recombination of excitons because the photon energy differs from the forbidden band width of ZnS by the binding energy of excitons, which is ~ 0.1 eV in ZnS [174, 175].

The activation energy of the thermal quenching process, deduced for the short-wavelength luminescence band of ZnS, is ~ 0.4 eV, which is in satisfactory agreement with the direct estimates of the binding energy of excitons in the ZnS lattice ($\sim 0.02-0.03$ eV) [175] and with the experimental data obtained for the hexagonal modification of ZnS [4] ($E_{ex} \sim 0.047$ eV). It must be stressed that the luminescence bands in the region of 321.5 and 327.5 nm, observed in the absorption spectra of the hexagonal and cubic phases of ZnS, are usually attributed to the free-exciton absorption.

At $T \leq 77°K$ the structure of the short-wavelength luminescence bands of ZnS, which agrees closely with the positions of the absorption bands, may be more complex: the half-width of $\sim 4kT$ at $T = 77°K$ suggests the presence of free and bound excitons [140, 176].

When the temperature is increased, the contribution of bound excitons decreases and the luminescence band becomes simpler [177]. In the quasiequilibrium case when the occupancy of the local levels is governed primarily by the thermal liberation of carriers,* the ratio of concentrations of free and bound excitons is

$$\frac{n_{free}}{n_{bound}} = \frac{N^*_{free}}{N^*_{bound}} \exp\left(-\frac{\Delta E_{bound}}{kT}\right).$$

If $\Delta E_{bound} \sim kT$, which is true in the range $T \geq 300°K$, the ratio of the exciton concentrations is governed solely by the densities of their states. The density of the free-exciton states is $N^*_{free} = 4.8 \times 10^{15} [(m^*/m)T]^{3/2} \sim 10^{19}$ cm^{-3} and the density of the bound-exciton states N^*_{bound} should in any case be below the residual impurity concentration ($\sim 10^{15}$ cm^{-3}) and the concentration of intrinsic lattice defects ($\sim 10^{17}$ cm^{-3}). Thus, in the high-temperature range the ratio of the concentrations is $n_{free}/n_{bound} \geq 10^2$ and the free-exciton luminescence predominates. Naturally, the intensities are governed not only by the concentrations of the free and bound excitons but also by the transition probabilities. Since at low temperatures the free- and bound-exciton luminescence intensities are comparable and since the oscillator strength of optical transitions is usually independent of temperature, this should not influence the above conclusion that the free-exciton luminescence predominates at high temperatures.

This is the first observation of the exciton luminescence at such very high temperatures. The short-wavelength luminescence observed in our study can also be attributed to the interband recombination in ZnS since in the range $E_{ex} \leq kT$ it is difficult to separate rigorously the exciton and interband luminescence. However, since the luminescence maximum shifts monotonically with rising temperature and it coincides with the exciton reflection peak of ZnS [171], it is more likely that the luminescence is of the exciton nature. This is in agreement with the existence of the exciton absorption at high temperatures [178].

The low efficiency of the short-wavelength luminescence indicates that the probability of nonradiative exciton annihilation is high and, in particular, this may happen on the surface of a single crystal. However, when exciting electrons are subjected to high accelerating voltages, a strong reabsorption of the short-wavelength luminescence is observed,† which is manifested clearly by a change in the ratio of the intensities of the edge and short-wavelength luminescence of ZnS (§ 4 in Chap. III).

Changes in the ratio of the intensities of these two types of luminescence in favor of the short-wavelength exciton luminescence are observed at high temperatures and also at high electron excitation densities [6]. Therefore, when we compare the data on the short-wavelength

---

*At sufficiently high temperatures and moderate electron excitation densities we can ignore the de-exciting action of the incident electrons [125].

†The depth of penetration of 20 keV electrons and of optical radiations of energies exceeding E_g is $\sim 10^{-4}$ cm, which is comparable with the reciprocal of the absorption coefficient in the $\sim 320–330$ nm range, i.e., it is comparable with the absorption coefficient of the exciton bands of the hexagonal and cubic phases of ZnS.

luminescence band of ZnS we must bear in mind not only the temperature at which measurements are carried out and the phase composition of the sample, but also the excitation rate.

Investigations of the ultraviolet luminescence of various ZnS powders and single crystals carried out in the wide temperature range of 77-625°K under identical excitation conditions have thus indicated the possibility of the emission of various types of luminescence differing in their nature and properties. We must stress particularly the considerable difference between the influence of temperature on the edge and short-wavelength luminescence bands (thermal shift, quenching, and half-width of the spectrum) and the similarity of the influence of temperature when a particular type of luminescence is considered but the objects are different (ZnS powders and single crystals).

When the temperature is raised from 77 to 293°K the intensity of the ultraviolet luminescence emitted by ZnS powders decreases by a factor of several thousands. As the temperature is increased, the edge luminescence spectrum gradually loses its structure and the short-wavelength series with $\lambda_{max} = 334.5$ nm becomes more important. At $T \approx 150°K$ the half-width of the luminescence spectrum is ~ 14 nm; when the structure is lost, the profile becomes asymmetric. The shift of the spectrum is slight: $dE/dt \approx -1.3 \times 10^{-4}$ eV/deg.

The short-wavelength ultraviolet luminescence of ZnS is affected much less by the temperature: the quenching is only by a factor of 5-6. Thus, in the case of ZnS powders the ratio of the intensities of the edge and short-wavelength luminescence bands is ~ 1 at $T = 293°K$ and this explains the complexity of the spectrum.

The short-wavelength ultraviolet luminescence spectra of ZnS crystals recorded at $T = 293°K$ are located in the same wavelength range as the edge luminescence at $T = 77°K$ (§ 3 in Chap. III). However, the ~ 337.5 nm luminescence band of a single crystal is considerably narrower than the edge luminescence spectrum recorded at low temperatures: the half-width of this band is ~ 0.05 eV, i.e., about $2kT$. This band corresponds to the short-wavelength band (~ 327.5 nm) if allowance is made for its thermal shift ($dE/dt \approx -5.2 \times 10^{-4}$ eV/deg), which is close to the temperature dependence of the forbidden band width.

Thus, whereas at 77°K the intensity of the short-wavelength luminescence of single crystals is approximately an order of magnitude less than the intensity of the edge luminescence, at 293°K the short-wavelength luminescence is much stronger than the edge luminescence.

When the temperature is increased still further (from 293 to $\sim 600°K$), both types of luminescence shift gradually (by ~ 0.3 eV) toward longer wavelengths at the same rate of $dE/dT \approx (5-6) \times 10^{-4}$ eV/deg. In this range of temperatures the intensities of both types of luminescence fall by more than an order of magnitude but the short-wavelength luminescence decreases more slowly. Throughout this range the short-wavelength luminescence spectrum of ZnS single crystals is almost elementary: its half-width is $\sim (2.4 \pm 0.2)kT$.

Investigations at high temperatures have enabled us to detect very deep traps in ZnS phosphors.

Luminescence of wavelengths shorter than the $\lambda_{max} \sim 321.5$ nm (hexagonal phase) and $\lambda_{max} \sim 327.5$ nm (cubic phase) bands of unactivated ZnS have not been observed, although some samples have spectra with a step in the range of frequencies corresponding to the direct interband transitions in ZnS. However, such luminescence is only slightly stronger than the background so that it is practically impossible to obtain any information on this feature.

§ 4.   Energy Structure of Local Levels in Unactivated Zinc Sulfide

Unactivated zinc sulfide is far from an ideal semiconductor and it is no less complex than ZnS containing high concentrations of heavy-metal or chloride dopants.

Investigations of various (including thermoluminescence) luminescence spectra have demonstrated that unactivated ZnS contains at least five different types of luminescence center and five different storage levels. Moreover, these local states need not be elementary. The number of nonequilibrium carriers localized at these discrete levels and the relative contribution of these carriers to the overall energy balance depend on the concentration of such centers and equally on the excitation conditions, particularly on temperature. When the temperature is increased, the role of the deeper levels becomes greater because the probability of liberation of nonequilibrium carriers is governed by an exponential function of $\Delta E_i/kT$.

It should be stressed that in the simplest case the kinetics of the luminescence emitted by an activated crystal phosphor is usually calculated employing a semiconductor energy band model with one type of radiative recombination center (usually Ag and Cu acceptors) and one type of nonradiative recombination center (usually donors in the form of halogens or "efficient" quenching impurities Co, Ni, and Fe). Compensated crystal phosphors are often prepared, i.e., we often have $n_d \approx n_a$. If the number of localized nonequilibrium carriers is considerably smaller than the number of the main donor and acceptor impurities, we can obtain a quasi-stationary solution of the balance (rate) equations for carriers which shows that $N^- (N^+) \ll \nu_d (\nu_a)$ and the quasineutrality conditions yield $n_d \approx n_a$ (equality of the number of localized carriers of both signs). Thus, a change in the luminescence efficiency with temperature or excitation rate reduces to a change in the ratio of the densities of free electrons and holes, whereas the corresponding ratio of localized electrons and holes remains unaltered, i.e., the number of the ionized luminescence centers remains equal to the number of localized electrons in spite of the difference between their energy depths which determine the concentration of free charges:

$$\eta = \frac{I_{lum}}{I_{lum} + I_{nr}} = \frac{1}{1 + \dfrac{I_{nr}}{I_{lum}}} = \frac{1}{1 + \dfrac{\beta_{nr}}{\beta_{lum}} \dfrac{N^+}{N^-}},$$

where β_{nr} is the probability of nonradiative recombination of free holes with quenching centers; β_{lum} is the probability of radiative recombination between free electrons and luminescence centers. This model is justified if the concentration of impurities of one kind (donors and acceptors) is considerably greater than the densities of other types of localized states. This condition is necessary but not sufficient: we must ensure that a more stringent conditions is observed which demands that the density of nonequilibrium carriers localized at such levels be considerably greater than the total density of nonequilibrium carriers at other (secondary) levels. Hence, it follows that $(\nu_{main}/\nu_{sec}) \exp[-(E_{main} - E_{sec})/kT]$ should be considerably greater than unity (here, "main" and "sec" represent the main and secondary impurities). This requirement reduces the range of temperatures in which the model is valid because we can always find a low but finite density of deeper localization levels of both types due to various structure defects and residual impurities. The concentration of residual impurities in the II–VI compounds purified by the currently available methods is at least 10^{14}–10^{15} cm^{-3} [159].

These residual impurities may influence the luminescence of activated ZnS phosphors and, particularly, the ultraviolet luminescence of unactivated zinc sulfide because of: 1) the low absolute number of the edge luminescence centers, which — according to the most optimistic estimates — does not exceed 10^{17} cm^{-3} and — according to the model of donor–acceptor pairs — is less than the number of single lattice defects and may be even comparable with the residual impurity concentration (10^{15} cm^{-3}); 2) the shallow depth of the levels involved in the principal radiative transitions in the ultraviolet edge luminescence region (according to the spectral data this depth does not exceed ~ 0.2 eV); 3) the relatively small difference between the densities of the various localization levels which can be of very different origin (single and associated defects, and also residual impurities), whose ratios do not exceed two order of magnitude, whereas in activated phosphors the activator concentrations are usually considerably greater than the concentrations of other defects and impurities.

Thus, the range of temperatures in which a given local level dominates the overall balance of localized and free carriers and the ratio of the radiative and nonradiative recombination events is much less than in the case of activated ZnS phosphors. Therefore, the kinetics of the process which occur in unactivated zinc sulfide should be considered stage by stage for each fairly narrow temperature range, allowing separately not only for the change in the ratio of the densities of free and bound carriers but also for changes in the number of nonequilibrium carriers localized at various levels. It follows from our investigation that unactivated zinc sulfide has a complex system of donor and acceptor levels whose depths are within a wide range from $\Delta E \leq 0.1$ eV to $\Delta E \geq 1$ eV. Depending on the population and thermal liberation of carriers of these levels, we have to consider at least five ranges of temperature (whose boundaries are naturally approximate and depend on the excitation rate) in discussing the luminescence (and excited conductivity) kinetics of zinc sulfide.

1. At very low temperatures ($4°K \leq T \leq 80°K$) all the deep and moderately deep levels are frozen-in and the thermal liberation is possible only from levels of depth below 0.1 eV. In this temperature range the intensity of the short-wavelength series of the edge luminescence rises relative to the intensity of the long-wavelength series when the temperature is increased and this is evidence of an increase in the density of free electrons relative to the density of electrons localized in a shallow donor level at ~ 0.03 eV. The total ultraviolet edge luminescence intensity remains almost constant in this temperature range, i.e., the probability of the liberation of carriers from acceptors of energies ≥ 0.1 eV is low. The efficiency of the ultraviolet edge luminescence is constant in the range $T \sim 40$-$50°K$ (a kink can be seen in the curve in Fig. 1b), which may be attributed to the liberation of electrons from ~ 0.07 eV traps, which are the principal nonradiative recombination centers at low temperatures.

2. In the low temperature range (~ 80-$110°K$), the ultraviolet edge luminescence is quenched strongly (by more than an order of magnitude) and at the same time the visible luminescence bands are enhanced. The infrared quenching of the edge luminescence and the stimulation of the blue luminescence can be observed only at the lower end of this range. At $T \geq 100°K$ the probability of thermal liberation of carriers from the acceptor levels at ~ 0.12-0.14 eV exceeds $\sim 10^5$ sec^{-1}, which is considerably higher than the probability of the optical liberation by very strong infrared radiation sources (ribbon lamps). However, the levels of depth ≥ 0.2 eV are hardly affected, i.e., we may assume that the transfer of carriers of these levels is the cause of a strong quenching of the ultraviolet luminescence in this range of temperatures.

In the case of CdS [75], the green edge luminescence is quenched and the photoconductivity decreases in this range of temperatures: this is due to the transfer of carriers to the quenching center levels characterized by a higher probability of nonradiative transitions, which reduce the free electron density.

3. Moderate temperatures (~ 100-$300°K$) correspond to the main thermoluminescence peaks of the unactivated and activated zinc sulfide (this follows from the results of our investigations and numerous published data). In this range the ultraviolet luminescence no longer makes any significant contribution to the overall energy balance but the rate of its quenching slows down somewhat and the most efficient electron traps of depth ~ 0.3 eV are emptied at $\sim 160°K$. It is this group of traps that stores the main part of a light sum. It is also possible that two types of level exist in this region because the profiles of the peaks exhibited by different phosphors are different and the estimates of the depths ΔE vary from 0.28 to 0.34 eV.

A low-temperature thermoluminescence peak observed in the 110-130°K range, which becomes a more effective storage level when excess Zn is introduced, is also complex and it is due to different charge states of the interstitial zinc atoms. A higher thermoluminescence peak at $\sim 220°K$ is evidently due to the presence of oxygen or other impurities.

4. At high temperatures (~ 300-$500°K$) the typical activator luminescence bands are strongly quenched and the higher levels of the known "poisons" (quenching centers) of the ac-

tivator luminescence (such as CO with $\Delta E \approx 0.44$ eV) become empty. The concentration of such poison or quenching centers in ZnS phosphors is at least 10^{16} cm^{-3}. At the lower end of this temperature range the densities of electrons and holes localized at levels of ~ 0.3 eV depth may even increase and this is accompanied by a further thermal quenching of the ultraviolet luminescence. At the upper end of this range the emptying of these levels slows down at the rate of quenching of the edge and short-wavelength (exciton) luminescence.

5. At very high temperatures ($500°K \leq T < 700°K$) the changes in the physicochemical properties of ZnS phosphors are still very slow. In this range the deepest electron (and probably hole) levels with energies of ~ 1 eV become emptied. Such emptying is accompanied by a strong enhancement of the visible green luminescence. The high activation energy of this process ($\Delta E \sim 1.2$ eV) is confirmed by indirect data (later stages of the afterglow) on the luminescence of ZnS phosphors and by the results of direct measurements of the temperature dependence of the dark conductivity [143]. An increase of the temperature to $\sim 700°K$ should establish comparable densities of free and localized carriers in ZnS phosphors. The ratio of the densities of the carriers localized at levels of different depth is practically independent of ΔE ($10 > e^{\Delta E/kT} \sim 1$), i.e., this ratio is equal to the ratio of the densities of the corresponding local levels irrespective of their depths. In this range of temperatures we may expect the external quenching to stop and there may be some recovery of the luminescence associated with shallow levels.

This is particularly important in the case of the exciton states whose density is close to the density of states of free carriers. Our studies indicate that the quenching of the short-wavelength exciton bands slows down or even stops at $T \sim 600°K$. Experimental difficulties prevented us from extending the detailed studies to higher temperatures but a qualitative confirmation of the ideas put forward above was obtained.

The region $T \sim 700°K$ lies completely outside the range of the reported investigations because the lattice itself becomes unstable. Recrystallization, annealing, and slow forming in activated phosphors all begin at these temperatures.

This raises doubts about the actual possibility of the determination of the thermal width of the forbidden band of ZnS in this range of temperatures (the thermal value of E_g can be determined only in this range because of the high impurity concentration and the high value of E_g).

Mixing of the electron and ion conduction processes in the lattice because of the motion of the lattice defects V_{Zn} and V_S is possible.

Moreover, in this range of temperatures the optical width of the forbidden band of ZnS is ~ 3.4 eV, which is practically identical with the thermal width of the forbidden band of ZnS [29].

If E_g is a linear function of the temperature, i.e., $E_g = E_{g_0} - \alpha T$, we should compare the optical and thermal forbidden band widths at the lowest possible temperatures. If the forbidden band width is a more complex function of the temperature, which is true of ZnS ($\alpha \sim 5 \times 10^{-4}$ eV/deg at $T > 80°K$ and $\alpha \sim 2 \times 10^{-4}$ eV/deg at $T < 80°K$), one can no longer compare reliably the thermal and optical widths of the forbidden band.

Therefore, the question of the relative and absolute values of the difference between the optical and thermal widths of the forbidden band of zinc sulfide requires further studies.

We can point out here that the thermal and optical depths of local levels in CdTe are the same [39] but the edge luminescence with properties similar to the ultraviolet edge luminescence of zinc sulfide is still observed [148].

CONCLUSIONS

1. The ultraviolet luminescence emitted by zinc sulfide is not a unique property of some samples prepared under special conditions or excited in a special way but is observed for a great variety of samples. If the rate of excitation of the host substance is sufficiently high ($\sim 10^{21}$ pairs $\cdot$ cm^{-3} $\cdot$ sec^{-1}), the ultraviolet luminescence is emitted by all unactivated ZnS powders and single crystals and even by samples with high concentrations with such activators as the heavy metals and rare-earth elements.

2. At 77°K all the investigated forms of unactivated ZnS emitted mainly three types of luminescence near the fundamental absorption edge: the short-wavelength luminescence, the edge luminescence discovered by Kröger, and the structure-free luminescence in the 360–390 nm range. The actual positions of these three types of luminescence are governed by the crystal phase (hexagonal or cubic) of the lattice and by the temperature of the sample.

3. The present paper reports the first determination of the optimal efficiency of the edge luminescence of ZnS ($\sim 10-15\%$), which is comparable with the efficiency of the activated phosphors. When particular conditions are used during the preparation of ZnS, the ultraviolet luminescence becomes the main radiative recombination channel.

The edge luminescence kinetics is basically the same as that of the activator luminescence. The thermal quenching is due to external heating. For the same volume excitation rates the kinetics of the thermal quenching of the cathodoluminescence is close to the corresponding kinetics of the photoluminescence. Up to $\sim 80\%$ of the recombination events occur at the quenching (nonradiative) centers.

The first observations are reported of the infrared quenching of the edge luminescence of unactivated ZnS: this quenching process has an activation energy of ~ 0.14 eV, which is close to the activation energy of the thermal quenching 0.12 eV.

4. The conditions during synthesis that ensure the optimal development of the ultraviolet edge luminescence of unactivated zinc sulfide are in agreement with the donor–acceptor model of the luminescence centers, each of which is composed of two lattice defects (V_{Zn}, V_S).

Freezing-in of the thermodynamic equilibrium by rapid cooling (quenching) increases the number of defects and raises the edge luminescence efficiency. The close agreement between the edge luminescence efficiencies of powders and single crystals shows that the luminescence centers are not grain boundaries. Moreover, the luminescence is not due to phase boundaries (there is no efficiency maximum at the phase transition temperature), surface levels (this is in conflict with the dependence on the accelerating voltage), or lattice strains. Single lattice defects (V_S, S_i, V_{Zn}, Zn_i) also cannot act as the edge luminescence centers in ZnS because the efficiency of such luminescence decreases on deviation from stoichiometry.

5. Approximate estimates of the concentration of the edge luminescence centers in ZnS obtained by various methods (from the infrared quenching regarded as a function of the excitation rate, from the quenching due to introduction of controlled amounts of the Ag and Sm activators, and from the data on the saturation at high electron excitation rates) suggest that the total number of these centers does not exceed $\sim 10^{16}$ pairs/cm^3 and the average distance between such centers is slightly greater than the mean free path of carriers.

The quenching of the edge luminescence of ZnS on introduction of impurities may be due to the formation of an additional recombination channel in the case of heavy metals (Ag) or due to a reduction in the number of the edge luminescence centers because of the formation of complexes from intrinsic lattice defects and impurities (Sm ions).

6. The high concentration of the associated edge luminescence centers in ZnS is possible only because of the existence of sufficiently "distant" ($\geq 20-30$ Å) donor–acceptor pairs. This

corresponds to an overlap of the wave functions of a donor and an acceptor at distances of the order of the sum of their Bohr orbit radii (estimated for ZnS using the effective mass theory).

The large internal distances in donor–acceptor pairs are confirmed also by a comparison of the experimental data on the spectral widths of the zero-phonon lines (~ 0.015 eV) with the theoretical calculations of Williams, which give the difference between the photon energies (~ 0.01 eV) for donor–acceptor pairs with internal distances amounting to about six lattice constants in ZnS, and with the experimental data on the shift of the afterglow spectra (~ 0.01 eV) and on the duration of afterglow τ.

7. The presence of two edge luminescence series and the changes in their intensities with rising temperature, as well as the thermal and optical (infrared) quenching are explained by a very simple model of the luminescence centers which is applied here for the first time to ZnS but which has been confirmed earlier for CdS and CdTe (without specification of the nature of the donor and acceptor in the pair).

According to this model a $\bar{V}_{Zn}$ acceptor, located at a depth of $\sim 0.12 \pm 0.02$ eV, first captures a hole from the valence band (this hole is generated by the external excitation of the lattice).

The radiative transitions in the first edge luminescence series are due to the direct recombination of free electrons with localized holes (free–bound transitions), whereas the transitions in the second series represent the recombination of electrons captured by a shallow (~ 0.03 eV) donor level V^+ holes localized at the deeper acceptor level (bound–bound transitions).

8. The ultraviolet luminescence spectra of powders differ somewhat from the spectra of ZnS single crystals. The difference is due to the existence of the two types of luminescence (short-wavelength edge), because the ratio of the intensities of these two types of luminescence depends strongly on the nature of the sample and on temperature. In comparisons of the luminescence spectra we must allow not only for the temperture but also for the phase composition of the ZnS lattice.

The energy shift of the ultraviolet bands (~ 0.07 eV) is in agreement with the difference between the values of E_g of the hexagonal and cubic phases of zinc sulfide.

The difference between the influence of temperature (shift, quenching, half-width) on the edge and short-wavelength ultraviolet luminescence and the similarity of the effects of temperature on the luminescence of a given kind emitted by different objects (ZnS powders and single crystals) indicate that the edge luminescence is of different origin from the short-wavelength luminescence. It is most likely that the short-wavelength band is due to exciton radiative transitions.

9. The properties of the edge and short-wavelength ultraviolet luminescence bands of ZnS are similar to the properties of the corresponding bands emitted by other II–VI compounds.

The published information confirms our interpretation of the edge luminescence of ZnS, which presumes that the radiative recombination occurs in associated lattice defects (donor–acceptor pairs) and it supports the proposed energy scheme of these transitions.

10. Unactivated ZnS contains many lattice defects: there are at least five types of storage centers and five types of luminescence centers.

The depths of local levels in unactivated ZnS extend from ~ 0.06–0.07 eV to ~ 1.0–1.2 eV. The shallow levels are the most important in the overall balance of carriers at low temperatures ($4°K \leq T < 100°K$), whereas the deep levels are important at high temperatures (~ 300–$600°K$). At still higher temperatures even the deepest levels gradually become empty and this is accompanied by the enhancement of the visible luminescence bands and the absence of further quenching of the ultraviolet bands.

11. The reported investigation has demonstrated that there is no basic limit to the efficiency of the edge luminescence compared with the luminescence of activated ZnS phosphors. This opens up the possibility of the development and manufacture of efficient narrow-band cathodoluminescence sources of near ultraviolet radiation for the excitation of the visible radiation of other phosphors and, possibly, for the pumping of laser crystals.

The author is deeply grateful to his Research Director V. L. Levshin for his continuing interest and daily help, to Yu. P. Timofeev for his constant interest and discussion of the results, to E. Ya. Arapova, V. V. Shchaenko, and N. A. Gorbacheva for the preparation of phosphors and growing of single crystals, and to the staff of the author's laboratory for their help and advice in the course of the investigation reported above.

LITERATURE CITED

1. F. A. Kröger, Physica, 7:1 (1940).
2. É. Ya. Arapova, Yu. V. Voronov, V. L. Levshin, V. A. Chikhacheva, and V. V. Shchaenko, Izv. Akad. Nauk SSSR Ser. Fiz., 30:1490 (1966).
3. O. V. Bogdankevich, M. N. Zverev, A. N. Pechenov, and L. A. Sysoev, Fiz. Tverd. Tela, 8:2547 (1966).
4. L. G. Suslina, Thesis for Candidate's Degree [in Russian], Leningrad State University (1966).
5. Yu. V. Bochkov, A. N. Georgobiani, A. S. Gershun, L. A. Sysoev, and G. S. Chilaya, Opt. Spektrosk., 20:183 (1966).
6. É. L. Nolle, V. S. Vavilov, G. P. Golubev, Yu. D. Dumarevskii, and V. S. Mashtakov, Opt. Spektrosk., 23:267 (1967).
7. S. A. Fridman and A. A. Cherepnev, Luminescent Compositions for Continuous and Temporary Use [in Russian], Izd. AN SSSR, Moscow (1946), p. 50.
8. K. V. Shalimova and N. K. Morozova, Opt. Spektrosk., 16:659 (1964).
9. A. Addamiano and M. Aven, J. Appl. Phys., 31:36 (1960).
10. M. Aven and J. A. Parodi, J. Phys. Chem. Solids, 13:56 (1960).
11. L. C. Green, D. C. Reynolds, S. I. Gzuzak, and W. M. Baker, J. Chem. Phys., 29:1375 (1958).
12. H. Samelson and V. A. Brophy, J. Electrochem. Soc., 108:150 (1961).
13. W. W. Piper, Phys. Rev., 92:23 (1953).
14. K. V. Shalimova, Dokl. Akad. Nauk SSSR, 80:587 (1951).
15. C. K. Coogan, Proc. Phys. Soc. Lond., 70B:845 (1957).
16. J. F. Hall, Jr., J. Opt. Soc. Am., 46:1013 (1956).
17. N. A. Vlasenko, Opt. Spektrosk., 7:511 (1959).
18. E. F. Gross and L. G. Suslina, Opt. Spektrosk., 6:115 (1959).
19. K. V. Shalimova, Doctoral Thesis [in Russian], Lebedev Physics Institute, Academy of Sciences of the USSR, Moscow (1952).
20. K. V. Shalimova and N. K. Morozova, Izv. Vyssh. Uchebn. Zaved. Fiz., No. 2, 98 (1964).
21. É. Ya. Arapova, Opt. Spektrosk., 13:416 (1962).
22. M. Cardona and G. Harbeke, Phys. Rev., 137:A1467 (1965).
23. J. L. Birman, Phys. Rev. 114:1490 (1959).
24. J. L. Birman, Proc. Intern. Conf. on Luminescence, Budapest, 1966 (ed. by G. Szigeti), Vol. 1, Akadémiai Kiadó, Budapest (1968), p. 919.
25. F. J. Low and A. R. Hoffman, Appl. Opt., 2:649 (1963).

26. A. Lempicki, D. R. Frankl, and V. A. Brophy, Phys. Rev., 107:1238 (1958).
27. B. D. Saksena, Phys. Rev., 81:1012 (1951).
28. P. F. Browne, J. Electron., 2:154 (1956).
29. M. V. Fok, Fiz. Tverd. Tela, 5:1489 (1963).
30. P. Zalm, Philips Res. Rep., 11:417 (1956).
31. J. C. Miklosz and R. G. Wheeler, Phys. Rev., 153:913 (1967).
32. J. J. Hopfield and D. G. Thomas, Phys. Rev., 122:35 (1961).
33. R. G. Wheeler and J. O. Dimmock, Phys. Rev., 125:1803 (1962).
34. C. Z. van Doorn, Physica, 20:1155 (1954).
35. M. N. Alentsev and E. I. Panasyuk, Opt. Spektrosk., 5:207 (1958).
36. T. S. Moss, Optical Properties of Semi-Conductors, Butterworths, London (1959).
37. E. F. Gross, L. G. Suslina, and K. F. Komarovskikh, Opt. Spektrosk., 8:516 (1960).
38. L. A. Vinokurov, Opt. Spektrosk., 1:901 (1956); L. A. Vinokurov and M. V. Fok, Opt. Spektrosk., 10:374 (1961).
39. E. N. Arkad'eva, Fiz. Tverd. Tela, 6:1034 (1964).
40. Yu. V. Bochkov, A. N. Georgobiani, and G. S. Chilaya, Fiz. Tverd. Tela, 8:1273 (1966); Izv. Akad. Nauk SSSR Ser. Fiz., 30:628 (1966).
41. I. Uchida, J. Phys. Soc. Jap., 19:670 (1964).
42. F. Seitz, Trans. Faraday Soc., 35:74 (1939).
43. J. S. Prener and D. J. Weil, J. Electrochem. Soc., 106: 409 (1959).
44. H. Baba, J. Electrochem. Soc., 110:79 (1963).
45. R. H. Bube, Photoconductivity of Solids, Wiley, New York (1960).
46. S. Rotschild, Trans. Faraday Soc., 42:635 (1946).
47. R. E. Shrader and S. Larach, Phys. Rev., 103:1899 (1956).
48. A. Addamiano, J. Chem. Phys., 23:1541 (1955).
49. É. Ya. Arapova, Izv. Akad. Nauk SSSR, Ser. Fiz., 25:324 (1961).
50. J. S. Prener and F. E. Williams, J. Chem. Phys., 25:361 (1956).
51. V. F. Tunitskaya, Zh. Prikl. Spektrosk., 10:1004 (1969).
52. E. E. Bukke, T. I. Voznesenskaya, N. P. Golubeva, N. A. Gorbacheva, Z. P. Kaleeva, E. I. Panasyuk, and M. V. Fok, Zh. Prikl. Spektrosk., 12:1047 (1970).
53. N. Riehl and H. Ortmann, J. Phys. Radium, 17:620 (1956).
54. N. Riehl and H. Ortmann, Dokl. Akad. Nauk SSSR, 46:613 (1954).
55. A. M. Gurvich, R. V. Katomina, and A. P. Nikiforova, Izv. Akad. Nauk SSSR, Ser. Fiz., 29:507 (1965).
56. F. Seitz, J. Chem. Phys., 6:150 (1938).
57. F. E. Williams, J. Chem. Phys., 19:457 (1951).
58. C. C. Klick and J. H. Schulman, J. Opt. Soc. Amer., 42:910 (1952).
59. C. Kittel and A. H. Mitchell, Phys. Rev., 96:1488 (1954); W. Kohn, Solid State Phys., 5:257 (1957).
60. V. L. Levshin and V. F. Tunitskaya, Opt. Spektrosk., 18:328 (1965).
61. N. Riehl, Proc. Intern. Conf. on Luminescence, Budapest, 1966 (ed. by G. Szigeti), Vol. 1, Akadémiai Kiadó, Budapest (1968), p. 974
62 N. Riehl, J. Phys. Chem. Solids, 29:1827 (1968).
63. H. M. James and K. Lark-Horovitz, Z. Phys. Chem. (Leipz.), 198:107 (1951).
64. C. C. Klick, J. Opt. Soc. Am., 41:816 (1951).
65. J. Ewles, Proc. R. Soc. A, 167:34 (1938).
66. F. Nicoll, Appl. Phys. Lett., 9:13 (1966).
67. J. J. Hopfield, J. Phys. Chem. Solids, 10:110 (1959).
68. S. I. Pekar, Investigations in the Electron Theory of Crystals [in Russian], Izd. AN SSSR, Moscow-Leningrad (1951); Usp. Fiz. Nauk, 50:197 (1953).
69. M. A. Krivoglaz, Opt. Spektrosk., 1:54 (1956).
70. K. Maeda, J. Phys. Chem. Solids, 26:1419 (1965).

71. D. G. Thomas and J. J. Hopfield, Phys. Rev., 116:573 (1959).
72. E. Grillot and M. Bancie-Grillot, Izv. Akad. Nauk SSSR, Ser. Fiz., 22:1356 (1958).
73. O. Goede and E. Gutsche, Phys. Status Solidi, 17:911 (1966).
74. C. Z. van Doorn, Philips Res. Rep., 21:163 (1966).
75. I. B. Ermolovich, A. V. Lyubchenko, and M. K. Sheinkman, Fiz. Tekh. Poluprovodn., 2:1639 (1968).
76. E. F. Gross and L. G. Suslina, Fiz. Tverd. Tela, 8:872 (1966).
77. H. Samelson and A. Lempicki, Phys. Rev., 125:901 (1962).
78. F. A. Kröger and H. J. G. Meyer, Physica, 20:1149 (1954).
79. C. C. Klick, Phys. Rev., 89:274 (1953).
80. V. L. Levshin, É. Ya. Arapova, Yu. V. Voronov, and Yu. P. Timofeev, Zh. Prikl. Spektrosk., 12:674 (1970).
81. D. Curie, Luminescence in Crystals, Methuen, London (1963).
82. E. F. Gross, B. S. Razbirin, and S. A. Permogorov, Fiz. Tverd. Tela, 7:558 (1965).
83. F. E. Williams, J. Phys. Chem. Solids, 12:265 (1960).
84. Yu. V. Voronov, Opt. Spektrosk., 24:957 (1968).
85. K. V. Shalimova, N. K. Morozova, O. I. Korolev, and M. M. Vaselkova, Proc. Intern. Conf. on Luminescence, Budapest, 1966 (ed. by G. Szigeti), Vol. 1, Akadémiai Kiado, Budapest (1968), p. 1190.
86. F. E. Williams, Phys. Status Solidi, 25:493 (1968).
87. K. Maeda, J. Phys. Chem. Solids, 26:595 (1965).
88. Yu. V. Voronov and Yu. P. Timofeev, Zh. Prikl. Spektrosk., 1:15 (1964).
89. W. Klein, J. Phys. Chem. Solids, 26:1517 (1965).
90. A. I. Dirochka, L. A. Kurbatov, and V. E. Mashchenko, Fiz. Tekh. Poluprovodn., 2:1682 (1968).
91. C. Benoit a la Guillaume and J. M. Debever, C. R. Acad. Sci. (Paris), 261:5428 (1965).
92. N. G. Basov, O. V. Bogdankevich, and Yu. M. Popov, Fiz. Tverd. Tela, 7:3289 (1965).
93. C. E. Hurwitz, Appl. Phys. Lett., 9:116 (1966).
94. A. G. Akmanov, V. S. Dneprovskii, A. I. Kovrigin, and A. N. Penin, Zh. Eksp. Teor. Fiz., 53:1293 (1967).
95. D. G. Thomas, M. Gershenzon, and F. A. Trumbore, Phys. Rev., 133:A269 (1964).
96. P. Debye and E. Hückel, Phys. Z., 24:185 (1923).
97. R. H. Fowler and E. A. Guggenheim, Statistical Thermodynamics, Cambridge University Press (1939), p. 409.
98. A. M. Gurvich, Usp. Khim., 35:1495 (1966).
99. H. Reiss, C. S. Fuller, and F. J. Morin, Bell. Syst. Tech. J., 35:535 (1956).
100. J. S. Prener and F. E. Williams, Phys. Rev., 101:1427 (1956).
101. E. F. Apple and F. E. Williams, J. Electrochem. Soc., 106:224 (1959).
102. S. Shionoya, K. Era, and Y. Washizawa, J. Phys. Soc. Jap., 21:1624 (1966).
103. I. Kh. Rammo and Kh. Yu. Voolaid, Zh. Prikl. Spektrosk., 10:79 (1969).
104. V. F. Tunitskaya, T. F. Filina, E. I. Panasyuk, and Z. P. Ilyukhina, Izv. Akad. Nauk SSSR, Ser. Fiz., 35:1437 (1971).
105. E. F. Gross and D. S. Nedzvetskii, Dokl. Akad. Nauk SSSR, 146, 1047 (1962); 152:309 (1963).
106. J. J. Hopfield, D. G. Thomas, and M. Gershenzon, Phys. Rev. Lett., 10:162 (1963).
107. M. V. Fok, Introduction to the Kinetics of the Luminescence of Crystal Phosphors [in Russian], Nauka, Moscow (1964).
108. V. V. Antonov-Romanovskii, Kinetics of the Photoluminescence of Crystals [in Russian], Nauka, Moscow (1966).
109. A. V. Moskvin, Cathodoluminescence [in Russian], Part 1, GITTL, Moscow (1948).
110. G. F. J. Garlick, Proc. IRE, 43:1907 (1955).
111. A. Bril and H. A. Klasens, Philips Res. Rep., 7:401 (1952).

112. V. L. Levshin, É. Ya. Arapova, A. I. Blazhevich, Yu. V. Voronov, I. G. Voronova, V. B. Gutan, A. V. Lavrov, Yu. M. Popov, S. A. Fridman, V. A. Chikhacheva, and V. V. Shchaenko, Tr. Fiz. Inst. Akad. Nauk SSSR, 23:64 (1963).

113. P. Lenard and S. Saeland, Ann. Phys. (Leipz.), 28:476 (1908).

114. A. Bril, Physica, 15:361 (1949).

115. V. L. Levshin, Photoluminescence of Liquid and Solid Substances [in Russian], GITTL, Moscow (1951).

116. G. F. J. Garlick, Brit. J. Appl. Phys., 13:540 (1962).

117. V. L. Levshin, Izv. Akad. Nauk SSSR Ser. Fiz., 29:346 (1965).

118. Yu. P. Timofeev, Diploma Thesis [in Russian], Moscow State University (1963).

119. S. I. Kataev, Television Tubes [in Russian], Znanie, Moscow (1936).

120. R. Dadding, J. Brit. Inst. Radio Eng., 11:445 (1951).

121. Yu. V. Voronov and Yu. P. Timofeev, Zh. Prikl. Spektrosk., 6:310 (1967).

122. A. A. Sokolov, Introduction to Quantum Electrodynamics, Fizmatgiz, Moscow (1955), p. 260.

123. G. Gergely, Acta Phys. Hung., 12:253 (1960).

124. W. Ehrenberg and J. Franks, Proc. Phys. Soc. B, 66:1057 (1953).

125. Yu. M. Popov, Opt. Spektrosk., 7:697 (1959).

126. A. Bril and F. A. Kröger, Philips Tech. Rev., 12:120 (1950).

127. Yu. M. Popov, Zh. Eksp. Teor. Fiz., 35:505 (1958); Yu. M. Popov and V. P. Shabanskii, Opt. Spektrosk., 6:769 (1959).

128. N. F. Malyuk, G. A. Fedorus, V. D. Fursenko, I. A. Shakh-Melikova, and M. K. Sheinkman, Fiz. Tverd. Tela, 8:3133 (1966).

129. A. I. Blazhevich, Fiz. Tekh. Poluprovodn., 1:1561 (1967).

130. B. I. Pochtarev, K. K. Raspletin, and D. V. Fetisov, Izv. Akad. Nauk SSSR, Ser. Fiz., 23:462 (1959).

131. B. I. Pochtarev, Izv. Akad. Nauk SSSR, Ser. Fiz., 25:514 (1961).

132. Yu. V. Voronov and A. G. Ovchinnikov, Prib. Tekh. Eksp., No. 3, 190 (1963).

133. Yu. V. Voronov and A. G. Ovchinnikov, Prib. Tekh. Eksp., No. 4, 238 (1965).

134. K. V. Shalimova and N. K. Morozova, Opt. Spektrosk., 16:659 (1964); N. K. Morozova and K. V. Shalimova, Opt. Spektrosk., 21:192 (1966).

135. A. Bril and H. A. Klasens, Philips Tech. Rev., 15:63 (1953).

136. E. F. Gross, L. G. Suslina, and V. G. Mokerov, Fiz. Tverd. Tela, 7:291 (1965).

137. V. L. Levshin, É. Ya. Arapova, Yu. V. Voronov, V. B. Gutan, and Yu. P. Timofeev, Izv. Akad. Nauk SSSR, Ser. Fiz., 33:944 (1969).

138. K. V. Shalimova and N. K. Morozova, Kristallografiya, 9:559 (1964).

139. C. Z. van Doorn, Physica, 20:1155 (1954).

140. E. F. Gross, L. G. Suslina, É. B. Shadrin, Fiz. Tverd. Tela, 10:1036 (1968).

141. V. L. Levshin, Yu. V. Voronov, and Yu. P. Timofeev, Opt. Spektrosk., 30:1063 (1971).

142. V. L. Levshin, Yu. V. Voronov, and Yu. P. Timofeev, Fiz. Tekh. Poluprovodn., 4:601 (1970).

143. G. S. Chilaya, Thesis for Candidate's Degree [in Russian], Lebedev Physics Institute, Academy of Sciences of the USSR, Moscow (1967).

144. V. S. Gavrilov, B. V. Novikov, and R. I. Shekhmamet'ev, Fiz. Tekh. Poluprovodn., 1:371 (1967).

145. M. Schön, Z. Naturforsch., 6a:287 (1951).

146. É. Ya. Arapova, V. L. Levshin, N. V. Mitrofanova, T. S. Reshetina, V. F. Tunitskaya, S. A. Fridman, and V. V. Shchaenko, Izv. Akad. Nauk SSSR, Ser. Fiz., 30:573 (1966).

147. Yu. V. Voronov and Yu. P. Timofeev, Izv. Akad. Nauk SSSR, Ser. Fiz., 33:951 (1969).

148. Zh. R. Panosyan, A. A. Gippius, and V. S. Vavilov, Abstracts of Papers presented at Second All-Union Conf. on Luminescence, Uzhgorod, 1969 [in Russian].

149. H. Gobrecht and D. Hofmann, J. Phys. Chem. Solids, 27:509 (1966).

150. Ch. B. Lushchik, Tr. Inst. Fiz. Astron. Akad. Nauk Est. SSSR, 3:3 (1955).
151. V. L. Levshin, Yu. V. Voronov, N. A. Gorbacheva, and Yu. P. Timofeev, Zh. Prikl. Spektrosk., 12:1061 (1970).
152. G. F. J. Garlick, Proc. Br. Ceram. Soc., 1:21 (1964).
153. A. M. Gurvich and R. V. Katomina, Zh. Fiz. Khim., 42:2199 (1968).
154. A. B. Lidiard, "Ionic conductivity," in: Handbuch der Physik (ed. by S. Flügge), Vol. 20, Springer Verlag, Berlin (1957), pp. 246-349.
155. A. B. Lidiard, Phys. Rev., 94:29 (1954).
156. N. Kh. Abrikosov, V. F. Bankina, L. V. Poretskaya, E. V. Skudnova, and L. E. Shelimova, Semiconductor Compounds, Their Preparation and Properties [in Russian], Nauka, Moscow (1967).
157. M. Kh. Karapet'yants and M. L. Karapet'yants, Tables of Some Thermodynamic Properties of Various Substances [in Russian], Nauka, Moscow (1961).
158. A. Rose, Concepts in Photoconductivity and Allied Problems, Interscience, New York (1963).
159. L. Ya. Markovskii, F. M. Pekerman, and L. N. Petoshina, Luminescent Phosphors [in Russian], Khimiya, Moscow-Leningrad (1966).
160. A. I. Blazhevich, A. V. Lavrov, and E. I. Panasyuk, Izv. Akad. Nauk SSSR Ser. Fiz., 33:980 (1969).
161. W. van Gool, Philips Res. Rep., 13:157 (1958).
162. J. J. Lambe, C. C. Klick, and D. L. Dexter, Phys. Rev., 103:1715 (1956).
163. E. T. Melamed, Phys. Rev., 107:1727 (1957).
164. V. L. Levshin, Yu. V. Voronov, V. B. Gutan, S. A. Fridman, and V. V. Shchaenko, Izv. Akad. Nauk SSSR, Ser. Fiz., 25:392 (1961).
165. A. Pfahnl, Proc. Fifth National Conf. on Advances in Electronic Tube Techniques, 1961, p. 204.
166. A. Pfahnl, Bell Syst. Tech. J., 42:181 (1963).
167. N. K. Morozova, Thesis for Candidate's Degree [in Russian], Moscow Power Institute (1964).
168. K. V. Shalimova, N. K. Morozova, and V. S. Soldatov, Kristallografiya, 8:461 (1963).
169. N. K. Morozova and K. V. Shalimova, Opt. Spektrosk., 21:192 (1966).
170. W. W. Piper, P. D. Johnson, and D. T. F. Marple, J. Phys. Chem. Solids, 8:457 (1959).
171. A. N. Georgobiani and Kh. Fridrikh, Fiz. Tverd. Tela, 12:1086 (1970).
172. A. M. Gurvich, Lectures on the Physical Chemistry of Crystalline Phosphors [in Russian], Moscow Institute of Electronic Machine Construction (1967).
173. R. J. Collins and J. J. Hopfield, Phys. Rev., 120:840 (1960).
174. I. A. Ansel'm, Introduction to the Theory of Semiconductors [in Russian], Fizmatgiz, Moscow-Leningrad (1962), p. 195.
175. C. Kittel, Introduction to Solid State Physics, 2nd ed., Wiley, New York (1956).
176. W. Klein, J. Phys. Chem. Solids, 27:1631 (1966).
177. S. Shionoya, Tech. Rep. Inst. Solid State Phys. Tokyo Univ. A, No. 376 (1969).
178. É. L. Nolle and V. B. Stopachinskii, in: Exciton Generation and Annihilation Processes [in Russian], Nauka, Moscow (1970).

IONIZATION DOMAINS IN STRONG FIELDS AND MOTION OF LUMINOUS REGIONS IN CRYSTALS

M. V. Fok and E. Yu. L'vova

The application of static voltages to manganese-activated sodium zincogermanate crystals produced luminous regions moving from the cathode to the anode at a velocity of about 4×10^{-2} cm/sec. These luminous regions were due to a new type of instability called ionization domains. An analysis was made of the processes of impact ionization in high fields, which can generate and maintain space-charge regions in a crystal. Quantitative estimates based on the experimentally determined values indicated that the field intensity in a double layer associated with such domains was of the order of 2×10^6 V/cm. The effective ionization cross section of donors was estimated to be of the order of 3×10^{-14} cm^2.

Sodium zincogermanate (Na_2ZnGeO_4) crystals were first synthesized a few years ago at the Institute of Crystallography of the USSR Academy of Sciences [1] and are relatively little known.[*] The greatest interest lies in manganese-activated crystals because they are capable of emitting strong luminescence when excited in various ways. The ability to emit electroluminescence in a static field was first reported in [2]; the luminescence appeared in $E \approx 1$ kV/cm at 40°C. However, the electroluminescence was not analyzed in detail in [2]. We found that this luminescence had interesting features which are described below.

Preliminary Experiments

First of all, we decided to determine whether the emitted radiation was electroluminescence because such a check was not carried out in [2]. The question was not trivial, because the luminescence appeared at voltages of the order of 1 kV applied to a crystal several millimeters long and was frequently accompanied by a clearly visible discharge. Therefore, one could not a priori exclude the possibility of the excitation of the luminescence by the ultraviolet light produced in such a discharge or even by hot electrons generated in its plasma.

The following experiments were carried out in order to prove that the luminescence appeared independently of any discharges in air.

1. A crystal which emitted radiation in an electric field was covered by a second crystal which was also capable of emitting radiation when a sufficiently high voltage was applied to its electrodes (Fig. 1). In this case the second crystal could only be excited by a discharge in air which appeared between the electrodes of the first crystal. No luminescence was emitted by the second crystal under these conditions although the first crystal emitted radiation in the usual way. This was repeated using several crystals including two halves of the same crystal and each time the same negative result was obtained. Hence, we concluded that the luminescence

---

[*]In the literature they are frequently referred to as the "phase D" crystals.

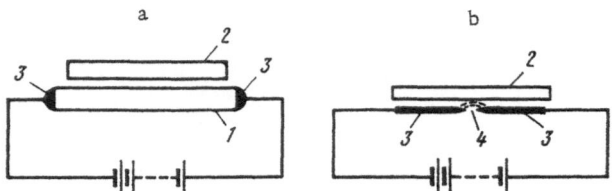

Fig. 1. Investigation of the role of discharges
in air in the excitation of the electroluminescence
of Na_2ZnGeO_4 crystals: a) experiment with two
crystals: b) experiment with a discharge gap;
1) luminous crystal; 2) investigated crystal; 3)
electrodes; 4) discharge.

of the second crystal, excited by the discharge in the air surrounding the first crystal, was
either nonexistent or represented only a small proportion of the total luminescence. However,
this experiment did not exclude the possibility that photoluminescence was excited by the light
generated in the discharge because such an effect could be masked by the scattered luminescence of the first crystal.

2. The investigated crystal (the second crystal in the first experiment) was placed above
a specially constructed discharge gap in which discharges were produced by applying a voltage
of about 2 kV. Again no luminescence was emitted by the investigated crystal. The light produced by the discharge did not interfere with the observations because it was of different spectral composition (the luminescence was green and the discharge was violet). Although one could
not say that the conditions in the discharge gap and at the electrodes of the crystal were identical, there was no reason to assume that the spectral composition of the radiation emitted by
the discharges in these two cases should be very different.

3. The discharge conditions in air were altered drastically by placing a crystal with its
electrodes inside an evacuated chamber. When the pressure was several hundredths of a millimeter of mercury no discharge was observed whereas the luminescence emitted by the crystal
did not differ from that observed at atmospheric pressure.

All these experiments showed that a discharge in air, if it occurred at all, played no significant role in the excitation of the luminescence of manganese-activated sodium zincogermanate crystals and, consequently, the radiation emitted by these crystals was the electroluminescence.

However, the radiation emitted by a discharge in air hindered the study of the electroluminescence. In order to eliminate this factor, we tried various electrodes. The first experiments were carried out using pressure electrodes. Subsequently, we tried Aquadag, indium,
silver paste, and a thermally stable electrically conducting enamel of the 19-102-69 type. The
most reproducible results were obtained when the electrodes were made of this enamel and in
this way most of the influence of discharges in air was eliminated. In the main part of the study
we used mainly the enamel electrodes.

Distribution of Luminescence

The electroluminescence of Mn-activated sodium zincogermanate crystals was excited in
the same way as in [2], using a static voltage applied at a temperature of about 50°C. In these
crystals the concentration of Mn was of the order of a few tenths or hundredths of a percent.
Unactivated crystals did not emit the electroluminescence. Some of the crystals were perfectly
transparent but many had visible internal defects.

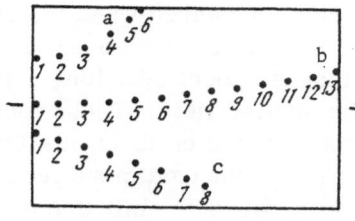

Fig. 2. Motion of luminous spots in crystals excited by a static voltage: a) -c) possible trajectories of a luminous spot; 1), 2), . . . are successive positions of a luminous spot.

We investigated several tens of crystals, which were prepared in different ways, had different activator concentrations, had many or few defects, and were oriented in different ways relative to the applied electric field. In all cases the basic luminescence pattern was the same (Fig. 2).

The application of several hundreds volts (the actual voltage depended on the nature of the sample and of the contacts) produced a weak luminescence throughout a crystal. This luminescence had no visible inhomogeneities. At voltages between 500 V and 1 kV, when the current across a crystal reached several microamperes, a bright yellow spot (almost a point) appeared on the cathode. This spot separated from the cathode and began to move slowly toward the cathode but it did not always follow the shortest path. When the spot appeared, the current became unstable. Sometimes the spot disappeared before reaching the anode and on other occasions it stopped inside the crystal near some visible defect. Still other spots reached the anode and disappeared there. The time taken by a spot to move from the anode to the cathode, which were 5-6 mm apart, was 15-20 sec. When the voltage was increased still further, several luminous spots appeared and some of them began to flicker so that it became very difficult to follow the motion of a single spot. At still higher voltages (about 2 kV) most of the crystal began to emit very strongly and luminous regions of fantastic shape were observed in the bulk of the crystal. The luminescence was highly unstable and it resembled the flow of a luminous liquid.

Comparison with Known Types of Instability

There is no doubt that the phenomenon described above is due to some kind of electric instability. However, the observations do not fit any of the known types of instability in semiconductors. The observed electroluminescence differs from the acoustoelectric instability by the extreme slowness: a "domain" does not move at the velocity of sound but rather at a snail's pace, of the order of 4×10^{-2} cm/sec. Clearly, this low velocity is due to the delay of the charge by some local levels.

Local levels are known to be involved in the recombination instability. However, this instability appears when carriers of both signs are generated rapidly throughout a semiconductor. In wide-gap semiconductors, characterized by a negligible intrinsic conductivity, such generation can only be produced by an external ionizing agent, such as light of sufficiently short wavelengths. In our case no such agent was required in spite of the fact that the forbidden band width of sodium zircogermanate, found in [3] by the electroreflection method, was 3.94 eV.

Deep and shallow local levels are also involved in the temperature—electric instability. This instability arises because of the thermal quenching of the photoconductivity, which is due to an enhanced evolution of the Joule heat in regions with accidentally lower conductivity. However, this instability requires continuous photoexcitation, which is not true of our experiments. The appearance of a high-field region and its slow motion during photoexcitation was observed experimentally by Böer et al. [4] in CdS single crystals. However, the motion of such region was not accompanied by the luminescence.

We found only one published description of a phenomenon similar to that observed by us. It was reported by Diemer who studied CdS single crystals [5]. He gave a description of the phenomenon which we paraphrased as follows: "Orange-red spots appeared when the current was increased. Their number and intensity increased. When the current exceeded 120 μA, a strange optical phenomenon was observed: tiny orange-red tongues about 20μ long separated from the bright spots and crept at an angle with respect to the applied field. These tongues moved at a velocity of about 100 μ/sec. They disappeared at a distance of about 50-100 μ from the anode. When the current was in excess of 200 μA, the tongues appeared more frequently and, finally, formed a continuous 'curtain' spreading from the anode. The emission of light and the current were unstable in this range." Diemer was of the opinion that the observed phenomenon was similar to a streamer discharge in gases but he gave no quantitative estimates. Since the differences between the conditions of motion of electrons in a gas and in a complex semiconductor are quantitatively miles apart, this analogy is not very apt. For example, the mobility of electrons in gases is 3-4 orders of magnitude higher simply because of the lower density. Moreover, we must bear in mind that although the description given above resembles our observations, it differs by the direction of motion of the spots which was from the anode to the cathode, i.e., opposite to that found by us for sodium zincogermanate.

Possible Nature of the Observed Phenomenon

Since our observations do not fit any of the known electric instabilities, we must seek a different mechanism. The appearance of a luminous spot inside a crystal means that inside the spot the field is high and the crystal is excited. The excitation may be due to collisions of hot electrons with the lattice atoms (and consequent ionization) or with the activator atoms (Mn). In the latter case the collisions may not cause ionization because the luminescence may appear also as a result of impact excitation of Mn. However, our experiments indicated that a fairly long (lasting about 10 min) afterglow was observed in the investigated crystals. This suggested a recombination mechanism of the luminescence. In this case a luminescence region does not coincide with a high-field region but is shifted relative to the latter in the direction of travel of the minority carriers.

We investigated the thermoelectric power in order to determine the sign of conduction in our crystals. It was found that the conduction of a freshly prepared surface was always n-type.[*] Consequently, in our crystals a luminous region was located on the "cathode" side of a high-field region.

It follows from elementary electrostatic considerations that an electric field may be concentrated in any region far from the electrodes provided only two layers of charge, one positive and one negative, appear either next to one another or separated by a short distance. Such double layers need not be strictly two-dimensional but may occupy a certain volume and the charges need not be free but may also be localized. If the charge-localization levels are sufficiently deep, a charged region of this kind can move only very slowly across a crystal. Electrons which traverse such a double layer are accelerated inside it and can cause impact ionization of all the centers inside the layer. Estimates given below show that, under certain conditions, both positive and negative layers may be formed and maintained by such ionization. Then, the luminescence is due to the recombination of holes which are generated by the impact ionization of the crystal lattice.

Thus, the main cause of the appearance of a high-field region (a domain) is the impact ionization. Therefore, such regions can be called ionization domains.

We shall now consider what are the conditions necessary for the appearance of ionization domains in a crystal and we shall obtain estimates showing to what extent these conditions are satisfied in our crystals.

---

[*]Crystals with an aged surface sometimes exhibited a local emf of the opposite sign. However, etching or grinding restored the original sign and magnitude of the thermoelectric power.

Estimate of the Space Charge

First of all, we shall estimate the densities of the positive and negative charges which may form a double layer in our crystals. This will allow us to find the field intensity in such a layer and to determine whether the flux of electrons crossing this layer can maintain both positive and negative charges.

The concentration of manganese used as the dopant in our sodium zincogermanate crystals was determined by the ESR method and by a spectroscopic analysis. The two methods gave results in order-of-magnitude agreement. The concentration of Mn could differ several fold from sample to sample but it was always of the order of 10^{19} cm^{-3}. According to the Goldschmidt rule, the radius of a Mn^{2+} ion (0.8 Å) is suitable for the replacement of Zn^{2+} (0.74 Å) or Na^+ (0.95 Å). However, manganese is more likely to replace zinc since the valences of Zn^{2+} and Mn^{2+} are the same, although manganese becomes electrically active only when it replaces sodium. In the latter case it acts as a donor, which is also confirmed by the sign of the thermoelectric power that corresponds to n-type conduction. The replacement of a germanium atom (0.53 Å) by manganese is unlikely because of the large difference between the ionic radii.

We did not know exactly what proportions of manganese replaced zinc and sodium, respectively. However, an analysis of the photoluminescence spectra of Na_2ZnGeO_4 : Mn crystals, carried out in [6] by the Alentsev method, demonstrated that the proportions of manganese replacing zinc and sodium were of the same order of magnitude. It was also found in [6] that the luminescence spectra of the investigated crystals always had three bands: two were narrow and green ($h\nu_{max}$ = 2.37 and 2.32 eV et 77°K) and one was wide and orange.* An investigation of the structure of the phase D crystals [7] indicated that the lattice of this compound consisted of tetrahedra formed from oxygen atoms and these tetrahedra contained zinc and germanium atoms and half the sodium atoms (Na_I^+). The remaining sodium atoms (Na_{II}^+) were located inside irregular octahedra, which were also formed by oxygen atoms.

It is known that the color of the luminescence emitted by the manganese-activated phosphors depends on the valence of Mn and on its oxygen coordination [8, 9]. Green luminescence is obtained if a divalent Mn^{2+} ion is surrounded by four oxygen atoms [10] and orange or red luminescence is obtained if such an ion is surrounded by six oxygen atoms [8, 11]. Since such a correlation between the color of the luminescence and the oxygen coordination of the Mn^{2+} ion is observed in different host lattices, it is clear that the color of the luminescence is governed primarily by the immediate environment of the impurity centers. Thus, we may assume that in our case the two green bands correspond to two possible positions of manganese in the tetrahedral configuration (replacing Zn^{2+} and Na_I^+), whereas the orange band corresponds to the octahedral environment (replacing Na_{II}^+). The band with a maximum at 2.37 eV is strongest in all the spectra whereas the relative intensities of the other two bands vary from sample to sample. This can be explained by the influence of the redox potential of the medium on the replacement of univalent sodium with divalent manganese. It also follows that at least one of the bands (the orange one) can be attributed definitely to manganese replacing Na_{II}^+ in the tetrahedral environment. The intensity of this band is at least several percent of the intensity of the main green band. Hence, it follows that at least several percent of manganese occupies the octahedral sodium positions. The concentration of manganese in the tetrahedral positions (replacing N_I^+) may be of the same order of magnitude although the weak green band with a maximum at 2.32 eV cannot easily be separated from the background of the strong band at 2.37 eV. Therefore,

---

*Moreover, the spectra of two crystals exhibited an additional band in the near infrared but this band was probably due to some other impurity, since it was not observed in the spectra of most of the other crystals.

the sodium sites are occupied by not more than 10% of the total number of manganese atoms.[*] It follows that the donor concentration in our crystals is of the order of 10^{18} cm^{-3}.

A less direct evidence that the concentrations of manganese replacing sodium and zinc are similar is provided by the results of an investigation of the ESR spectra of Mn^{2+} in our crystals. An exceptionally large number of lines is found in these ESR spectra.[†] Sodium zincogermanate has only one symmetry element (a plane of symmetry). The existence of this element is manifested by the pairwise merging of the ESR lines when the magnetic field is directed along the plane of symmetry. However, even then about 100 lines remain in the ESR spectrum. Such spectra are characteristic of Mn^{2+} because unactivated crystals do not exhibit these spectra. Therefore, obviously all these lines belong to manganese. The very large number of lines indicates that there are at least three different positions of manganese ions in the investigated crystals. This can be compared with the three possible ways of incorporation of manganese in the lattice mentioned above. However, until the ESR spectrum is interpreted (this is a very difficult task because of the large number of lines), such an attribution can be regarded only as a tentative hypothesis. Nevertheless, the complexity of the ESR spectrum is itself an indirect confirmation that manganese replaces not only zinc but also sodium because otherwise the ESR spectrum would be much simpler.

The absolute intensities of the lines in the ESR spectrum differ by not more than one order of magnitude. Therefore, if we assume that even the weakest lines in this spectrum are due to manganese at the sodium sites, it follows that if the total concentration of manganese is 10^{19} cm^{-3}, the donor concentration is $N_d \approx 10^{18}$ cm^{-3}. This is in agreement with an estimate of the total donor concentration deduced from the photoluminescence spectra.

The highest density of the positive space charge should be somewhat lower than the donor concentration because a crystal contains also acceptors whose concentration may be comparable with the concentration of manganese occupying the sodium lattice sites. Hence, it follows that the density of a positive space charge is of the order of 10^{17} cm^{-3}. A more detailed calculation shows that this density is 2×10^{17} cm^{-3}. The density of a negative charge in a double layer may be of the same order of magnitude.[‡]

Estimate of the Field Intensity in a Double Layer

The field intensity in a space charge layer can be found from

$$\mathscr{E} = \sqrt{\frac{8\pi\rho}{\varepsilon} V_0}. \tag{1}$$

Here, ρ is the charge density, assumed to be uniform in the layer. The electric field $\mathscr{E}$ acting in the double layer can be estimated if we know the permittivity of the crystal ε and the potential drop V_0 across the double layer.

The static permittivity ε of sodium zincogermanate is not known. All that is known is its refractive index n = 1.67 [12]. The permittivity should naturally be higher than $n^2 = 2.8$ because of a considerable contribution of the ionic mechanism to the binding in crystals as complex as sodium zincogermanate. On the other hand, if we exclude ferroelectrics, we can say

---

[*]We must bear in mind that Mn replacing Na acts as a donor in an n-type semiconductor and has a ratio of the electron- and hole-capture cross sections unfavorable for a large recombination flux. Therefore, our estimates of the relative concentration of Mn occupying Na can only be underestimated.

[†]The authors are grateful to G. E. Arkhangel'skii for a special study of the ESR spectra of our crystals.

[‡]Here and later the unit of charge is the charge of a free electron.

that the permittivity of inorganic crystals rarely exceeds 10-12. Therefore, we shall take the value corresponding to the mid-point of this range and assume that $\varepsilon = 6$. The error in the value of $\mathscr{E}$ cannot exceed a factor of 1.5 because it follows from Eq. (1) that $\mathscr{E}$ is proportional to $\varepsilon^{1/2}$.

The potential drop V_0 across the double layer is more difficult to estimate. If the double layer crosses the whole crystal, almost all the applied potential may be concentrated in it provided the resistivity of the crystal is not too high. All the electrons escaping from the cathode would then have to cross such a double layer on their way to the anode. However, our observations show that the luminous spots and, consequently, the associated double layers are only a few tenths of a millimeter in diameter, i.e., their areas are much smaller than the cross-sectional area of a crystal. Therefore, electrons traveling along a crystal may simply bypass such double layers. This is particularly likely because the field on one side of a double layer is stronger than the field in the region between the layer and the electrodes. To obtain information on this point, we calculated the distribution of the potential in a crystal on the basis of the following assumptions.

1. A crystal is homogeneous and isotropic in respect of its electrical properties.

2. Electrodes on the surface of a crystal are plane and so large that in the part of crystal of interest to us the edge effects can be ignored.

3. A double layer is located half-way between the electrodes (Fig. 3I). The densities in the component positive and negative layers are equal in their absolute values, and the thicknesses of the component layers are also equal.

4. The charge layers are bounded laterally by a cylindrical surface whose diameter a is considerably less than the distance between the electrodes b, but much greater than the thickness of the charge layers c.

The results of our calculations are shown in Fig. 3II. This figure shows the distribution of the potential along the symmetry axis of a double layer, which is denoted by the line AB in Fig. 3I. The curves depend on the fraction of the total potential V which is concentrated in the double layer. We can see that if the fraction V_0 is large, a maximum and minimum of the po-

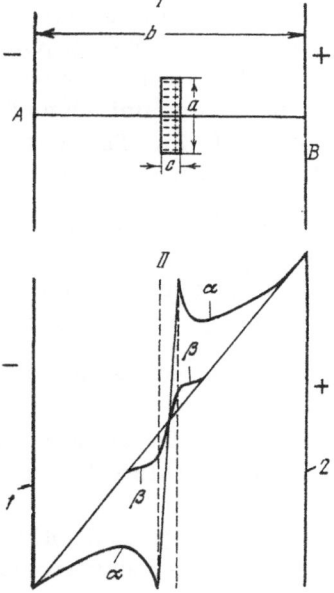

Fig. 3. Model of a high-field region (I) and distribution of the potential along the symmetry axis of a double layer (II) a) diameter of a high-field region; b) distance between electrodes; c) thickness of space-charge layer; 1), 2) electrodes; α is the potential in the case when $V_0 = V$; β is the potential when $V_0 = Va/b$.

tential are formed. Therefore, electrons arriving from the cathode cannot enter the double layer. Calculations show that these extrema disappear if

$$V_0 \leqslant V \, a/b, \tag{2}$$

where V is the potential difference applied to the whole crystal. A considerable reduction in the field near the double layer is maintained until the potential difference falls to less than half the limit stated above. Therefore, in real situations the value of V_0 is half the limit predicted by Eq. (2).

In our case the ratio a/b was of the order of 0.1. Consequently, about $V_0 = 50$ V should be concentrated in a double layer when the applied potential is V = 1 kV. If the positive and negative charge densities are about 2×10^{17} cm^{-3}, the field $\mathscr{E}$ in such a layer is of the order of 2×10^6 V/cm.

Estimate of the Electron Energy in a Strong Field

We shall now consider the energy distribution of electrons in a high-field region. This distribution governs the relative rates of the processes which occur under the influence of a strong field. Electrons moving in such a field should rapidly acquire an energy sufficient for the ionization of centers of any depth and of the host lattice. In fact, in order to acquire an energy of 10 eV, an electron need only travel a distance of 500 Å in a field $\mathscr{E} = 2 \times 10^6$ V/cm without a collision. This energy an electron can lose only by ionizing the host lattice.

The actual probability of a collision between an electron and an impurity in a path of this length is extremely low even if impurities are present in high concentrations. The average distance l_{imp}, which an electron may travel before colliding with an impurity atom, is

$$l_{imp} = 1/SN. \tag{3}$$

Here, S is the cross section for the interaction with an impurity and N is the impurity concentration. If $N = 10^{19}$ cm^{-3} and $S = 10^{-15}$ cm^2, we find that $l_{imp} = 10^4$ Å. This means that in a distance of 500 Å only 5% of all the electrons may collide with impurity atoms and thus lose their energy. It follows that the probability of the collisions with impurities is very low and, therefore, the loss of energy due to such collisions can be ignored.

Collisions with phonons are more frequent but in each such collision an electron loses much less energy than in the interaction with an impurity. The greatest loss of energy results from the emission of an optical phonon by an electron. If the electron momentum is high, the collision with a phonon changes only slightly the direction of motion of the electron. The energy lost by electrons in the emission of phonons is, in the first approximation, equivalent to the action of a retarding field. The magnitude of this field is governed by the energy of the emitted phonons and by the mean free path of an electron l_0 between two phonon-emission events.

The energy of optical phonons in the lattice of sodium zincogermanate is deduced in [6] from the temperature dependence of the width of the green luminescence band (2.37 eV). It is assumed that the frequency of local vibrations with which a luminescence center interacts is close to the frequency of optical phonons. The energy of optical phonons is thus found to be 0.02 eV.

The mean free path of an electron l_0 is usually about 100 Å or more.* Even if we assume

---

*If it is assumed that the effective mass of an electron in the conduction band is equal to the free-electron mass, this value of l_0 corresponds to an electron mobility of about 150 cm^2 · V^{-1}· sec^{-1}. The mobility of electrons in wide-band semiconductors is usually of the order of several hundreds of the conventional units (cm^2 · V^{-1}· sec^{-1}) so that our estimate of l_0 is close to the smallest values found experimentally. Clearly, a "hot" electron undergoes a much smaller number of collisions. Moreover, not each collision is accompanied by the emission of

that l_0 = 50 Å, the retarding field is found to be $\mathscr{E}_r = 4 \times 10^4$ V/cm. This is almost two orders of magnitude less than the acting field $\mathscr{E}$.

Thus, in a high-field region the electron energy is limited not by the interaction of electrons with impurities or with phonons but by the impact ionization of the crystal lattice.

We can now find approximately the electron energy distribution function on the assumption that the conduction band is parabolic. The upper limit of this distribution lies above the lattice ionization threshold E_{th}. This is due to the fact that the ionization cross section is a finite quantity and, consequently, having reached the ionization threshold an electron must travel a further short distance before the ionization takes place. In this distance an electron may acquire an additional energy ΔE from the strong field. This energy, $\Delta E = E - E_{th}$, can be estimated from the dependence of the ionization cross section S on the electron energy E, which can be represented by the expression

$$S = S_0 \left(\frac{E - E_{th}}{E_{th}} \right)^j ,$$ (4)

where $S_0 = 10^{-14}$–10^{-16} cm^2, and the exponent j can assume the values 1, 2, or 3 [13].

Let us estimate E_{th}. In the case of semiconductors with direct energy gaps and equal effective electron and hole masses, we have $E_{th} = 1.5 E_g$, where E_g is the forbidden band width (band gap); however, if the effective mass of holes is much greater than the effective mass of electrons, we find that $E_{th} = E_g$ [14]. According to [3], the forbidden band width is E_g = 3.94 eV. Consequently, E_{th} lies between 3.9 and 5.9 eV.

If we substitute the average values of the parameters $S_0 = 10^{-15}$ cm^2 and j = 2 in Eq. (4), we find that before an impact ionization event an electron acquires an additional energy ΔE = 2.8 eV if $E_{th} = 1.5 E_g$ and ΔE = 2.1 eV if $E_{th} = E_g$. Consequently, in the former case the average electron energy reaches 8.7 eV and in the latter case it reaches 6 eV. Then, an electron loses a considerable part of its energy by impact ionization. In subsequent estimates we shall assume that the maximum energy reached by electrons is 7.4 eV, which corresponds to the midpoint of the range 6–8.7 eV.

The lower limit of the electron energy distribution function can be found by estimating the electron energy directly after an impact ionization event, because after such an event an electron remains in a high-field region and its energy continues to rise. The ionization requires an energy which we shall assume to be equal to the forbidden band width and the rest of the energy is distributed between three particles, which are two electrons and one hole. If the effective masses of these particles are equal, they share this energy equally and, consequently each particle acquires about 1.6 eV. If the effective mass of holes is greater, only the electrons share the energy. Therefore, the energy of electrons after ionization is about 1 eV. Thus, there are practically no thermal electrons in a high-field region and there are very few electrons even with an energy which is tens of times greater than the thermal energy. In fact, it follows from our estimates that the electrons in a high-field region are highly energetic and the minimum electron energy in such a region is 1.3 eV.

In the range 1.3–7.4 eV the electron energy distribution can be assumed, at least in the first approximation, to be uniform because in this range electrons simply acquire energy and lose a very small proportion of it.

---

a phonon and there are some processes which may result in the absorption of energy. These circumstances reduce the average energy transferred by an electron to the lattice as a result of its interaction with phonons. Therefore, the adopted value of the electron mean free path l_0 = 50 Å can only be an underestimate.

Formation of a Positive Space Charge in a Double Layer

Obviously, a positive space charge layer may appear because of the impact ionization of that proportion of donors which remain nonionized in the equilibrium state. Since the depth of the donor levels does not exceed 1.3 eV (this follows from our experiments), electrons of energy in excess of 1.3 eV can ionize these levels.

The depth of donors can be estimated from the photoconductivity spectra. It is clear from these spectra that photons of energy below 1.2 eV produce a noticeable photoconductivity in our crystals, whereas photons of energy below 0.7 eV produce no photoconductivity even when the energy density reaching a crystal is made the same. This equality of the energy density in both cases allows us to reject with assurance the hypothesis that the increase in the conductivity due to illumination with light of shorter wavelengths results from the heating of a crystal. On the other hand, since the number of photons reaching a crystal illuminated with light of longer wavelengths is greater, the absence of a significant photoconductivity suggests that the photosensitivity edge and, consequently, the depth of the donor levels lie at higher energies. In subsequent estimates we shall assume that $E_d = 1$ eV. This corresponds to the midpoint of the range from 0.7 to 1.2 eV. A donor level separated by 1.0 eV from the conduction band was found more recently [3] by the electroabsorption method.

Hence, it follows that practically all free electrons in a high-field region can ionize donor centers if, of course, these centers contain localized electrons.

In a high-field region all the donors should be ionized. However, hot electrons may escape from this region and ionize donors also in the neutral part of the crystal adjoining this region on the anode side, since they may travel a considerable distance before losing the energy needed for the ionization. In fact, electrons of energy below the ionization threshold of the lattice can dissipate their energy only in two ways, either by the ionization of impurities or by the emission of phonons. As mentioned earlier, the energy losses can be represented by a retarding field whose intensity is about 4×10^4 V/cm. This means that outside a high-field region electrons lose about 4 eV (per 1μ of path) in the emission of phonons. Electrons of energies considerably greater than the ionization threshold of the lattice rapidly lose this energy after such ionization and other electrons can travel several tenths of a micron retaining an energy sufficient for the ionization of donors. We shall approximate the complex dependence of the impact ionization cross section on the electron energy by a step-like function and we shall assume that beginning from an energy exceeding the threshold by one-third, the cross section is constant but below this energy it is zero. We then find that the highest energy of electrons emerging from a high-field region and capable of ionizing the lattice (in the case $E_{th} = 1.5 E_g$) is 7.9 eV and the average energy of such electrons is above 4 eV (this value corresponds to the midpoint of the range from 1.3 to 7.4 eV). Consequently, these electrons can lose 2-3 eV and still retain the ability to ionize donors. They can lose this energy by the emission of phonons in traveling a distance $l_{hot} = 0.5\mu$.

If the neutral donor concentration is 2×10^{17} cm^{-3} and the impact ionization cross section is 10^{-15} cm^2, every hundredth electron can ionize a donor outside a high-field region in a distance $l_{hot} = 0.5\mu$. Moreover, electrons naturally ionize these few nonionized donors which remain in the high-field region or which have appeared there as a result of electron capture. Thus, electrons crossing the whole positive space-charge layer not only maintain this layer but can also create a new positive layer in a region about 0.5μ thick adjoining the existing layer on the anode side.

These ideas can be checked by estimating the cross section for the ionization of a donor S_I by a hot electron knowing the experimentally determined velocity of a luminous spot v and the current density i. This can be done on the basis of the following considerations. Under electron impact conditions the lifetime of a donor electron τ is

$$\tau = \frac{1}{iS_i}. \tag{5}$$

On the other hand, we know that the ionization of donors occurs in a layer l_{hot} thick. The time needed to ionize all the donors in this layer is

$$\tau = \frac{l_{hot}}{v}, \tag{6}$$

since a luminous spot moves a distance l_{hot} in this time.

Equating the right-hand sides of Eqs. (5) and (6), we can express S_i in terms of known quantities:

$$S_i = \frac{v}{il_{hot}}. \tag{7}$$

The velocity of a luminous spot is $v = 4 \times 10^{-2}$ cm/sec and $l_{hot} = 0.5 \mu$. Luminous spots begin to appear when the current through a crystal is about 10μA. In the first approximation we may assume that almost all the current passes through a luminous spot because impact ionization occurs in this spot and each electron reaching the spot produces a large number of free carriers. The diameter of a luminous spot is about 0.5 mm. Hence, the current density i, expressed as an electron flux density, is 2.5×10^{16} electrons $\cdot$ sec$^{-1} \cdot$ cm^{-2}. Substituting these quantities in Eq. (7), we find that $S_i = 3 \times 10^{-14}$ cm^2. Since the value of l_{hot} may actually be several times smaller and since i is known only to within an order of magnitude, we may conclude that this value of the impact ionization cross section of donors is in full agreement with the values found by other means (10^{-14}-10^{-16} cm^2).

It must be stressed that in estimating the quantities which occur in Eq. (7) we have not made any assumptions about the magnitude of the impact ionization cross section of donors. Therefore, the agreement reported above can be regarded as an independent proof of the correctness of the proposed mechanism of the formation of a positive space charge layer.

Formation of a Negative Space Charge in a Double Layer

The mechanism of formation of a negative space charge is somewhat more complex. It can be described as follows. In a high-field region electrons may not only ionize the crystal lattice and donors but they may also deionize donors, i.e., they may transfer electrons from the valence band to an empty donor level (Fig. 4). The threshold energy of this process can be estimated if the depth of the donor level is subtracted from the forbidden band width. In this way we obtain about 3 eV, which is slightly less than the average energy of electrons in a

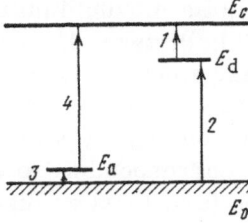

Fig. 4. Electron transitions in a strong field. Here, E_d and E_a are the donor and acceptor levels; 1) ionization of a donor; 2) deionization of a donor by electron impact; 3) tunnel liberation of a hole from an acceptor; 4) liberation of an electron from an acceptor.

high-field region. Applying the same approximation to the dependence of the ionization cross section on the electron energy, we find that the probability of ionization of donors by hot electrons in a high-field region is only 1.5 times greater than the probability of their deionization. According to our estimates, electrons in a high-field region have an energy which is one-third higher than the threshold energy for the ionization of donors, and consequently, they can ionize them with an equal probability. The energy needed for the deionization of donors is, according to our approximation, about 4 eV (the threshold energy is 3 eV). This or higher energy is possessed only by about two-thirds of electrons. However, this is sufficient for a significant deionization process. In fact, the role of the impact deionization of donors is probably even greater because at energies several times greater than the threshold value the probability of the interaction between a hot electron and an impurity decreases. Most of the electrons in a high-field region lose an energy which is several times greater than the donor ionization threshold. Consequently, the donor ionization should be affected by a reduction in the ionization cross section at high energies. On the other hand, the energy of the hottest electrons corresponds approximately to the maximum of the cross section for the donor deionization.

If the probability of the tunnel ionization of donors is less than the probability of their impact ionization, about two-fifths of all the donors are filled with electrons under the conditions discussed above. If the acceptor concentration is so close to the donor concentration that the degree of electron occupancy of the donors is even less in the equilibrium state, the net charge in a high-field region may become negative under the influence of hot electrons. This happens when a further condition is satisfied, namely that practically all the acceptors are occupied by electrons in and out of the high-field region. This condition is satisfied if the acceptor levels are sufficiently shallow to prevent the capture of holes by the acceptors in a high-field region so that holes are transferred by the tunnel effect to the valence band.

We shall now consider to what extent the conditions for the appearance of a negative space charge are satisfied in a high-field region. The probability of the tunnel liberation of electrons can be estimated from the following formula [15]:

$$w = \frac{e\mathscr{E}d}{2\pi\hbar} \exp\left(-\frac{\pi\sqrt{2m^*}}{2e\hbar\mathscr{E}} E_d^{3/2}\right), \tag{8}$$

where $\mathscr{E}$ is the electric field intensity; E_d is the depth of a donor level; m^* is the effective mass of electrons, assumed to be equal to the free-electron value; d is the lattice constant. For a level at a depth of 1 eV in a field of 2×10^6 V/cm acting in a lattice with a constant of the order of 10 Å, this probability is of the order of 10^{-3} sec^{-1}. It should be compared with the lifetime of a donor electron under electron impact conditions, which can be estimated from the velocity of a luminous spot v.

As established above, impact ionization outside a positive space charge region occurs in a layer about $0.5\,\mu$ thick. If the velocity of a moving luminous spot and, therefore, of a space charge region is $v \sim 4 \times 10^{-2}$ cm/sec, this distance is crossed in 1.25×10^{-3} sec. During this time the donors should become completely empty because there are practically no electrons in the donor levels in a positive space charge region. This means that the lifetime of an electron in a donor level under the impact of electrons flying out of a high-field region is about 10^{-3} sec, which is much smaller than the lifetime corresponding to the tunnel ionization probability. Since the probability of impact ionization is at least as high in a high-field region, it follows that the first condition is satisfied with some margin even at a point where the field in the double layer is highest.

The second condition is satisfied for the assumed degree of compensation* of 0.8 (under these conditions the density of the positive space charge is equal to the density of the negative

---

*We recall that the concentration of uncompensated donors is assumed to be 2×10^{17} cm^{-3} for $N_d = 1 \times 10^{18}$ cm^{-3}.

charge, as assumed in estimating the acting field). The concentration of acceptors in sodium zincogermanate can be found from the composition and compensation of impurities deduced from a spectroscopic analysis. According to this analysis, the impurity content of our crystals varies strongly from sample to sample but there are always impurities which can form acceptor centers if they occupy a suitable lattice site (for example, Cu^+ occupying the Zn^{2+} sites or Al^{3+} occupying the Ge^{4+} sites). The concentration of these impurities is several times samller than the concentration of magnanese. However, if we bear in mind that not all manganese atoms become donors, it becomes clear that the degree of compensation of our crystals may be quite high. Therefore, our assumption on the degree of compensation is likely to be correct.

Finally, the last condition may be satisfied if the depth of the acceptor level is less than 0.6 eV. A strong dependence of the probability of the tunnel liberation on the depth of this level follows from Eq. (8), where the level depth occurs in the argument of the exponential function and is present in this argument in a power of 3/2. Nothing is known about the depth of the acceptor levels in sodium zincogermanate. All that we can say is that the range from zero to 0.6 eV is sufficiently wide for an acceptor level to be located there with a high probability.

Appearance of Ionization Domains

Thus, we have established that once a double electric layer and an associated high-field region are formed, they are maintained for a long time and can move toward the anode, as found in our experiments. We shall now consider how such a double layer may appear. We have mentioned at the beginning of the paper that a luminous spot appears in a crystal near the cathode surface. Therefore, we may assume that this spot appears because of the formation of a positive space charge layer.

The contacts with our crystals were nonohmic since no special care was taken to produce ohmic contacts. The nonohmic nature of the contacts is also indicated by the nonlinearity of the current-voltage characteristic even in the range of voltages too low for the appearance of a luminous spot. All these observations indicate the existence of a barrier between the electrode and the crystal or in the cathode region of the crystal. This barrier hinders the penetration of electrons from the cathode into the crystal. Therefore, when a sufficiently high voltage is applied, free electrons are removed from the cathode region but their loss is not compensated by a corresponding flow of electrons from the cathode. Consequently, a positive space charge appears near the cathode and a considerable fraction of the applied potential may be concentrated in that region. Gradually, the density of the accumulated charge rises because more and more deep donors are emptied. Since the applied potential is constant (at least its sign does not change), eventually all the donors in the cathode region lose their electrons. The field in this part of the crystal may reach a value sufficient for the onset of the processes discussed above.

Since the contacts and even the crystals themselves are not fully homogeneous, a space charge layer may be denser at some points or it may form a little earlier. At these points the impact ionization and deionization of donors begins earlier. Since a negative space charge begins to form in the adjoining high-field region, the whole double layer begins to move toward the anode. The luminescence appears soon after the formation of a negative space charge layer because on the side of this layer facing the cathode the electric field is weaker and, therefore, the conditions are more favorable for the recombination of the minority carriers (holes) generated by the impact ionization of the lattice. As the negative charge density increases, the luminescence brightness becomes greater. When a double layer is fully formed and begins to move toward the anode, the luminous region follows it extending over a distance equal to the diffusion length of free holes.

Since a double layer that breaks away from the cathode is subject to a small proportion of the total potential, the conditions in the cathode region occupied initially by the double layer differ little from the conditions elsewhere in the cathode region as soon as the double layer moves away a distance of the order of its diameter. Therefore, a new spot may appear at a different or at the same place near the cathode. It is found experimentally that there is no correlation between the points of appearance of the first and subsequent luminous spots.

It follows from our discussion that the appearance of luminous spots can be explained in a natural maner by the same processes which are responsible for their subsequent motion. The only additional assumption which is needed (the existence of a barrier at the cathode) is also very likely to be true although it has not yet been proved experimentally because of the relatively high resistivity of the investigated crystals.

Conclusions

We can now formulate the conditions under which ionization domains may appear in a crystal if it is subjected to a sufficiently high voltage. It follows from our analysis that an ionization domain is kept stable in an electric field in the same way as a whirlpool in a river is maintained by the flow of the river. Therefore, we can speak of the stability conditions for ionization domains.

Ionization domains may appear in a semiconductor if it has a sufficiently wide forbidden band. Then, the upper half of this band contains levels from which electrons are liberated thermally in negligible amounts compared with the liberation by impact ionization. (At low temperatures this condition can be satisfied also by semiconductors with narrower forbidden bands but a visible luminescence may not appear because the photon energy is always less than the forbidden band width.) Moreover, a semiconductor should contain both donors and acceptors in sufficiently high concentrations but below the level at which hopping conduction becomes significant. In this respect more favorable conditions are found in semiconductors with heterodesmic lattices in which impurity atoms are surrounded by oxygen tetrahedra since in this situation impurities are separated by large potential barriers resulting from the presence of oxygen atoms between them (sodium zincogermanate is a crystal of this kind). The difference between the donor and acceptor concentrations should be sufficiently large in the absolute sense to ensure a sufficiently high positive space charge formed as a result of complete ionization of the donors and to ensure also a sufficiently high electric field associated with this charge. On the other hand, the ratio of these concentrations should be quite close to unity to ensure the appearance of a negative space charge.

The donor levels should be sufficiently deep so that electrons are not liberated from donors by the tunnel effect in strong fields. On the other hand, these levels should be several times closer to the conduction than to the valence band so that the threshold energies for the impact ionization and deionization of the donor levels are quite different. This facilitates the formation of a negative space charge. The levels contributed by acceptors should be as shallow as possible so that holes cannot be captured by these levels in strong fields but are transferred by tunnel effect to the valence band.

The mobilities of electrons and holes should differ as much as possible because otherwise holes also cause impact ionization and a double layer loses its stability because of the development of avalanche processes. (In our crystals these processes begin at voltages above 1.2 kV, when luminous spots begin to flicker, and this is indirect evidence of the instability of the space charge layer.)

The appearance of an ionization domain is accompanied by the luminescence if all the above conditions are satisfied and a semiconductor contains also centers capable of emitting light as a result of recombination of the minority carriers.

Each of the above conditions is satisfied quite often but the necessary combination, unless deliberately attempted, can only arise because of a happy coincidence. Obviously, this is the case in Na_2ZnGeO_4:Mn crystals. The discovery of ionization domains and a study of the conditions for their appearance may now lead to the discovery of similar phenomena in other wide-gap semiconductors.

LITERATURE CITED

1. I. P. Kuz'mina, O. K. Mel'nikov, and B. P. Litvin, in: Hydrothermal Synthesis of Crystals (ed. by A. N. Lobachev), Consultants Bureau, New York (1971), p. 99.
2. K. A. Verkhovskaya, I. P. Kuz'mina, A. N. Lobachev, and V. M. Fridkin, Fiz. Tverd. Tela, 10:1906 (1968).
3. V. S. Vavilov, V. B. Stonachinskii, and Fan ba Nyan, Kratk. Soobshch. Fiz., No. 4, 3 (1972).
4. K. W. Böer, H. J. Hänsch, and U. Kümmel, Z. Phys., 155:170 (1959).
5. G. Diemer, Philips Res. Rep., 9:109 (1954).
6. G. E. Arkhangel'skii, E. Yu. L'vova, and M. V. Fok, Zh. Prikl. Spektrosk., 14:97 (1971).
7. É. A. Kuz'min, V. V. Ilyukhin, and N. V. Belov, Kristallografiya, 13:976 (1968).
8. M. A. Kosntantinova-Shlezinger, Zh. Eksp. Teor. Fiz., 21:252 (1951).
9. V. V. Osiko and G. V. Maksimova, Opt. Spektrosk., 9:478 (1960).
10. M. A. Konstantinova-Shlezinger, Izv. Akad. Nauk SSSR Ser. Fiz., 30:707 (1966).
11. N. A. Gorbacheva and A. I. Kabakova, Zk. Prikl. Spektrosk., 6:478 (1967).
12. I. P. Kuz'mina, Thesis for Candidate's Degree [in Russian], Institute of Crystallography, Academy of Sciences of the USSR, Moscow (1968).
13. V. A. Chuenkov, in: Injection Electroluminescence [in Russian], Tartu State University (1968), p. 116.
14. J. R. Hauser, J. Appl. Phys., 37:507 (1966).
15. K. B. McAfee, E. J. Ryder, W. Shockley, and M. Sparks, Phys. Rev., 83:650 (1951).

INVESTIGATION OF THE AMBIPOLAR DIFFUSION OF FREE CARRIERS IN INHOMOGENEOUSLY PHOTOEXCITED ZINC SULFIDE CRYSTALS

N. N. Grigor'ev and M. V. Fok

Several methods, based on an investigation of the luminescence of a crystal, were developed for the determination of the transport processes and their range. The luminescence contour method was used to show that the range of ambipolar diffusion was 5-10 μ, whereas the bulk of the crystal was excited by the reabsorption of the luminescence. A new transport parameter α_0, representing the change in the recombination rate under diffusion conditions and the spatial distribution of the nonequilibrium carrier density, was introduced in a theoretical justification of this method for semiconductors with the quadratic recombination kinetics. Investigations of the influence of the excitation rate, size of the excitation region, long-wavelength background radiation, and an alternating electric field on the spatial distribution of the luminescence excited in a crystal by a narrow strip of light yielded results in good agreement with the theory.

Introduction

The processes of transport of energy and charge in semiconductors with relatively narrow forbidden bands have been thoroughly investigated and have already found practical applications (for example, thermoelements and solar cells). However, the situation is not as favorable in the case of semiconductors with wider forbidden bands. Relatively little work has been done on the transport phenomena in such semiconductors. This is due to the fact that the wide-gap semiconductors have a number of features which make it difficult to study these phenomena.

Theoretical descriptions of the transport phenomena are usually based on the concept of the average lifetime of nonequilibrium carriers [1-3]. This concept is very fruitful in those cases when the average lifetime is constant. The transport equations then become linear and can be solved quite easily. This is always true in those cases when the recombination alters significantly the density of only one of the particles participating in the recombination processes. In the case of wide-gap semiconductors, this condition is not always satisfied. Such semiconductors have deep donors and acceptors present in almost equal concentrations. Consequently, the equilibrium densities of free carriers are usually so low that even a weak excitation increases the densities of free carriers of both signs by a large factor and this alters the charge states of the recombination centers. Consequently, the average free-carrier lifetime before recombination begins to fall rapidly with rising excitation rate and this gives rise to a quadratic rather than linear recombination term [4]. This is particularly true of zinc sulfide [5].

Experimental difficulties encountered in investigations of transport phenomena in wide-gap semiconductors are primarily due to their high resistivity which makes it difficult to

apply the usual well-developed electrical investigation methods and sometimes renders these
methods completely unsuitable [6]. On the other hand, most of the wide-gap semiconductors
emit visible luminescence and this makes it possible to study the transport phenomena employing the luminescence method developed for homogeneously excited crystals. However,
this method has hardly been used in studies of such phenomena. This is unfortunate because
the transport phenomena in wide-gap semiconductors are very important and can probably be
used as successfully as the phenomena in narrow-gap semiconductors. The main difficulty
is that the available information is insufficient.

Several different mechanisms of energy transfer may take place in semiconductor compounds. These include the scattering and reabsorption of luminescence, inductive-resonance
mechanism, transfer of energy by excitons, and ambipolar diffusion of free carriers. These
mechanisms may act simultaneously although under certain conditions one usually predominates.
For example, the resonance mechanism and reabsorption are favored by the overlap of the
absorption and recombination radiation spectra of the luminescence centers. If the concentration of such centers is low, the reabsorption process predominates in the transport of energy
over long distances. On the other hand, the resonant energy transfer is favored by an increase in the concentration of the luminescence centers because the probability of such transfer
is inversely proportional to the sixth power of the distance between the centers. The
transfer of energy by excitons is important at low temperatures and low impurity and defect
concentrations because, under these conditions, the exciton lifetime is long and an increase
in the excitation intensity raises strongly the number of excitons. Such an increase in the
excitation intensity is undesirable if the energy is transferred by ambipolar diffusion because
this increase reduces the lifetime of free carriers.

If follows from this discussion that the question of the energy transfer mechanism is
not trivial. The purpose of our present paper is to report a theoretical and experimental investigation of the photoexcitation energy transfer in ZnS crystals under strongly inhomogeneous
excitation conditions.

The prime aim was to establish the nature of the mechanism responsible for the transfer
of the photoexcitation energy and to estimate the parameters of this mechanism. Since the
investigated material is zinc sulfide, characterized by a high energy efficiency of the luminescence,
we can use the luminescence method for investigating the energy transfer processes developed
by us. However, the methods used to calculate the transport phenomena on the basis of ambipolar diffusion are general and can be applied to the diffusion of nonequilibrium carriers in
any semiconductor compound with a quadratic kinetics of the recombination processes.

1. Energy Transfer over Long Distances

The energy transfer was detected by the following method. The crystal was illuminated
with light of $\lambda < 300$ nm, which excited directly only a thin surface layer. The thickness of
this layer, estimated from the absorption coefficient, was $0.1-0.3\,\mu$. Next, the excitation was
stopped, a layer of a certain thickness was removed from the surface by etching, and the
light sum in the remainder of the crystal was measured. This light sum was deduced from
the luminescence flash induced by the rapid heating of a crystal for different periods after
the end excitation. As a check, in some cases a crystal was excited not with ultraviolet light
but with light from an electroluminescent capacitor operated in such a way that its spectrum
was identical with the photoluminescence spectrum of the investigated crystal.

These experiments showed [7] that the bulk of the crystal was excited by the reabsorption of the luminescence because the light sum was distributed homogeneously throughout
the crystal and it was independent of whether the crystal was excited with ultraviolet light
or visible light produced by an electroluminescent capacitor. Moreover, it was found that,
under short-wavelength ($\lambda < 300$ nm) excitation conditions, there was a directly excited layer

as well as a second layer about 5 μ thick where the excitation did not penetrate by reabsorption because this layer was excited very inhomogeneously and more strongly than the rest of the crystal.

Since a large number of free-carrier pairs was produced when ZnS was excited in the fundamental absorption region, these carriers could participate in the transfer of energy over the distances involved. Moreover, the participation of excitons in this process was equally likely [8] because they were generated by such excitation. However, the etching method was not sufficiently precise to study the nature of the energy transfer over such distances. Therefore, we developed a different technique [9], known as the luminescence contour method, in which we determined the distribution of the luminescence brightness along a crystal excited by a narrow ultraviolet beam, producing a brightly luminous transverse strip. This strongly inhomogeneous photoexcitation produced a carrier-density gradient and this gave rise to an ambipolar diffusion whose effective range was governed by the recombination of carriers and by their diffusion and drift parameters.

When a substance was capable of emitting luminescence, such carrier diffusion was manifested by a broadening of the luminescence region compared with the excitation region. A similar broadening appeared in the case of energy transfer by excitons but the distribution of the brightness in the luminescence region and its dependence on the excitation conditions were different. Therefore, the problem of the energy transfer mechanism required a calculation of the kinetics of the processes in a crystal emitting luminescence as a result of a strongly inhomogeneous excitation and a comparison of the results of the calculations with experiments.

2. Theory of the Luminescence Contour Method

Let us assume that a homogeneous semi-infinite crystal is excited by a narrow strip of light which produces a strongly inhomogeneous distribution of the densities of free carriers of both signs. This gives rise to an equalizing diffusion flux of the carriers. The simultaneous or ambipolar motion of free carriers of both signs may transport the excitation energy into that part of the crystal which is not subject to direct optical excitation. We shall consider a very simple model of a crystal phosphor with two local levels. Figure 1 shows the energy band scheme of a crystal under consideration, where the transitions of free electrons are indicated. The transition 1 is accompanied by luminescence. The problem under consideration can be solved using the transport equations discussed in [5].

Under steady-state conditions, the equations which describe the balance of the particles in the allowed bands and in the local levels have the following form when the occupancy of the local levels is low and carrier diffusion takes place:

$$I(x) - \beta n(x) N_-(x) - \delta_1 N_-(x) + w_1 n_1(x) + \frac{1}{q} \operatorname{div} j_{N_-}(x) = 0, \tag{1a}$$

$$I(x) - \beta_1 n_1(x) N_+(x) - \delta N_+(x) + w n(x) - \frac{1}{q} \operatorname{div} j_{N_+}(x) = 0, \tag{1b}$$

$$- w n(x) + \delta N_+(x) - \beta n(x) N_-(x) = 0, \tag{1c}$$

$$- w_1 n_1(x) + \delta_1 N_-(x) - \beta_1 n_1(x) N_+(x) = 0. \tag{1d}$$

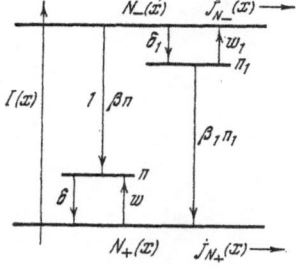

Fig. 1. Electron transitions in a crystal with two levels. The vertical arrows represent the transitions and the symbols alongside give the transition probabilities. The transition denoted by 1 is radiative.

In the above system, $I(x)$ represents the distribution function of the number of ionization events in the lattice along the crystal; $N_-(x)$ and $N_+(x)$ are the densities of free electrons and holes; $n(x)$ is the concentration of the ionized luminescence centers; $n_1(x)$ is the density of the trapped electrons; β, β_1, δ, δ_1, w, and w_1 are the recombination coefficients and the probabilities of the transitions shown in Fig. 1; q is the electronic charge; $j_{N_-}(x)$ and $j_{N_+}(x)$ are the densities of the free-carrier currents. We shall ignore the direct recombination of free carriers and the exciton formation process.

The system given above is complex and cannot be solved in a finite form without additional assumptions. In our case, these assumptions are the conditions

$$\beta_1 N_+ \ll w_1, \ \beta N_- \ll w, \tag{2}$$

which imply that a thermal equilibrium is established rapidly between the levels and the nearest bands.

Expressions of the type $(\delta N + wn)$ in Eqs. (1a) and (1b) can be replaced with expressions of the type βNn, in accordance with Eqs. (1d). Next, substituting the expressions $n = \delta N_+/w$ and $n_1 = \delta_1 N_-/w_1$ obtained from Eqs. (1d) and (1c) into the equations obtained in this way, we find that — subject to the assumption (2) — the system (1a)-(1d) is replaced with two equations

$$I(x) - \left(\beta_1 \frac{\delta_1}{w_1} + \beta \frac{\delta}{w}\right) N_+(x) N_-(x) - \frac{1}{q} \operatorname{div} j_{N_+}(x) = 0, \tag{3a}$$

$$I(x) - \left(\beta_1 \frac{\delta_1}{w_1} + \beta \frac{\delta}{w}\right) N_+(x) N_-(x) + \frac{1}{q} \operatorname{div} j_{N_-}(x) = 0. \tag{3b}$$

In order to solve the system (3a) and (3b), we shall assume that the conditions are quasineutral in any sufficiently small volume of a crystal, i.e., we shall assume that the density of the space charge resulting from the diffusion is small compared with the densities of the nonequilibrium charges of each kind. In other words, we shall assume that the diffusing charges of opposite sign do not spread far apart from one another and, therefore, they neutralize almost completely the space charge that tends to appear. Nevertheless, in two neighboring regions of a crystal, the densities of free and localized carriers of each sign may differ. However, the important factor is that, at each point, the total densities of the localized and free carriers of opposite signs are almost equal, i. e.,

$$\frac{n + N_+}{n_1 + N_-} - 1 \ll 1. \tag{4}$$

The quasineutrality condition does not mean that there is no electric field in a crystal. On the contrary, if the densities and mobilities of the diffusing electrons and holes are different, the quasineutrality condition can be obeyed only because the space charge is partly neutralized: the field of this charge accelerates the lagging carriers and decelerates those that have shot ahead.

According to Eqs. (1c), the conditions (4) and (2) lead to a proportionality between the free-carrier densities N_- and N_+:

$$N_-(x) = \frac{1 + \delta/w}{1 + \delta_1/w_1} N_+(x). \tag{5}$$

Next, we shall assume that

$$\operatorname{div} j_{N_+} = \mathscr{E} \operatorname{div} \sigma_+ + \sigma_+ \operatorname{div} \mathscr{E} - qD_+ \frac{d^2 N_+}{dx^2},$$

$$\operatorname{div} j_{N_-} = \mathscr{E} \operatorname{div} \sigma_- + \sigma_- \operatorname{div} \mathscr{E} + qD_- \frac{d^2 N_-}{dx^2}, \tag{6}$$

where $\mathscr{E}$ is the space-charge field which appears because the quasineutrality condition is not obeyed exactly;

$$\sigma_+ = q\mu_+ N_+, \ \sigma_- = q\mu_- N_- \tag{7}$$

are the conductivities; μ_- and μ_+ are the mobilities; D_- and D_+ are the diffusion coefficients of free electrons and holes related by the Einstein equation $\mu_\pm = (q/kT)\, D_\pm$. The system (3) can then be reduced [4] to

$$I(x) - \nu N_+^2(x) + D_+' \frac{d^2 N_+(x)}{dx} = 0, \tag{8}$$

where

$$\nu = \left(\beta \frac{\delta}{w} + \beta_1 \frac{\delta_1}{w_1}\right) \frac{1 + \delta/w}{1 + \delta/w_1} \tag{9}$$

and

$$D_+' = \frac{2D_+}{1 + \dfrac{\mu_+}{\mu_-} \cdot \dfrac{1 + \delta_1/w_1}{1 + \delta/w}} \ . \tag{10}$$

The above results are obtained by multiplying Eq. (3a) by σ_- and Eq. (3b) by σ_+, adding the results, and applying the relationships (6). In Eq. (8), the quantity D_+' is the ambipolar diffusion coefficient of holes. Replacing the free-hole density in Eq. (8) with the free-electron density $N_+(x)$, we find — in accordance with Eq. (5) — that the diffusion of free carriers under the conditions in question is described by

$$I(x) - \nu' N_-^2(x) + D_-' \frac{d^2 N_-(x)}{dx^2} = 0, \tag{11}$$

where

$$\nu' = \left(\beta \frac{\delta}{w} + \beta_1 \frac{\delta_1}{w_1}\right) \frac{1 + \delta_1/w_1}{1 + \delta/w} \tag{12}$$

and

$$D_-' = \frac{2D_-}{1 + \dfrac{\mu_-}{\mu_+} \cdot \dfrac{1 + \delta/w}{1 + \delta_1/w_1}} \ . \tag{13}$$

In this case, D_-' is the ambipolar diffusion coefficient of electrons and ν' is the quadratic recombination coefficient of electrons..

Introducing

$$\tau_+(x) = [\nu N_+(x)]^{-1} \ \text{ and } \ \tau_-(x) = [\nu' N_-(x)]^{-1} \tag{14}$$

as the lifetimes of electrons and holes in the appropriate bands, we can define formally the ambipolar diffusion length L for an inhomogeneously illuminated crystal. This length is given by

$$L^2(x) = D_+' \tau_+(x) = D_-' \tau_-(x) = D_+' \nu^{-1} N_+^{-1}(x). \tag{15}$$

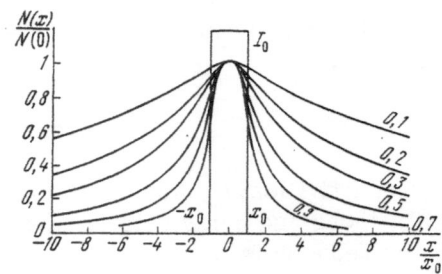

Fig. 2. Single-parameter family of curves representing a normalized distribution of the free-carrier density in the case of a Π-shaped distribution of the excitation (I_0). The numbers alongside the curves give the values of the parameter α_0.

As expected, the ambipolar diffusion length is not a constant of the process but depends on the absolute value of the carrier density which varies along the crystal under consideration. Hence, the concept of the diffusion length, which is essentially the derivative of the lifetime, has a very limited validity in the case of processes characterized by the quadratic recombination.

Equation (8) or (11) can be solved for $N_+(x)$ or $N_-(x)$, if we specify the exact form of the excitation function $I(x)$, for example, the Π-shaped distribution of the intensity of the exciting light (Fig. 2) in the form of a strip $2x_0$ wide: $I(x) = I_0$ when $|x| \leq x_0$ and $I(x) = 0$ when $|x| > x_0$. The origin of the coordinates is assumed to be located in the middle of the strip. In this case, in the excitation region, i.e., in the region where $|x| \leq x_0$, Eq. (8) is

$$\frac{d^2 N_+(x)}{dx^2} - \frac{v}{D'_+} N_+^2(x) + \frac{I_0}{D'_+} = 0. \tag{16}$$

We shall seek the solution of this equation in the form of a series:

$$N_+(x) = \sum_0^\infty C_n x^{2n}. \tag{17}$$

Substituting this series in Eq. (16), equating the coefficients of the terms with identical powers of x, and rejecting terms higher than x^6, we obtain the following solution:

$$N_+(x) = N_+(0)\left[1 + \frac{\alpha_0 - 1}{2}\,\gamma_0 x^2 + \frac{\alpha_0(\alpha_0 - 1)}{12}\,\gamma_0^2 x^4 + \frac{\alpha_0(\alpha_0 - 1)(5\alpha_0 - 3)}{360}\,\gamma_0^3 x^6\right], \tag{18}$$

where $N_+(0)$ is the density of carriers in the middle of the excited strip, and α_0 and γ_0 are the parameters defined by

$$\alpha_0 = \frac{v N_+^2(0)}{I_0} \tag{19}$$

and

$$\gamma_0 = \frac{I_0}{D'_+ N_+(0)}. \tag{20}$$

The parameter α_0 has a fairly clear physical meaning. It represents the ratio of the real rate of recombination at the center of the excited strip under diffusion conditions to the rate of recombination which would have been obtained in the absence of diffusion, i.e., in the case of a homogeneous excitation of the whole crystal. Apart from the excitation rate and the width of the illuminated strip, the quantity α_0 also depends on the properties of the crystal. The more intense the diffusion process, the greater the number of recombination events outside the excited strip and the smaller is the numerical value of α_0. In the absence of free-carrier diffusion, we have $\alpha_0 = 1$. We shall show later that α_0 can be determined quite readily in experiments. The parameter γ_0 does not have such a clear physical meaning.

It should be noted that Eq. (16) can also be solved exactly by an analytic method, employing the elliptic Weierstrass functions. This is pointed out by Blakemore, who refers to a private communication of Nomura [10] and Kamke's handbook [11]. Such a solution is obtained in [12]. However, its form is very inconvenient in any comparison with the experimental results. A representation of the solution as a series (17) has enabled us to obtain a form which can be compared with the experimental data because of the introduction of a new transport parameter α_0, which represents the ambipolar diffusion process in the quadratic recombination processes in the same way as the diffusion length represents the diffusion in the case of linear recombination processes.

In the part of the crystal not excited directly by incident light, i.e., in the region where $|x| > x_0$, Eq. (8) becomes

$$\frac{d^2 N_+(x)}{dx^2} - \frac{\nu}{D'_+} N_+^2(x) = 0. \tag{21}$$

The solution of this equation can easily be obtained by applying the boundary conditions $N_+(x) \to 0$ and $dN_+(x)/dx \to 0$ corresponding to $x \to \pm\infty$. This solution is

$$N_+(x) = \frac{6D'_+}{\nu(x + C_0)^2} = \frac{6N_+(0)}{\alpha_0 \gamma_0 (x + C_0)^2}, \tag{22}$$

where C_0 is a new parameter. We can show that the parameters γ_0 and C_0 depend explicitly only on α_0 and x_0 but are independent of the choice of the values of I_0, ν, D'_+, and $N_+(0)$, which simplifies greatly the calculation of the distribution of the carrier density and the comparison of the theory with experiment. Applying the conditions of continuity of $N_+(x)$ and $dN_+(x)/dx$ at the boundary of the excitation strip $x = x_0$, we obtain two equations:

$$\left[(\alpha_0 - 1)\gamma_0 x_0 + \frac{\alpha_0(\alpha_0 - 1)}{3}\gamma_0^2 x_0^3 + \frac{1}{60}\alpha_0(\alpha_0 - 1)(5\alpha_0 - 3)\gamma_0^3 x_0^5\right]^2 = \frac{2}{3}\alpha_0 \gamma_0$$
$$\times \left[1 + \frac{1}{2}(\alpha_0 - 1)\gamma_0 x_0^2 + \frac{1}{12}\alpha_0(\alpha_0 - 1)\gamma_0^2 x_0^4 + \frac{1}{360}\alpha_0(\alpha_0 - 1)(5\alpha_0 - 3)\gamma_0^3 x_0^6\right]^3, \tag{23}$$

$$C_0 = \frac{-2\left[1 + \frac{1}{2}(\alpha_0 - 1)\gamma_0 x_0^2 + \frac{1}{12}\alpha_0(\alpha_0 - 1)\gamma_0^2 x_0^4 + \frac{1}{360}\alpha_0(\alpha_0 - 1)(5\alpha_0 - 3)\gamma_0^3 x_0^6\right]}{(\alpha_0 - 1)\gamma_0 x_0 + \frac{1}{3}\alpha_0(\alpha_0 - 1)\gamma_0^2 x_0^3 + \frac{1}{60}\alpha_0(\alpha_0 - 1)(5\alpha_0 - 3)\gamma_0^3 x_0^5} - x_0, \tag{24}$$

the first of which allows us to find $\gamma_0 = f_1(\alpha_0, x_0)$, and the second $C_0 = f_2(\alpha_0, x_0)$. We recall that $\alpha_0 = f(I_0, x_0, \nu, D'_+)$.

In this way, we obtain the solution of the second-order nonlinear differential equation which describes the diffusion of free carriers in an inhomogeneously photoexcited crystal in which the recombination kinetics is quadratic. The solution allows for the capture of carriers by traps and their liberation. It is obtained in the form of a distribution of the density of free holes. The distribution of free electrons can easily be obtained by applying Eq. (5).

Figure 2 shows graphically the solutions obtained. Since x_0 is the parameter governed by the experimental conditions, it follows that $N(x)/N(0)$ is a single-parameter family of functions of the transport parameter α_0. Calculations show that the fall of $N(x)$ at $|x| \to \infty$ slows down with decreasing value of the parameter α_0, i.e., the distribution curve of the carrier density becomes flatter (Fig. 2).

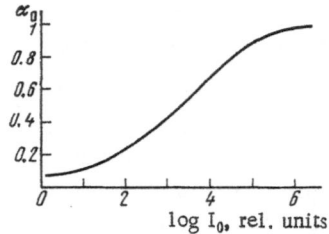

Fig. 3. Dependence of the transport parameter α_0 on the excitation rate $E = kI_0$, calculated in accordance with Eq. (25) for $\nu/(D'_+)^2 = 1$.

The relationships (19) and (20) allow us to determine the dependence of the parameter α_0 on I_0 and, therefore, on the intensity of the exciting light E. We find that

$$\alpha_0 \gamma_0^2 = \frac{\nu}{(D'_+)^2} I_0. \tag{25}$$

Since ν and D'_+ are independent of the excitation intensity (the de-exciting action of the incident light is ignored) and since γ_0 is, in accordance with Eq. (23), a function of only α_0 and x_0, and calculations show that this function rises with α_0, it follows that the parameter α_0 increases with I_0 (Fig. 3). Thus, as I_0 rises, the rate of fall N(x) accelerates and this results in a broadening of the initial distribution of the carrier density. This prediction can be explained by assuming that, when the intensity of the exciting light is increased, the probability of recombination of free carriers increases because it is proportional to the concentration of the ionized centers and the density of free carriers, so that the lifetime of free carriers decreases and, consequently, the distance to which these carriers can diffuse becomes shorter [see Eq. (15)].

The dependence of α_0 on the dimensions of the excitation region, i.e., on the value of x_0, can be found using the same relationship (25). Since the right-hand side of this equation is independent of x_0, we can find the function $\alpha_0(x_0)$ by plotting a family of the dependences of $\alpha_0 \gamma_0^2$ on α_0 for several values of x_0 and intersecting the family by the straight line $[\nu/(D'_+)^2]$ I_0 = const. Figure 4 shows the results obtained. We can see that as x_0 rises, so does α_0 and, in the case of higher values of $\nu I_0/(D'_+)^2$, the curves $\alpha_0(x_0)$ are located higher.

Knowing the dependence $\alpha_0(x_0, I_0)$, we can use Eq. (19) to show that the carrier density in the middle of the excited strip N(0) rises with increasing α_0, I_0, and x_0.

Since we are assuming that the radiative recombination occurs between a free electron and a localized (at a luminescence center) hole, it follows from Eq. (1a) that the term describing the luminescence is of the form $\beta n(x) N_-(x)$. Having found the value of n(x) from Eq. (1b) subject to the condition (2) and having applied Eq. (5), we obtain (to within a constant factor) the luminescence brightness

$$B(x) = \beta \frac{\delta}{w} \cdot \frac{1 + \delta/w}{1 + \delta_1/w_1} N_+^2(x). \tag{26}$$

Fig. 4. Dependence of the transport parameter α_0 on the size of the excitation region x_0 for several values of the quantity $\log(\nu I_0/(D'_+)^2)$ (given alongside the curves).

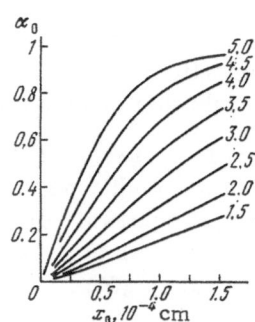

Thus, introducing the dimensionless quantities $X = x/x_0$, $C = C_0/x_0$, and $\gamma = \gamma_0 x_0^2$ with the aid of Eqs. (18) and (22), we find that, in the excitation region (where $|X| \leq 1$), the luminescence brightness is

$$B(X) = \beta \frac{\delta}{w} \left(\frac{1 + \delta/w}{1 + \delta_1/w_1} \right) N_+^2(0) \left[1 + \frac{\alpha_0 - 1}{2} \gamma X^2 + \right.$$
$$\left. + \frac{\alpha_0(\alpha_0 - 1)}{12} \gamma^2 X^4 + \frac{\alpha_0(\alpha_0 - 1)(5\alpha_0 - 3)}{360} \gamma^3 X^6 \right]^2 \tag{27}$$

and outside the excitation region (where $|X| > 1$) this brightness is

$$B(X) = \beta \frac{\delta}{w} \left(\frac{1 + \delta/w}{1 + \delta_1/w_1} \right) N_+^2(0) \frac{36}{\alpha_0^2 \gamma^2 (X + C)^4}. \tag{28}$$

It is clear from Eq. (25) that although B(X) is a stronger function of the coordinate than $N_+(x)$, the qualitative conclusions reached earlier on the dependences of $N_+(x)$ on α_0, x_0, and I_0 also remain valid for B(X). It follows that smaller values of α_0 correspond to less steep dependences B(X) and, correspondingly, as I_0 rises, the fall of B(X) becomes steeper.

In comparison of experimental results with the theory, it is convenient to use the normalized distribution of the brightness, which we have defined as the luminescence contour, i.e., it is convenient to use the function

$$Q(X) = B(X)/B(0). \tag{29}$$

Following Eqs. (26) and (27), we obtain an expression for the luminescence profile in the excitation region ($|X| \leq 1$):

$$Q(X) = \left[1 + \frac{1}{2}(\alpha_0 - 1)\gamma X^2 + \frac{1}{12}\alpha_0(\alpha_0 - 1)\gamma^2 X^4 + \right.$$
$$\left. + \frac{1}{360}\alpha_0(\alpha_0 - 1)(5\alpha_0 - 3)\gamma^3 X^6 \right]^2 \tag{30}$$

and outside the excitation region ($|X| > 1$):

$$Q(X) = \frac{36}{\alpha_0^2 \gamma^2 (X + C)^4}. \tag{31}$$

Figure 5 shows a family of the luminescence contours calculated for different values of α_0. We can clearly see the dependence of the steepness of the luminescence contours on the value of α_0.

Having determined experimentally the luminescence contour for given values of I_0 and x_0 and having established that it agrees with one of the theoretical contours, i.e., that it is

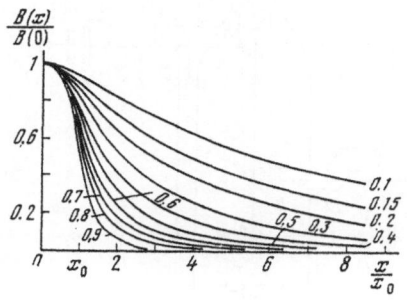

Fig. 5. Family of normalized luminescence contours, calculated in accordance with Eqs. (30) and (31) for several values of the transport parameter α_0 (given alongside the curves).

characterized by a definite value of the parameter α_0, we can apply Eq. (25) and determine the quantity

$$\frac{\nu}{(D'_+)^2} = \frac{\alpha_0 \gamma_0^2}{I_0} \,. \tag{32}$$

Next, we can estimate ν employing Eq. (9) and then find the ambipolar diffusion coefficient of holes

$$D'_+ = \sqrt{\frac{I_0}{\alpha_0 \gamma_0^2}\left(\beta \frac{\delta}{w} + \beta_1 \frac{\delta_1}{w_1}\right)\frac{1+\delta/w}{1+\delta_1/w_1}} \tag{33}$$

and the hole mobility μ_+ deduced from Eq. (10):

$$\mu_+ = \frac{D'_+ \mu_-}{\dfrac{2kT}{q}\mu_- - D'_+ \dfrac{1+\delta_1/w_1}{1+\delta/w}} \,, \tag{34}$$

as well as the density of free holes at the center of the excited strip $N_+(0)$, which — according to Eqs. (19) and (20) — is

$$N_+(0) = \sqrt{\frac{\alpha_0 I_0}{\nu}} = \frac{I_0}{D'_+ \gamma_0} \,. \tag{35}$$

3. Description of Apparatus

As mentioned earlier, the distances over which the excitation energy can be transferred by free carriers in ZnS crystals does not exceeed several microns. Therefore, to investigate the luminescence contours, we built apparatus that made it possible to record the luminescence of a crystal emerging from a very small volume and, at the same time, to record the coordinate of the region being investigated. Special measures were taken to ensure the maximum possible precision of the measurements of the luminescence intensity and coordinate of the luminous region.

Figure 6 is a schematic diagram of the apparatus. The apparatus consisted of two main units: one was used for the excitation and based on an MIM-7 microscope and the other was

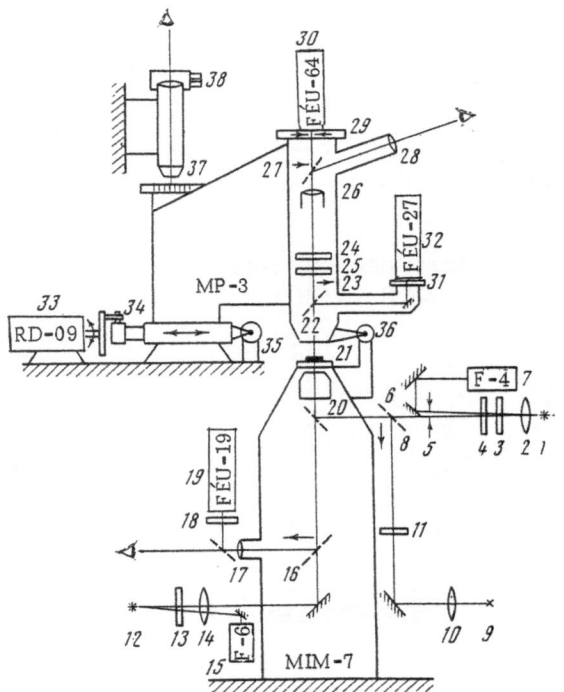

Fig. 6. Apparatus used in measurements of the luminescence contours.

an observation unit composed of two microscopes, one acting as a photometer and based on an MP-3 microscope and the other being a MIS-11 microscope which was used to study the surface of a crystal (this microscope is not shown in Fig. 6). The observation unit was driven relative to the excitation unit by a RD-09 motor. The coordinate of the relative positions of the two units was indicated by a mechanotron (electronic micrometer) [13] bolted rigidly to the excitation unit.

The excitation unit included a source of ultraviolet radiation (1), which was a DRSh-250 mercury discharge lamp, and an OK-40x quartz objective (20), which projected a reduced image of a slit (5) on the lower surface of a crystal (21) placed on the stage of the MIM-7 microscope. A quartz lens (2) made it possible to illuminate uniformly the slit (5). A small mirror (6) deflected some of the light flux to an F-4 photocell (7) to check the constancy of the ultraviolet radiation flux. The spectral composition of this flux was selected with filters (3 and 4). An additional illumination of the whole crystal (or some part of it) with a selected wavelength was produced by an incandescent lamp (12), a combination of filters (13) for the selection of the spectral range 0.7-0.8 or 1.1 μ, a lens (14) and the same OK-40x objective (20). A F-6 photocell (15) was used to measure the intensity and to monitor the constancy of this additional (background) illumination. A visual observation system (16, 17) of the MIM-7 microscope was used in recording the integrated brightness of the luminescence with the aid of a photomultiplier (19). A solution of sodium nitrite (16) absorbed the reflected and scattered ultraviolet exciting radiation.

A glass (20× or 40×) microscope objective (22) together with a 15× ocular (26) produced, in the plane of a second slit (29), a magnified image of the region emitting luminescence. This slit (29) was oriented parallel to the luminescence-emitting strip and it selected from the image a region corresponding to a strip of specified width on the surface of the crystal (this width was usually 0.5-0.7 μ, i.e., the strip had the dimensions which matched the resolving power of the microscope). Since this slit was located next to the photocathode of an FÉU-64 photomultiplier (30), practically all the light flux from the selected part of the image reached the photomultiplier. A sodium nitrite solution (25) was used to prevent the scattered ultraviolet radiation from reaching the photomultiplier and a cell containing a solution of copper sulfate removed the background red illumination. Moreover, the spectral range of the luminescence could be selected by placing suitable filters in front of the photomultiplier slit (these filters are not shown in Fig. 6). An ocular (28) and a tilting prism (27), taken from an MFN-2 microphotometer, were used to orient the crystal, select the appropriate region, and check that the selected region emitted luminescence homogeneously (this was done using a polarizer 11 and an analyzer 24). A dark-field illuminator with a deflecting prism (23) directed some of the luminescence to an FÉU-27 photomultiplier (32) and the filters in front of it (31); this photomultiplier recorded the luminescence from the whole light-emitting region of the crystal. This system for recording the integrated luminescence usually operated only when the additional illumination system (12-14) was employed because the recording system (16-19) was then disconnected. The former recording system could not be used in other situations because the dark-field illuminator always took away some of the light flux from the main recording channel.

The second microscope (MIS-11) in the observation unit was used to measure the thickness of the selected part of the crystal and to check the optical quality of its surface to within 0.5 μ; simple rotation replaced the photometric microscope (MP-3) with the MIS-11 microscope. After the necessary operations, the photometric microscope was returned to its original position.

The observation unit traveled at a velocity of 0.056 μ/sec relative to an immobile crystal excited with a narrow strip of light and this motion enabled us to record the distribution of the luminescence brightness at right angles to the excited strip. The motion was measured with

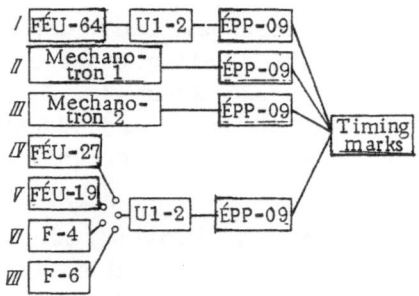

Fig. 7. Recording system used in measurements of the luminescence contours. The recording channels are as follows: I) luminescence contour; II) displacement of the photometric microscope; III) vibrations of the microscope objective; IV) integrated luminescence in the presence of background illumination; V) integrated luminescence in the absence
of background illumination; VI) excitation intensity; VII) background illumination intensity.

mechanotrons (35 and 36) of the 6MKh1S type which allowed us to determine displacements of
50-70 μ to within ±0.5 μ. These mechanotrons were calibrated with a measuring microscope
(35, 38).

Timing marks were used to synchronize the recording of the luminescence contour,
motion of the photometric microscope, integrated intensity of the luminescence, and constancy
of the excitation and background illumination conditions; these timing marks were applied to
all the recording systems (Fig. 7).

4. Method Used in Measurements and Comparison of Luminescence Contours

Since the measured contours were very narrow, special attention was paid to the distortions which could be introduced by the diffraction of light. The width of the slit placed in
front of the photomultiplier cathode was such (0.15-0.2 mm) that it selected from the image
a region which corresponded to a strip 0.5-0.7 μ wide, which was limited by the resolving
power of the microscope. Therefore, the distortions due to diffraction were small although
one would expect some spreading (within a tenth of a micron) of the observed relative to the
true contour.

The distribution of the intensity of the exciting light was difficult to measure directly
because light of the excitation wavelengths did not pass through the crystal. Therefore, we
measured instead the distribution of light of λ = 578 nm wavelength, which was not absorbed in
the crystal and did not excite luminescence. This light was selected, with suitable filters,
from the radiation emitted by the mercury lamp. Care was taken not to introduce additional
changes in the path of light between the slit and the crystal. The width of the slit was also
kept constant.

It is clear from the contour shown in Fig. 8 that, although the distribution is not rectangular, the width of the contour Δ = 1.3 μ is approximately equal to the expected width of the
image of the slit (1.2 μ). The deviation of the contour from the rectangular shape could only
be partly attributed to diffraction. A considerable contribution to the deviation was made
simply by the finite width of the slit. This meant that the actual contour was even closer to
the rectangular shape than the curve in Fig. 8. It is clear from this figure that the intensity

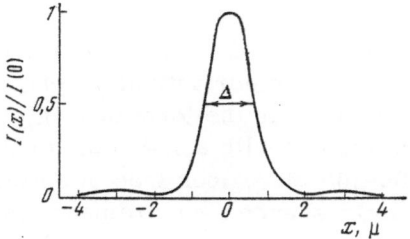

Fig. 8. Shape of a contour obtained by illuminating a ZnS:I crystal ~50 μ thick with light
of λ = 578 nm wavelength.

of the diffraction maxima on both sides of the main peak was slight and could not affect greatly the shape of the luminescence contour except for a slight increase in intensity in the wings.

The linearity of the whole recording channel was checked to within 2% in each of the ranges used in recording the luminescence contours. The absence of distortions associated with the speed of response of the system was checked by reversing the direction of motion of the observation unit.

The homogeneity of the emitted luminescence was checked by illuminating a crystal with a wide (100 μ) strip. In this case, a special check was made to ensure that the intensity of the orange light in the excitation strip was uniform in a slit of this width. This method was used to select parts of the crystal which emitted luminescence homogeneously.

To avoid photolysis, which could occur because of the high intensity of the exciting light and humidity of the atmosphere, we ensured a constant flow of dry nitrogen through the chamber containing the crystal. In this situation, we found no blackening of the crystal or change in the brightness of its luminescence even during prolonged illumination.

The defocusing of the image of the excitation slit on the crystal surface as a result of thermal effects was avoided by ensuring a steady temperature throughout the excitation unit and in the investigated crystal. The defocusing of the crystal image in the object plane was due to vibrations of the photometric microscope. Therefore, we decided to select only the best records of the luminescence contours. Before each recording session, the focusing of the image of the excitation slit on the crystal's surface and of the crystal image itself in the plane of the recording slit was checked visually. About 30 records were obtained under constant experimental conditions (this constancy was checked as described above). When various corrections were made for the width of the contour using the mechanotron readings, we selected only the narrowest contours, usually between three and five contours, because we assumed that the narrowest contours were obtained under the best focusing conditions. The results deduced from the selected contours were averaged in the usual manner.

An analysis of the errors in recording the luminescence contours was made allowing for all the possible sources of error and it indicated that we could distinguish reliably luminescence contours differing by 0.2 μ in half-width at mid-amplitude. It was not surprising that this error was less than the width of the region on the crystal (0.5-0.7 μ) selected by the observation slit. The point was that a luminescence contour was usually several microns wide, i.e., it was considerably wider than the observation slit. A simple geometrical calculation indicated that, in this case, the contour was reproduced very accurately and the smallest error corresponded to the case when the contour could be approximated by linear segments. In our case, the linear region corresponded to the middle part of the contour.

It was not surprising that the error was less than the diffraction limit of the resolving power of the microscope. We were dealing with the resolution of two closely spaced lines but with a comparison of the widths of two relatively smoothly falling contours which were only slightly distorted by diffraction. In the luminescence contour wings the error could rise to 1 μ.

5. Shape of Luminescence Contours

We investigated crystals of two types, ZnS:Cl and ZnS:I. The ZnS:Cl crystals were grown from the vapor phase in the presence of chlorine without any other activator [14]. They emitted bright blue luminescence at room temperature. They were in the form of long plates, several tens of microns thick; their structure was mainly hexagonal with some admixture of the cubic phase. We selected crystals with fairly large (200-500 μ) regions free of optical inhomogeneities, which were checked by polarization and luminescence measurements carried out as described above. The surface of these crystals was not treated in any way. The ZnS:I crystals were grown by the iodine transport reaction method [15], also without introduction of any other activator. These crystals were of bulk shape and emitted a bright blue luminescence. Their structure was cubic. We selected those crystals which had at least one perfect face. The side of the crystal opposite to this face was ground and polished until plane-parallel surfaces were obtained separated by a thickness of 40-100 μ. The homogeneity of the optical properties of the luminescence was then checked. Crystals of this kind were excited from the natural-face side.

Samples were bonded to a mica substrate with an aperture of 1 mm diameter for the excitation and were placed in a light-tight chamber. The chamber was fitted with three screws for the alignment of the crystal surface at right angles to the optical axes of the photometric and excitation microscopes; the axes of these microscopes were made parallel in an earlier operation.

Both types of crystal were photoexcited with light of λ = 313 nm, corresponding to the fundamental absorption of ZnS [16]. The spectral composition of the excited light was checked photographically and the exposure was varied by a factor of 10^4. This indicated that at the highest excitation intensity an admixture of the weakly absorbed light (λ = 365 nm) represented about 0.1% of the intensity of the strongly absorbed light (λ = 313 nm); the admixture was less than 1% at the lowest intensity. Since only the light with λ = 365 nm could reach the opposite face of the crystal, we concluded that the intensity of the excited light reflected from the rear surface was negligible and did not distort the luminescence contour.

Our measurements indicated that the luminescence contours obtained for ZnS:I were 3 μ wider at mid-amplitude than the excitation contour and, in the case of ZnS:Cl, the difference was 0.3 μ (Fig. 9). These values were greater than the maximum possible error (0.2 μ) in the determination of the shape of the luminescence contours at mid-amplitude. Since the wavelength at which the excitation contour was measured (578 nm) was different from the lu-

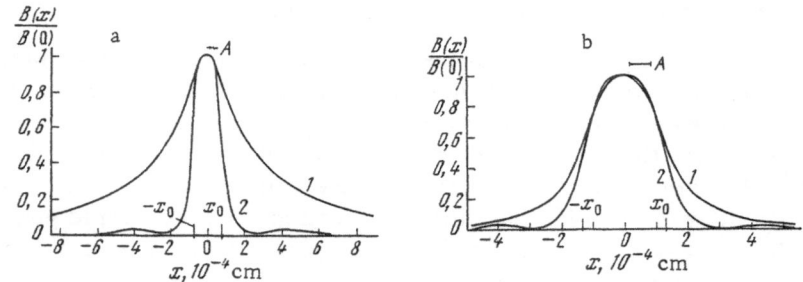

Fig. 9. Broadening of a normalized luminescence contour (1) compared with a contour of the exciting light of λ = 313 nm wavelength (2); A is the size of the recording slit. a) ZnS:I, $x_0' = 0.65 \times 10^{-4}$ cm, $E = 1.5 \times 10^{18}$ photons $\cdot$ cm^{-3} $\cdot$ sec^{-1}; b) ZnS:Cl, $x_0' = 1.3 \times 10^{-4}$ cm, $E = 6 \times 10^{18}$ photons $\cdot$ cm^{-3} $\cdot$ sec^{-1}

Fig. 10. Comparison of measured luminescence contours with a family of theoretical
curves calculated for the ambipolar diffusion mechanism and different values of α_0
(given alongside the curves): 1) ZnS:I ($\alpha_0 =$
0.2, $x_0 = 0.75 \times 10^{-4}$ cm); 2) ZnS:Cl ($\alpha_0 =$
0.75 $x_0 = 1.3 \ 10^{-4}$ cm).

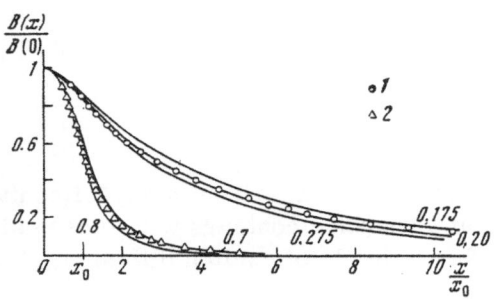

minescence wavelength, we compared the shape of the scattering contours using a nonluminescent material (a mica crystal, 40 μ thick) at the same two wavelengths. The difference
between the two scattering contours after corrections for the differences between the magnification was found to be within the limits of experimental error and evidently due to diffraction
broadening. Since the excitation light was of $\lambda = 313$ nm wavelength and its diffraction broadening
should be even less, the width of the excitation strip was less than the width of the measured
scattering contours. To determine the mechanism of energy transfer, we compared the shapes
of the luminescence contours with the shapes of those deduced theoretically on the assumption
of the electron—hole transfer mechanism. Figure 10 shows a family of theoretical contours
calculated in accordance with Eqs. (30) and (31) for different values of α_0; it also includes
the experimentally obtained luminescence contours of ZnS:Cl and ZnS:I crystals. We can
see that the contours agree well, particularly in the middle part. At the edges of the contours, where the experimental points lie somewhat higher than the theoretical curves, the
diffraction effects due to the excitation slit or the reabsorption transfer mechanism play some
role [7].

According to these results, the luminescence contour of ZnS:I corresponded to $\alpha_0 = 0.2$
and the contour of ZnS:Cl corresponded to $\alpha_0 = 0.75$. The values of x_0, compared with the
half-width of the excitation strip x_0', were also varied within a narrow range to ensure the
best agreement between the theoretical and experimental contours. The values obtained in
this way were as follows: $x_0 = 0.75 \times 10^{-4}$ cm for a half-width of the scattering contour $x_0' =$
0.65×10^{-4} cm in the case of ZnS:I; $x_0 = 1.3 \times 10^{-4}$ cm for $x_0' = 1.3 \times 10^{-4}$ cm in the case of
ZnS:Cl. The difference between x_0 and x_0' in the two cases was within the limits of experimental
error, which was again evidence of the good agreement between the theoretical and experimental
contours.

This agreement between the experimental and theoretical contours suggests that the
electron—hole energy transfer mechanism predominates and the small value of α_0 (0.2 in the
case of ZnS:I) indicates that this mechanism is very efficient because a considerable number
of recombination events occurs outside the excitation region.

We shall now consider to what extent the observed luminescence contours agree with
those calculated from the theory of exciton energy transfer. Since at the excitation rates
employed the exciton concentration cannot be high, we can assume that the lifetime of excitons is independent of their concentration. Then, the exciton concentration and, consequently,
the luminescence brightness decrease exponentially away from the excitation region. However,
Fig. 11 shows that the observed luminescence contours do not fit the exponential law. Therefore,
we may conclude that the role of excitons in the observed transfer phenomena is small.

The predominance of the ambipolar diffusion in the energy transfer processes was confirmed by recording the luminescence contours using excitation slits of different widths and
comparing the experimentally obtained dependences with the theoretical predictions. The

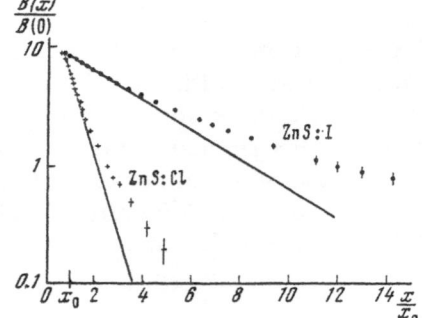

Fig. 11. Comparison of experimental (Fig. 9) luminescence contours with those calculated theoretically of exciton diffusion (straight lines).

TABLE 1

$x_0' \cdot 10^4$, cm	$x_0 \cdot 10^4$, cm	α_0^{cont}	α_0^{calc}	
			x_0	x_0'
ZnS:I, $E = 1.5 \cdot 10^{18}$ photons $\cdot$ cm$^{-3} \cdot$ sec^{-1}				
0.65	0.75	0.200	0.23	0.20
0.85	0.90	0.275	0.29	0.27
1.00	1.05	0.350	0.34	0.32
1.30	1.33	0.450	0.43	0.42
1.50	1.52	0.500	0.49	0.47
ZnS:Cl, $E = 6 \cdot 10^{18}$ photons $\cdot$ cm$^{-3} \cdot$ sec^{-1}				
0.65	0.75	0.45	0.50	0.43
0.85	0.95	0.55	0.60	0.55
1.00	1.05	0.65	0.65	0.63
1.30	1.30	0.75	0.75	0.75

results of measurements are presented in Table 1, where the second column gives the values of x_0' deduced from the measured scattering contours, whereas the third and fourth columns give the values of x_0 and α_0^{cont} corresponding to the theoretical contours which agreed best with the experimental curves. The last column gives the values of α_0^{calc}, calculated from x_0' and x_0. We can see that the values of α_0^{cont} and α_0^{calc} agree satisfactorily. It is also clear from Table 1 and Fig. 12 that, as x_0' rises, the value of α_0 increases, in agreement with the theoretical predictions based on an analysis of the function $\alpha_0 = f(x_0)$, i.e., as x_0' increases, the contour becomes steeper. This is evidence of the smallness of the errors due to the departure from the Π shape in the distribution of the exciting light intensity, particularly for $x_0' \geq 10^{-4}$ cm.

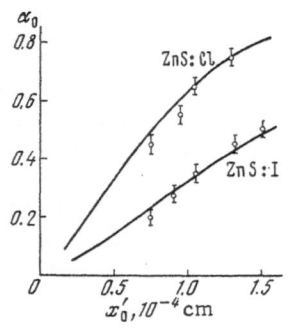

Fig. 12. Dependences of the transport parameter α_0 on the size of the excitation region x_0'. The theoretical curves are continuous and the experimental results are taken from Table 1.

TABLE 2

$E, \dfrac{\text{photons}}{\text{cm}^3 \cdot \text{sec}}$	$\alpha_0{}^{\text{cont}}$	$\alpha_0{}^* = \dfrac{B(0)}{B_0}$	$\alpha_0{}^{**} = k_0 \dfrac{B(0)}{E}$
Linear dependence of B_0 on E			
$5.7 \cdot 10^{17}$	0.275	0.25	0.30
$0.5 \cdot 10^{17}$	0.275	0.28	0.30
$1.5 \cdot 10^{18}$	0.35	0.33	0.32
$2.9 \cdot 10^{18}$	—	0.42	0.39
$6.0 \cdot 10^{18}$	0.45	0.48	0.46
$8.0 \cdot 10^{18}$	0.50	0.51	0.51
Nonlinear dependence of B_0 on E			
$1.5 \cdot 10^{19}$	—	0.57	0.56
$2.3 \cdot 10^{19}$	0.55	0.60	0.57
$5.5 \cdot 10^{19}$	0.60	0.62	0.52
$1.3 \cdot 10^{20}$	0.60	0.64	0.45

6. Influence of the Excitation Rate on the Shape of Luminescence Contours

Since the characteristic lengths of the energy transfer by ambipolar diffusion are governed by the time spent by free carriers in the allowed bands and since, in the bimolecular kinetics case, this time is inversely proportional to the concentration of ionized centers, it follows that at high excitation intensities, i.e., when the number of ionization events is high, the contour should broaden less than at low excitation intensities [17]. In other words, as the excitation rate rises, the parameter α_0 should also increase and (as reported later) this is found experimentally. On the other hand, if we assume the exciton energy transfer mechanism, we cannot expect a narrowing of the luminescence contour with rising excitation rate because the exciton lifetime does not decrease. On the contrary, when the excitation rate is increased, the number of excitons rises rapidly. Therefore, if the energy is transferred by excitons, the luminescence contours should broaden with rising excitation rate.

The second column in Table 2 gives the values of $\alpha_0{}^{\text{cont}}$ found for different values of the excitation rate* E for ZnS:I crystals. We can clearly see the rise of $\alpha_0{}^{\text{cont}}$ with increasing E.

All the values of α_0 given in Table 2 were obtained for $x_0' = 10^{-4}$ cm and $x_0 = 1.05 \times 10^{-4}$ cm.

The numerical values of α_0 corresponding to different excitation rates could also be determined independently: α_0 should be equal to the ratio of the luminescence brightness at the maximum of the contour B(0) to the brightness B_0 obtained as a result of uniform excitation of a crystal, i.e., in the absence of diffusion but for the same excitation rate.

If the excitation rates are the same, the total number of ionization events per unit volume (I_0) is the same in both types of excitation. Since in the uniform excitation case we have $d^2N_+(x)/dx^2 = 0$, it follows from Eq. (8) that

$$I_0 = \nu (N_+')^2, \tag{36}$$

---

*The maximum excitation rate was measured in absolute units, employing a thermopile and the Red-460 phosphor, known to have a constant efficiency in the spectral range under consideration [8]. The degree of attenuation on reflection from a thick phosphor layer was determined. The absorption coefficient of ZnS at $\lambda = 313$ nm was taken to be 4×10^4 cm^{-1} [16].

where N'_+ is the density of free holes under uniform excitation conditions and, consequently, we find that

$$\alpha_0^* \equiv \frac{B(0)}{B_0} = \frac{N_+^2(0)}{(N'_+)^2} \,.$$ (37)

Therefore, combining Eqs. (19) and (36), we find that

$$\alpha_0 \equiv \alpha_0^*.$$ (38)

We measured the values of B_0 and $B(0)$ for ZnS:I crystals employing different values of E. The values of α_0^* obtained in this way are given in Table 2. They are very close to α_0^{cont}. Since this agreement was obtained without normalization, it should be regarded as confirmation of the correctness of our reasoning.

The value of α_0 can also be determined by a third way, in which the ratio $B(0)/E$ is measured experimentally. Since $I_0 = k_1 E$ is satisfied quite accurately, because the excitation occurs within the fundamental absorption region, we obtain

$$\alpha_0 = k_0 \frac{B(0)}{E} \,.$$ (39)

The last column of Table 2 gives the experimentally obtained values of α_0^{**}, normalized for $E = 8 \times 10^{18}$ photons $\cdot$ cm$^{-3} \cdot$ sec^{-1} to the value of α_0^*. We can see from Table 2 that the values of α_0^{**} are close to the values of α_0^{cont} and α_0^* at all excitation rates E, except for the highest, at which the dependence $B_0(E)$ becomes sublinear because of the de-exciting action of the incident light, so that $k_0 \neq$ const.

The theoretical dependence $\alpha_0(E)$, calculated by us earlier and shown in Fig. 13, is also in good agreement with the results obtained in the linear range of the dependence $B_0(E)$. The reference or normalization point of all the calculated curves corresponding to different values of x'_0 is the point corresponding to $\alpha_0 = 0.5$ and $x'_0 = 10^{-4}$ cm. Since the relative positions of the members of a family of calculated dependences $\alpha_0(I_0)$ corresponding to different values

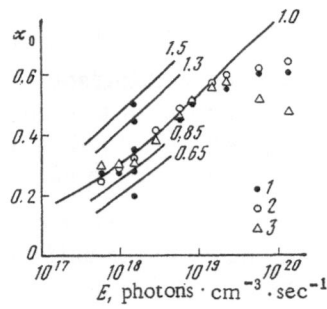

Fig. 13. Dependences of the transport parameter α_0 on the excitation rate E (λ = 313 nm) of a ZnS:I crystal: 1) α_0^{cont}; 2) α_0^*; 3) α_0^{**}. The continuous curves represent a theoretical family of dependences plotted for different values of $x'_0 \times 10^4$ cm (given alongside the curves); the results are normalized to $E = 8 \times 10^{18}$ photons $\cdot$ cm$^{-3} \cdot$ sec^{-1}.

TABLE 3

$x_0' = 10^4$, cm	Experiment		Theory
	$\alpha_0{}^{cont}$	$\alpha_0{}^* = B(0)/B_0$	$\alpha_0{}^{calc}{}_{(x_0')}$
1.30 *	0.75	0.75	0.75
1.00	0.65	0.68	0.63
0.85	0.55	0.53	0.55
0.65	0.46	0.42	0.43

*Normalization point.

of x_0' are not arbitrary but are predicted by the theory, the agreement between these curves and the corresponding experimental points is evidence of a quantitative match between the theory and experiment.

The somewhat slower rise of $\alpha_0{}^{cont}$ and $\alpha_0{}^*$ in the range of high values of E can be explained by the de-exciting action of the incident light. Since, in this range, the rise of the concentration of ionized centers slows down and the rate of recombination of free with localized carriers also slows down, the narrowing of the luminescence contours with increasing excitation rate becomes slower.

The reported experimental results were obtained for ZnS:I crystals. Similar measurements on ZnS:Cl crystals were limited to the determination of the values of the parameter α_0 corresponding to four widths of the excitation slit listed in Table 3, keeping constant the excitation rate $E = 6 \times 10^{18}$ photons $\cdot$ cm$^{-3} \cdot$ sec^{-1}. The value of α_0 was estimated in two ways: from the luminescence contour ($\alpha_0{}^{cont}$) and from the ratio $B(0)/B_0$. The results obtained are given in Table 3.

It is worth noting the good agreement between the values of $\alpha_0{}^{cont}$ and $\alpha_0{}^*$ obtained in the linear region of the dependence $B_0(E)$, i.e., in the region where there is still no interference because of the de-exciting action of the incident light. The last column in Table 3 gives the values of $\alpha_0{}^{calc}(x_0')$ calculated as a function of x_0' using Eq. (25). In this case, the normalization or reference point corresponds to $\alpha_0 = 0.75$ and $x_0' = 1.30 \times 10^{-4}$ cm. Once again, the theory and experiment agree. Thus, we can conclude that the ambipolar diffusion mechanism dominates the transfer of the excitation energy also in ZnS:Cl crystals.

7. Influence of Long-Wavelength Background Illumination on the Shape of Luminescence Contours

Illumination of ZnS-type crystal phosphors with long-wavelength light liberates the localized carriers and reduces the stored light sum [5]. This reduces the rate of recombination of free with localized carriers and increases the time spent by carriers in an allowed band, i.e., it facilitates diffusion and, consequently, broadens the luminescence profile. The influence of background illumination should increase with the depth of localization from which the carriers are liberated.

We investigated ZnS:Cl crystals because the traps were somewhat deeper than in ZnS:I crystals. (This was indicated by the longer afterglow, greater stored light sum, and finally — larger value of α_0.) Figure 14 shows the observed broadening of the luminescence contour as a result of additional background illumination with radiation of wavelengths exceeding 700 nm, generated by an incandescent lamp in combination with suitable filters. The principal photoexcitation was the same as in the absence of background illumination. The quenching effect of the background illumination was weak. When the exciting light strip was 15 μ wide, so that the instability of the measurements and the diffusion in the middle of the strip were negligible, the intensity of the luminescence in the presence of the additional background illumination was 0.95 of the intensity of the luminescence obtained originally. This ratio was not

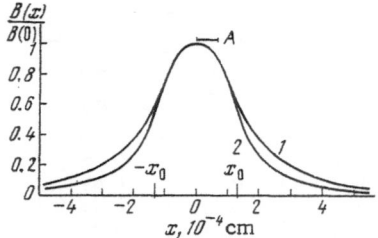

Fig. 14. Broadening of a normalized luminescence contour of a ZnS:Cl crystal subject to an overall background illumination
with radiation of $\lambda >$ 700 nm and a local excitation with light of λ = 313 nm: 1) contour
in the presence of background illumination;
2) without background illumination. Here,
A is the size of the recording slit.

affected when the principal excitation rate was reduced by an order of magnitude. Thus, the
influence of the quenching caused by the background illumination on the luminescence contours
could be ignored.

Figure 15 shows a family of theoretical luminescence profiles calculated for several
values of α_0 but for the same value of $x_0 = 1.3 \times 10^{-4}$ cm, obtained by us earlier for the luminescence contour of a ZnS:Cl crystal in the absence of background illumination (Fig. 10).
We can see from Fig. 15 that the luminescence contour in the presence of background illumination fits well the curve with $\alpha_0{}^{cont}$ = 0.6. The error in the determination of α_{cont}^0 is
about 0.4.

The value of the transport parameter α_0 in the presence of background illumination (b.i.)
could also be found from the ratio

$$\frac{B^{b.i.}(0)}{B_0^{b.i.}} \equiv \alpha_0^{*\,b.i.}, \tag{40}$$

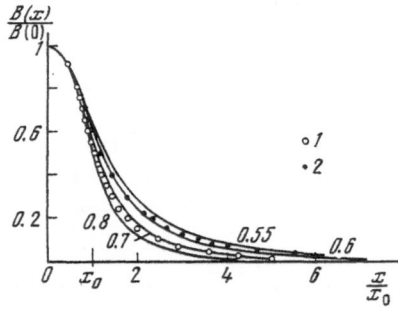

Fig. 15. Comparison of measured normalized luminescence contours of a ZnS:Cl crystal with a family of theoretical curves calculated for different values of α_0 (given alongside the curves): 1) in the absence of background illumination; 2) in the presence of
background illumination; $x_0' = 1.3 \times 10^{-4}$ cm;
E = 6×10^{18} photons·cm^{-3}·sec^{-1}.

whose experimental value was 0.69 ± 0.03, because the error in the determination of the luminescence brightness B was approximately 2 %. Thus, the values of $\alpha_0^{b.l.}$ could be regarded as being in reasonable agreement.

Unfortunately, similar effects could not be studied in ZnS:I crystals because they were practically unaffected by long-wavelength illumination. This was in agreement with the prediction that at room temperature the charges in ZnS:I should be localized in shallower levels.*

Thus, although in the case of ZnS:I and ZnS:Cl crystals we had to use somewhat different methods for the determination of the mechanism of energy transfer over distances of the order of several microns, the mechanism was found to be the same in both cases (ambipolar diffusion). This was supported by the qualitative and quantitative agreement between the experimental results and the theoretical ideas on the transfer of energy by free carriers of both signs.

8. Theory of Luminescence Contours in an External Alternating Electric Field

If it is true that the energy transfer in ZnS crystals takes place mainly via free carriers, the application of an electric field should alter considerably the shape of a luminescence contour. Therefore, the discovery of such an influence would provide an additional proof of the importance of the ambipolar diffusion in the transfer of energy under inhomogeneous photoexcitation conditions.

Zinc sulfide becomes strongly polarized in a static electric field. A large space charge which appears under these conditions and the associated internal electric field may screen quite appreciably the external electric field. This may distort the influence of the external field on the free-carrier distribution which determines the shape of a luminescence contour. On the other hand, an alternating electric field prevents the formation of a space charge. Moreover, since an alternating field extracts alternately (in both directions) electrons and holes from the excitation region, nonequilibrium electrons and holes which mutually neutralize one another may appear outside the excitation region. Such neutralization may reduce considerably the space-charge field and this makes it easier to study the role of the ambipolar diffusion.

The use of an alternating field is also convenient because, in measuring the luminescence contours by the method described above, we are dealing with a time-averaged distribution of the luminescence over a crystal. Therefore, the use of a sinusoidal electric field instead of pulses makes it possible to employ the measuring system described above.

Much work has been done on the influence of electric fields on recombination processes in photoexcited crystal phospors. For example, the problems of the kinetics of photoluminescence are discussed very extensively by Vereshchagin et al. [19]. However, this discussion is mainly concerned with homogeneously excited semiconductors. We are be interested in the problems of diffusion under inhomogeneous excitation conditions which have not yet been considered within the framework of the quadratic recombination kinetics [20].

Let us go back to the model adopted above. Since a crystal is subjected to an external alternating electric field, the problem is no longer of the steady-state type and we must allow for the time dependences of free and localized carriers of both signs. We shall assume that the external field is not too strong to alter the transport parameters of the system (such as the capture and recombination cross sections) or the probabilities of liberation of localized carriers.

---

*We were unable to check this hypothesis directly by a study of the thermoluminescence spectra because the light sum stored in these crystals was small.

In this case, we can follow the notation used in Fig. 1 and obtain the following rate or balance equations for the numbers of particles:

$$\frac{\partial N_-(x, t)}{\partial t} = I(x) - \beta n(x, t) N_+(x, t) - \delta_1 N_-(x, t) + w_1 n_1(x, t) + \frac{1}{q} \operatorname{div} j_{N_-}(x,t); \tag{41a}$$

$$\frac{\partial N_+(x, t)}{\partial t} = I(x) - \beta_1 n_1(x, t) N_-(x, t) - \delta N_+(x, t) + w n(x, t) - \frac{1}{q} \operatorname{div} j_{N_+}(x,t); \tag{41b}$$

$$\frac{\partial n(x, t)}{\partial t} = - w n(x, t) + \delta N_+(x, t) - \beta n(x, t) N_-(x, t); \tag{41c}$$

$$\frac{\partial n_1(x, t)}{\partial t} = - w_1 n_1(x, t) + \delta_1 N_-(x, t) - \beta_1 n_1(x, t) N_+(x, t). \tag{41d}$$

Here, all the concentrations and densities are, in contrast to the system (1), functions not only of the coordinate x but also of the time t.

We shall assume that

$$\frac{1}{n} \cdot \frac{\partial n}{\partial t} \ll w, \qquad \frac{1}{n_1} \cdot \frac{\partial n_1}{\partial t} \ll w_1. \tag{42}$$

This means that the density of localized charges does not vary greatly during the period of the applied field. In other words, it is assumed that a carrier which leaves a trap travels only a short distance to the next trap or, more exactly, it moves to a region where the concentration of filled traps or recombination centers differs only a little from the corresponding concentration at the point of liberation. The estimates given below confirm the validity of this assumption. If the trap concentration N_t is of the order of 10^{18} cm^{-3} and the effective capture cross section σ_c is 10^{-16} cm^2, a free carrier should, on the average, travel a distance $l = 10^{-2}$ cm before it is captured. If the thermal velocity is u = 10^7 cm/sec, this distance is traveled in a time t = 10^{-9} sec. If the field in a crystal is of the order of 10^2 V/cm and the carrier mobility is 100 cm$^2 \cdot$V$^{-1} \cdot$ sec^{-1}, the drift velocity is of the order of 10^4 cm/sec. This means that in 10^{-9} sec a free carrier drifts a distance of only 0.1 μ. In a weaker field or in the case of a lower mobility (this is true of holes), the path traveled is even shorter ($\sim$0.01 μ). This means that our assumption is valid if the concentration of localized charges does not vary greatly in a distance of the order of several hundredths of a micron.

Using the inequalities (42), we find from Eqs. (41c) and (41d) that

$$\begin{aligned} - w n + \delta N_+ &= \beta n N_-, \\ - w_1 n_1 + \delta_1 N_- &= \beta_1 n_1 N_+. \end{aligned} \tag{43}$$

Substituting the above expressions into Eqs. (41a) and (41b), respectively, we obtain the following equations for the densities of free electrons and holes

$$\left. \begin{aligned} \frac{\partial N_-(x, t)}{\partial t} &= I(x) - \left(\frac{\beta \delta}{\beta N_- + w} + \frac{\beta_1 \delta_1}{\beta_1 N_+ + w_1} \right) N_+(x, t) N_-(x, t) + \frac{1}{q} \operatorname{div} j_{N_-}(x,t), \\ \frac{\partial N_+(x, t)}{\partial t} &= I(x) - \left(\frac{\beta \delta}{\beta N_- + w} + \frac{\beta_1 \delta_1}{\beta_1 N_+ + w_1} \right) N_+(x, t) N_-(x, t) - \frac{1}{q} \operatorname{div} j_{N_+}(x,t). \end{aligned} \right\} \tag{44}$$

If we now assume, as before, that the conditions (2) are satisfied, i.e., if we postulate that the density of localized charges is governed primarily by the exchange with the relevant allowed band and not by the recombination processes, we find that the above equations simplify to

$$\left.\begin{array}{l}\dfrac{\partial N_-}{\partial t} = I(x) - \left(\beta\dfrac{\delta}{w} + \beta_1\dfrac{\delta_1}{w_1}\right) N_+(x,t)\,N_-(x,t) + \dfrac{1}{q}\,\mathrm{div}\,j_{N_-}(x,t), \\[3mm] \dfrac{\partial N_+}{\partial t} = I(x) - \left(\beta\dfrac{\delta}{w} + \beta_1\dfrac{\delta_1}{w_1}\right) N_+(x,t)\,N_-(x,t) - \dfrac{1}{q}\,\mathrm{div}\,j_{N_+}(x,t).\end{array}\right\} \tag{45}$$

Then, using the relationships for the current densities

$$\left.\begin{array}{l}j_{N_-} = q\mu_-\mathscr{E}N_- + qD_-\,\mathrm{grad}\,N_-, \\[3mm] j_{N_+} = q\mu_+\mathscr{E}N_+ - qD_+\,\mathrm{grad}\,N_+,\end{array}\right\} \tag{46}$$

we obtain, in the one-dimensional case,

$$\left.\begin{array}{l}\dfrac{\partial N_-}{\partial t} = I(x) - \left(\beta\dfrac{\delta}{w} + \beta_1\dfrac{\delta_1}{w_1}\right) N_-N_+ + \mu_-\mathscr{E}\dfrac{\partial N_-}{\partial x} + \mu_-N_-\,\mathrm{div}\,\mathscr{E} + D_-\dfrac{\partial^2 N_-}{\partial x^2}, \\[3mm] \dfrac{\partial N_+}{\partial t} = I(x) - \left(\beta\dfrac{\delta}{w} + \beta_1\dfrac{\delta_1}{w_1}\right) N_-N_+ - \mu_+\mathscr{E}\dfrac{\partial N_+}{\partial x} - \mu_+N_+\,\mathrm{div}\,\mathscr{E} + D_+\dfrac{\partial^2 N_+}{\partial x^2},\end{array}\right\} \tag{47}$$

where $\mathscr{E}$ is a field acting inside the crystal. The intial conditions for these equations can be replaced by the conditions of periodicity, which will be discussed in detail later. The boundary conditions can be the same as before, i.e., the densities and their derivatives with respect to x tend to zero for x → ± ∞.

We shall assume that an external field is much stronger than the internal field resulting from the presence of space charges. An estimate obtained below shows that this assumption was fully justified under the conditions in our experiments. In fact, a total of 9×10^{18} photons/sec reached a crystal in our case. Even if we assumed that each generated an electron – hole pair and that the charge of each pair, generated during the time when the applied voltage retained its sign, reached the edge of a crystal, the field due to such charges would have been only 15 V/cm (in these calculations, we assumed that the permittivity of the ZnS was of the order of 8 [21], the distance between the electrodes was 0.6 mm, and the applied voltage was of 3 kHz frequency). The external voltage was of the order of 10^2 V/cm. Clearly, the actual internal field was much lower than the estimated 15 V/cm because the charge generated in one half-period was partly neutralized by the charge dispersed by the field during the next half-period. Thus, we could ignore the field of the internal charges and drop the terms with div $\mathscr{E}$ from the system (47).

We shall now consider the solution of the equations obtained. Let us assume that the field applied to a crystal is

$$\mathscr{E} = \mathscr{E}_0 \sin \omega t. \tag{48}$$

We shall seek the solution of the system (47) for the free-carrier densities in the form of a Fourier series for electrons

$$N_-(x,t) = \sum_{k=0}^{\infty} [a_k^-(x) \sin k\omega t + b_k^-(x) \cos k\omega t] \tag{49}$$

and a corresponding series for holes

$$N_+(x, t) = \sum_{k=0}^{\infty} [a_\kappa^+(x) \sin k\omega t + b_\kappa^+(x) \cos k\omega t]. \tag{50}$$

Since we are dealing with an alternating electric field, it follows that, after one half-period, the free-carrier densities should be equal at points symmetric relative to the middle line of the excitation strip, i.e.,

$$N_-(x, t) = N_-\left(-x, t + \frac{\pi}{\omega}\right), \quad N_+(x, t) = N_+\left(-x, t + \frac{\pi}{\omega}\right). \tag{51}$$

Hence, we obtain the conditions for the amplitudes of the Fourier series representing the hole density

$$a_\kappa^+(-x) = (-1)^k a_\kappa^+(x), \quad b_\kappa^+(-x) = (-1)^k b_\kappa^+(x) \tag{52}$$

and, correspondingly, for electrons

$$a_\kappa^-(-x) = (-1)^k a_\kappa^-(x), \quad b_\kappa^-(-x) = (-1)^k b_\kappa^-(x). \tag{53}$$

Substituting now the series (49) and (50) into the system (47) and averaging over a period of the applied field $T = 2\pi/\omega$, we find that, in the case of electrons,

$$\frac{1}{T} \int_0^T \frac{dN_-}{dt} dt \equiv 0 = I(x) - v_1 \left[b_0^+(x) b_0^-(x) + \frac{1}{2} \sum_1^\infty (a_\kappa^+(x) a_k^-(x) + \right.$$

$$\left. + b_\kappa^+(x) b_\kappa^-(x)) \right] + D_+ \frac{d^2 b_0^+(x)}{dx^2} - \frac{1}{2} \mu_+ \mathscr{E}_0 \frac{da_1^+(x)}{dx}, \tag{54}$$

and, in the case of holes,

$$\frac{1}{T} \int_0^T \frac{dN_+}{dt} dt \equiv 0 = I(x) - v_1 \left[b_0^+(x) b_0^-(x) + \frac{1}{2} \sum_1^\infty (a_\kappa^+(x) a_\kappa^-(x) + \right.$$

$$\left. + b_\kappa^+(x) b_\kappa^-(x)) \right] + D_- \frac{d^2 b_0^-(x)}{dx^2} + \frac{1}{2} \mu_- \mathscr{E}_0 \frac{da_1^-(x)}{dx} \tag{55}$$

where

$$v_1 = \beta\delta/w + \beta_1\delta_1/w_1. \tag{56}$$

The first terms in these equations describe the generation of carriers, the second terms the recombination of carriers, the third terms the diffusion of carriers, and the last terms the drift in an electric field. The structure of these equations is such that we can separate the processes occurring in a crystal in the presence and absence of a field. If $\mathscr{E} = 0$, we substitute k = 0 and thus remove all the terms associated with the periodicity of the electric field, so that we obtain two equations:

$$I(x) - v_1 b_0^-(x) b_0^+(x) + D_+ \frac{d^2 b_0^-(x)}{dx^2} = 0,$$

$$\tag{57}$$

$$I(x) - v_1 b_0^-(x) b_0^+(x) + D_- \frac{d^2 b_0^+(x)}{dx^2} = 0,$$

which describe the processes under consideration in the absence of an electric field. In this case, term $\nu_1 b_0^-(x) b_0^+(x)$ describes the process of recombination in a field-free crystal. Bearing this point in mind, we find that Eqs. (54) and (55) yield

$$\mu_{\mp} \mathscr{E}_0 \frac{da_1^{\mp}(x)}{dx} = \pm \nu_1 \sum_{k=1}^{\infty} [a_k^+(x)\, a_k^-(x) + b_k^+(x)\, b_k^-(x)], \tag{58}$$

which equates the recombination term that appears additionally in the presence of a field to the first derivative (with respect to x) of the amplitude of the first term in the Fourier series expansion describing the free-carrier density. Hence, it follows that, to find the deformation of a luminescence contour due to the application of an alternating electric field, it is sufficient to find the quantities $da_1^{\mp}(x)/dx$.

We shall make a further assumption, which is only a very rough approximation, to solve this problem. This approximation is essential to obtain even a general idea of the influence of an electric field on a luminescence contour. According to this assumption, all the parameters of the electron traps and the electrons themselves are equal to the corresponding parameters of hole traps and of holes. This allows us to assume that

$$N_-(x, t) = N_+(-x, t). \tag{59}$$

Hence, using Eqs. (49), (50), (52), and (53), we obtain

$$a_k^-(x) = (-1)^k a_k^+(x), \quad b_k^-(x) = (-1)^k b_k^+(x), \tag{60}$$

which allows us to drop the plus and minus signs of a_k and b_k and thus combine the two equations of the system (57) into one:

$$I(x) - \nu_1 b_0^2(x) + D \frac{d^2 b_0(x)}{dx^2} = 0. \tag{61}$$

In order to find $da_1(x)/dx$, we shall return to the system (47) and carry out the following operations: first, we shall subtract one equation from the other, then replace N_- and N_+ using Eqs. (49), (50), and (60), and finally equate the coefficients of identical terms. Then, considering only the first harmonics of the series (the terms with sin ωt and cos ωt), we obtain the following system of equations for finding the function $a_1(x)$:

$$\left.\begin{array}{l} -\omega b_1(x) = D \dfrac{d^2 a_1(x)}{dx^2} + \mu \mathscr{E}_0 \dfrac{db_0(x)}{dx} - \dfrac{1}{2} \mu \mathscr{E}_0 \dfrac{db_2(x)}{dx} \\[2mm] \omega a_1(x) = D \dfrac{d^2 b_1(x)}{dx^2} + \dfrac{1}{2} \mu \mathscr{E}_0 \dfrac{da_2(x)}{dx}. \end{array}\right\} \tag{62}$$

Ignoring the terms $db_2(x)/dx$ and $da_2(x)/dx$, i.e., assuming that the amplitudes of the second terms in the series vary with the coordinate x much more slowly than $b_1(x)$ and $a_1(x)$, we obtain the following system which can be used to find $da_1(x)/dx$:

$$\frac{db_0(x)}{dx} + \frac{\omega}{\mu \mathscr{E}_0} b_1(x) + \frac{D}{\mu \mathscr{E}_0} \cdot \frac{d^2 a_1(x)}{dx^2} = 0, \tag{63}$$

$$\frac{d^2 b_1(x)}{dx^2} - \frac{\omega}{D} a_1(x) = 0. \tag{64}$$

We have to know $b_0(x)$ to solve this system. We shall assume that this coefficient is independent of the value of $\mathscr{E}_0$ and that it can be found from Eq. (61). It is not necessary to solve this equation because an analogous equation (8) has already been solved and we can use the results

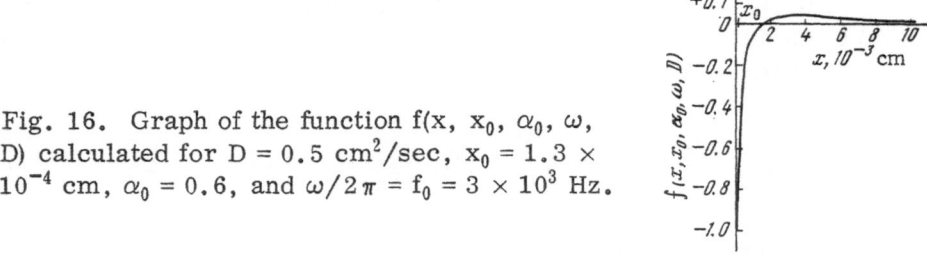

Fig. 16. Graph of the function f(x, x_0, α_0, ω, D) calculated for D = 0.5 cm^2/sec, x_0 = 1.3 × 10^{-4} cm, α_0 = 0.6, and $\omega/2\pi$ = f_0 = 3 × 10^3 Hz.

obtained earlier simply substituting $\delta = \delta_1$, $w = w_1$, $\beta = \beta_1$, and $\mu_- = \mu_+ = \mu$. Having found $b_1(x)$ from Eq. (63) and substituted it into Eq. (64), we obtain one fourth–order equation

$$\frac{d^4 a_1(x)}{dx^4} + \frac{\omega^2}{D^2} a_1(x) = -\frac{\mu \mathscr{E}_0}{D} \cdot \frac{d^3 b_0(x)}{dx^3},\qquad (65)$$

whose solution can be expressed in terms of elementary functions [11]. According to Eqs. (22) and (18), outside the excitation region, i.e., in the range $|x| > x_0$, we have

$$\frac{d^3 b_1(x)}{dx^3} = -\frac{144 D}{\nu_1 (x + C_0)^5}\qquad (66)$$

and outside the excitation region, where $|x| \le x_0$, we obtain

$$\frac{d^3 b_0(x)}{dx^3} = b_0(0)\,\alpha_0\,(\varkappa_0 - 1)\,\gamma_0^2 x \left[2 + \left(\frac{5}{3}\,\alpha_0 - 1 \right) \gamma_0 x^2 \right].\qquad (67)$$

Having found $a_1(x)$ and $b_1(x)$ inside and outside the excitation region, we match the solutions obtained by selecting the undetermined coefficients in the solutions of a homogeneous equation corresponding to Eq. (65) in such a way that $a_1(x)$ and $b_1(x)$ and their first derivatives are continuous. We obtain at the end the required expression for $da_1(x)/dx$, which is so long that we shall not give it here. We shall simply note that it is of the form

$$\frac{da_1(x)}{dx} = \mu \mathscr{E}_0 I_0 f(x, x_0, \alpha_0, \omega, D).\qquad (68)$$

We can see that the quantities I_0 and $\mathscr{E}_0$ occur only as the coefficients which do not affect the nature of the dependence of da_1/dx on the other parameters. Thus, the function $f(x, \alpha_0, x_0, D, \omega)$ depends on just one parameter which cannot be found from independent experiments and this parameter is D. The other quantities are either set by the experimental conditions (ω) or can be determined from the measured luminescence profiles in the absence of a field (α_0 and x_0).

Figure 16 shows the graph of the function $f(x, x_0, \alpha_0, \omega, D)$, calculated for the case closest to the experimental conditions.[*] The parameters α_0 = 0.6 and x_0 = 1.3 × 10^{-4} cm correspond to a luminescence contour in the absence of a field and D = 0.5 cm^2/sec is close to estimates obtained from measurements in the absence of a field. We shall show later that studies of the field–induced deformation of the luminescence contours confirm the correctness of this choice of the parameter D. The value of ω is 3 kHz.

It is clear from Fig. 16 that in the excitation region we have $f(x, x_0, \alpha_0, D, \omega) < 0$. This means that the recombination rate in an alternating field and, therefore, the luminescence brightness at the center of the excited region decrease in the absolute sense. Away from the

---

[*]The complexity of the problem made it necessary to carry out the calculations on a computer. The authors are deeply grateful to Z. A. Fedorova for her help in these calculations.

center of the excitation region, this fall becomes gradually weaker and is eventually replaced by a rise in the recombination rate. However, the rise begins far from the boundary of the excitation region and represents only one-twentieth of the field effect at the center of the luminescence contour.

These results can be used to predict the effect of the field on the shape of a normalized luminescence contour. This shape is governed by the ratio of the rates of fall of $da_1(x)/dx$ and $N^2(x)$. We can show that if $|x| > x_0$, the derivative is $|d^2a_1(x)/dx^2| < |dN^2(x)/dx|$, i.e., that $da_1(x)/dx$ falls more slowly than $N^2(x)$. This means that, as we go away from the center of the luminescence contour, the role of the former term increases and the normalized contour should become narrower. This is not a trivial result because qualitative considerations would lead us to expect broadening of the contour. In fact, such broadening does occur but well outside the investigated parts of the contours, where the brightness is so low that it cannot be measured.

In order to obtain a quantitative estimate of the expected effect of an electric field, we shall calculate the quantity

$$\Delta Q_{\text{theor}}(X) = Q(X) - Q_f(X),\tag{69}$$

where Q and Q_f are the normalized luminescence contours in the absence and presence of a field. Since the luminescence brightness in the absence of a field (in the case of constant transport parameters) is

$$B(x) \equiv v_{\text{lum}}\, b_0^2(x) = \frac{v_1}{2}\, b_0^2(x),\tag{70}$$

it follows that

$$\frac{1}{2} v_{\text{lum}} \sum_1^\infty (-1)^k (a_\kappa^2 + b_\kappa^2) = \frac{v_1}{4} \sum_1^\infty (-1)^k (a_\kappa^2 + b_k^2) = \frac{1}{4}\,\mu\mathscr{E}_0 a_1'(x),\tag{71}$$

where a_1' is the derivative of $a_1(x)$ with respect to x. In this case, the luminescence brightness in the presence of a field is given by

$$B_f(x) = \frac{v_1}{2}\, b_0^2(x) + \frac{1}{4}\,\mu\mathscr{E}_0 a_1'(x).\tag{72}$$

Then,

$$\Delta Q_{\text{theor}}(x) = \frac{b_0^2(x)}{b_0^2(0)} - \frac{2v_1 b_0^2(x) + \mu\mathscr{E}_0 a_1'(x)}{2v_1 b_0^2(0) + \mu\mathscr{E}_0 a_1'(0)}.\tag{73}$$

The above expression can be transformed to

$$\Delta Q_{\text{theor}}(x) = \frac{1}{1 + \xi\, \dfrac{b_0^2(0)}{a_1'(0)}}\, \varphi(x),\tag{74}$$

where

$$\varphi(x) = \frac{b_0^2(x)}{b_0^2(0)} - \frac{a_1'(x)}{a_1'(0)}, \qquad \xi = \frac{2v_1}{\mu\mathscr{E}_0}.\tag{75}$$

It is clear from Eq. (74) that the dependence ΔQ_{theor} (x) is governed entirely by the
the function φ(x) which depends on the theoretical parameters α_0 and D and on the experimental
conditions (x_0, ω). Since the value of α_0 is found from the shape of a luminescence contour
in the absence of a field, the only parameter of the theory of luminescence contours in the
presence of a field is the quantity D, which is an effective ambipolar diffusion coefficient.
It will become clear from Fig. 18 that the value of D governs the position and amplitude of
the maximum of the function φ(x) and this makes it possible to determine D from the experimental results without assuming the values of I_0 and $\mathscr{E}_0$.

9. Experimental Investigation of the Influence of an Electric Field on the Shape of Luminescence Contours

As mentioned earlier, the experiments involving the use of an alternating electric
field were carried out using the same apparatus as before. An audiofrequency electric field
was produced in a capacitor with electrodes ~0.6 mm apart. Since the permittivity of ZnS is
of the order of 8 [21], the field inside a crystal was considerably less than the external field.

The applied field was sufficiently high to observe the predicted effect and to ignore the
internal space-charge field. On the other hand, it was sufficiently low not to influence the
photoluminescence under homogeneous excitation conditions or the constant component of the
free-carrier density b_0(x). Bearing these points in mind and the value of the permittivity of
ZnS, we selected an external field of the order of several hundreds of volts per centimeter.
The absence of field-induced enhancement or quenching of the luminescence was checked experimentally for a crystal illuminated uniformly with light intensity exactly equal to that used
in the measurements of the luminescence contours and of intensity an order of magnitude lower.

Since the applied field was not very high, the deformation of the luminescence contour
due to the drift of carriers in an electric field was detected under conditions which facilitated
carrier diffusion, i.e., in the presence of an additional background illumination with light
of λ > 700 nm. Moreover, the use of such background illumination, which destroyed the dif-

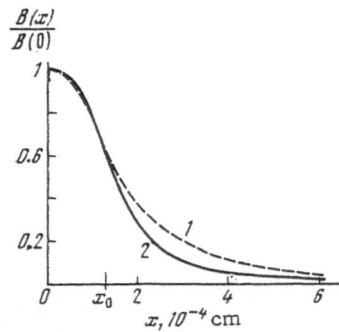

Fig. 17. Deformation of a normalized luminescence contour of a ZnS:Cl crystal caused
by the application of an alternating electric
field of f_0 = 3 kHz frequency. The crystal
was excited with a narrow (x_0' = 1.3 × 10^{-4} cm)
strip of λ = 313 nm light and the whole crystal was subjected to background illumination
with λ > 700 nm wavelength. 1) results obtained in the absence of a field; 2) in the presence of a field. Excitation rate E = 6 × 10^{18}
photons·cm^{-3}·sec^{-1}.

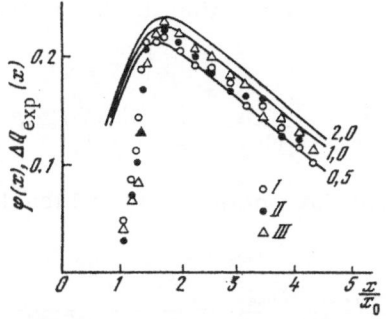

Fig. 18. Comparison of the measured values of $\Delta Q_{exp}(x)$ with the theoretical curve $\varphi(x)$. The continuous curves represent theoretical dependences for several values of Δ (given alongside the curves), whereas the points are the experimental results obtained in different external fields of 400, 600, and 800 V/cm (I, II, and III, respectively) of 3 kHz frequency normalized to the maximum of the function $\varphi(x)$ ($\alpha_0 = 0.6$, $x_0 = 1.3 \times 10^{-4}$ cm).

ferences between the depths of the luminescence centers ε_1, made the conditions approach more closely the conditions assumed in the calculations of the expected field effect.

We used thin ($\sim 50\ \mu$) ZnS:Cl crystals of the same type as those employed in a study of the luminescence contours in the absence of a field. As before, the principal excitation was of $\lambda = 313$ nm wavelength and the width of the illuminated strip was $2x_0' \sim 2.6 \times 10^{-4}$ cm. Figure 17 shows the experimental data obtained in an external field of 800 V/cm at a frequency of 3 kHz. We can see some narrowing of the luminescence contour in the presence of a field (Q_f), compared with the contour determined in the absence of the field (Q). This behavior of the contour in an alternating electric field was predicted theoretically. A similar effect was also observed in weaker fields of 600 and 400 V/cm. As expected, the effects in weaker fields were smaller.

Figure 18 compares the results of calculations of the quantity $\varphi(x)$ with experimental data. A family of theoretical curves was plotted for several values of D using Eq. (74) and the parameters $\alpha_0 = 0.6$ and $x_0 = 1.3 \times 10^{-4}$ cm, determined from the shape of the luminescence contour in the absence of a field but in the presence of an additional long-wavelength illumination (Fig. 15). We also assumed that the electric field frequency was 3 kHz. The experimental results were obtained in external fields of 800, 600, and 400 V/cm intensity. We can see that in the region of the maxima and to their right the theoretical curves agree with the experimental results. The discrepancy in the region of $x/x_0 = 1$ is due to some secondary phenomenon. This may be a slight vibration of a crystal under the action of an alternating electric field, which would cause some broadening of the contour that would be most pronounced in the steepest region, i.e., near $x/x_0 = 1$.

The agreement between the calculated and experimental curves indicates that the ambipolar diffusion coefficient D lies within the range 0.5-1 cm^2/sec.

Since the coincidence of the theoretical curves $\varphi(x)$ and of the experimental dependences $\Delta Q(x)$ is obtained by introducing a constant factor $\varkappa$ which is defined — in accordance with Eq. (74) — by the condition

$$\Delta Q_{exp}(x)|_{max} \approx \varkappa \varphi(x)_{max}, \qquad (76)$$

we can use the absolute value of this factor to find the internal field $\mathscr{E}_0$. According to Eqs. (24) and (68), we have

$$\frac{2\nu_1}{\mu\mathscr{E}_0} \cdot \frac{b_0^2(0)}{a_1'(0)} = \frac{2\alpha_0}{\mathscr{E}_0^2 D^2}\left(\frac{kT}{q}\right)^2 f_{x=0}^{-1}, \tag{77}$$

where $f_{x=0}$ is the calculated value of the function $a_1'(0)$, which does not depend explicitly on $\mathscr{E}_0$ and I_0, so that

$$\varkappa = \frac{\Delta Q_{exp}(x)|_{max}}{\varphi(x)|_{max}} = \frac{1}{1 + \dfrac{2\alpha_0 (kT/q)^2}{\mathscr{E}_0^2 D^2} f_{x=0}^{-1}} \tag{78}$$

and, consequently,

$$\mathscr{E}_0^2 = \frac{2\alpha_0 (kT/Dq)^2}{\dfrac{\varphi(x)}{\Delta Q_{exp}(x)}\bigg|_{max} - 1} f_{x=0}^{-1}. \tag{79}$$

It should be noted that we do not have to know the absolute value of the number of ionization events I_0 needed to determine $\mathscr{E}_0$. Table 4 gives the values of $\mathscr{E}_0$ obtained for $D = 0.75$ cm^2/sec and $\varphi_{max} = 0.223$.

Since the permittivity of ZnS can be taken as 8.3 [21], the agreement between the values of of $\mathscr{E}_0$ and $(1/8.3)\,\mathscr{E}_{ext}$, which correspond to different intensities of the external field, suggests that the theory of the deformation of luminescence contours by an electric field describes satisfactorily the magnitude of the effect and its dependence on $\mathscr{E}_{ext}$. The value of D which is then obtained is considerably closer to the diffusion coefficient of holes D_+' than to the diffusion coefficient of electrons D_-' since holes are less mobile in ZnS and the diffusion processes are governed primarily by their behavior.

The agreement between the theory and experiment can be demonstrated also by a different and independent method. We can use theoretical considerations and calculate the ratio of the values of the luminescence brightness at the maximum of the contour in a field $B_f(0)$ and in the absence of a field $B(0)$, and then compare this ratio with the experimental value.

It follows from Eqs. (72) and (70) that

$$\frac{B_f(0)}{B(0)} = 1 + \frac{1}{2} \cdot \frac{\mu\mathscr{E}_0}{\nu_1} \cdot \frac{a_1'(0)}{b_0^2(0)} = 1 + \frac{\mathscr{E}_0^2 D^2}{2\left(\dfrac{kT}{q}\right)^2 \alpha_0} f_{x=0}. \tag{80}$$

Taking the value of $\mathscr{E}_0$ from Table 4 and the values of the parameters α_0, D, and the function $f_{x=0}$, we can calculate $B_f(0)/B(0)$. The last two columns in Table 4 give the results of such

TABLE 4

$\mathscr{E}_{ext}$, V/cm	ΔQ_{max}	$\mathscr{E}_0$, V/cm	$\dfrac{\mathscr{E}_{ext}}{8.3}$	$B_f(0)/B(0)$	
				experiment	theory
800	0.040	90	96	0,90±0,05	0,85
600	0.025	74	72	0,92±0,05	0,90
400	0,015	58	48	1.00±0,05	0.94

calculations alongside the experimental value of the ratio obtained in different external fields $\mathscr{E}_{ext}$. We can see once again a satisfactory agreement between the theory and experiment.

Thus, an investigation of the influence of an external alternating electric field on the transfer of the photoexcitation energy gives a further and independent confirmation of the dominant role of the ambipolar diffusion. Moreover, such investigations make it possible to estimate the effective diffusion coefficient of carriers and the magnitude of the internal field.

10. Estimates of the Transport Properties of Zinc Sulfide

We have already shown that the main mechanism responsible for the transfer of the initial photoexcitation energy in zinc sulfide crystals for distances of the order of several microns is the ambipolar diffusion of free carriers, which is characterized by the transport parameter α_0 and the diffusion coefficients D'_+. We shall use the numerical values of these quantities and estimate various transport properties of zinc sulfide.

We shall need to know the numerical values of the ratios δ/w and δ_1/w_1. These values can be expressed quite easily in terms of other better known properties of ZnS. Under thermodynamic equilibrium conditions, the reciprocals of these ratios are given by

$$\frac{w}{\delta} = \frac{N_v}{N_a} e^{-\varepsilon/kT} \text{ and } \frac{w}{\delta_1} = \frac{N_c}{N_d} e^{-\varepsilon_1/kT}, \tag{81}$$

where N_c and N_v are the densities of states in the conduction and valence bands (at room temperature these densities are of the order of 10^{19} cm^{-3}); $N_a = 10^{K_a}$ cm^{-3} and $N_d = 10^{K_d}$ cm^{-3} are the concentrations of acceptor and donor defects which have levels at depths ε and ε_1, respectively. Since we are dealing with n-type crystals, it follows that $K_d \geq k_a$. On the other hand, since ZnS is an almost totally compensated semiconductor [22], it follows that $\Delta K = K_d - K_a$ satisfies the condition $0 \leq \Delta K < 1$. Moreover, in the case of ZnS-type phosphors which luminesce strongly, the values of N_d and N_a are usually such that K_d and K_a lie in the range 17-18. Therefore, in the case of ZnS crystals, for which ε and ε_1 do not usually exceeed 0.2 eV, we find that the required ratios are

$$\frac{\delta_1}{w_1} = 10^{K_d-19} e^{\varepsilon_1/kT} \gg 1, \frac{\delta}{w} = 10^{K_a-19} e^{\varepsilon/kT} \gg 1. \tag{82}$$

It should be noted that, in our case, $\varepsilon > \varepsilon_1$ because, in the case of a homogeneously excited ZnS:I or ZnS:Cl crystal, the luminescence brightness decreases when the temperature is raised. Moreover, in the case of "self-activated" ZnS:Cl crystals exhibiting a bright blue luminescence at room temperature, the energy of one of the levels is $\varepsilon_1 = 0.25$ eV [23]. The values of ε for the same crystals are known less accurately. However, it is reported in [24] that, in the case of the centers responsible for the green luminescence of ZnS:Cu:Cl, the depth of the activator level is 0.35 eV. The depth of the centers responsible for the blue luminescence should be less, as indicated by the nature of the recombination interaction in which these centers are involved. Hence, it follows that $\Delta\varepsilon = \varepsilon - \varepsilon_1 = 0.05-0.1$ eV. We shall estimate the transport properties of zinc sulfide bearing in mind these limits of variation of $\Delta\varepsilon$ and ΔK.

Estimate of the Hole Mobility μ_+

The mobility of minority carriers μ_+ in ZnS:Cl crystals can be estimated from Eq. (34). In combination with Eq. (82), we obtain

$$\mu_+ = \frac{D'_+\mu_-}{\frac{2kT}{q}\mu_- - D'_+\frac{\delta_1}{w_1} \cdot \frac{w}{\delta}}. \tag{83}$$

Our study of the deformation of the luminescence contours by an alternating electric field yielded the effective ambipolar diffusion coefficient D = 0.5-1.0 cm^2/sec. As mentioned earlier, we shall assume that D $\approx$ D$'_+$. Since the value of the coefficient D was found in the presence of a long-wavelength background illumination, the value $(\delta_1/\omega_1) \cdot (\omega/\delta)$ should also be taken in the presence of such a background. This value could be determined from the influence of the background illumination on the luminescence efficiency. Since the thermal quenching curves indicated that the luminescence of our ZnS:Cl crystals was partly quenched at T = 300°K, the change in the efficiency was governed [5, p. 77] by the value of $[1 + (\delta_1/\omega_1) \cdot (\omega/\delta)]^{-1}$. Under the experimental conditions, the background illumination altered the luminescence brightness by just 5%. When the expression just given was used, we found that the value of $(\delta_1/\omega_1) \cdot (\omega/\delta)$ was practically unaffected by the background illumination. In this case, we could estimate μ_+ from Eq. (83), rewriting it, with the aid of Eq. (81), in the form

$$\mu_+ = \frac{D\mu_-}{\frac{2kT}{q}\mu_- - D \cdot 10^{\Delta K} e^{-\Delta\varepsilon/kT}} . \qquad (84)$$

Table 5 gives the results of calculation of μ_+ for ZnS:Cl crystals and several values of ΔK and $\Delta\varepsilon$. These values of μ_+ were calculated on the assumption that $\mu_- = 140$ cm$^2 \cdot$ V$^{-1} \cdot$ sec^{-1}, which was obtained experimentally at T = 300°K [25] for samples prepared by a method similar to that employed in our study. However, for lower values of μ_- and higher values of ΔK ($\Delta K > 0$), the value of μ_+ would increase. The results given in Table 5 indicate that the minimum values of the hole mobility should be $\mu_+ = 9-18$ cm$^2 \cdot$ V$^{-1} \cdot$ sec^{-1} (these values correspond to a reasonable range of $\Delta\varepsilon$ and D and to a reasonable ratio of N_d to N_a). The mobility corresponding to the most probable values D = 0.7-0.8 cm^2/sec, $\Delta K = 0$, and $\Delta\varepsilon = 0.05$ eV was $\mu_+ = 15$ cm$^2 \cdot$ V$^{-1} \cdot$ sec^{-1}. We shall show later that, in the case of ZnS:I crystals, the value μ_+ is of the same order of magnitude (~ 10 cm$^2 \cdot$ V$^{-1} \cdot$ sec^{-1}).

It should be noted that the published values of the hole mobility in ZnS are very contradictory. For example, $\mu_+ = 5$ cm$^2 \cdot$ V$^{-1} \cdot$ sec^{-1} at 700°K [25] and $\mu_+ = 0.8$ cm$^2 \cdot$ V$^{-1} \cdot$ sec^{-1} at room temperature [27]. According to other workers, $\mu_+ = 5 \times 10^{-4}$ cm$^2 \cdot$ V$^{-1} \cdot$ sec^{-1} [28] and $\mu_+ = 10^{-4}$ cm$^2 \cdot$ V$^{-1} \cdot$ sec^{-1} [29].

Our values of the hole mobility (10-15 cm$^2 \cdot$ V$^{-1} \cdot$ sec^{-1}) are in good agreement with the empirical ratio $\mu_-/\mu_+ = 10-15$, which is known to be characteristic of the majority of the II-VI compounds. This follows from Table 6, which gives the 300°K mobilities in several II-VI compounds. Our values make it possible to join zinc sulfide into one series with other II-VI compounds.

Our value of μ_+ is obviously only an estimate because its exact value can be found only if we know the parameters δ, w, and β of both levels. However, the method adopted above can be used in those cases when direct measurements of the hole mobility are difficult.

TABLE 5

D, cm^2/sec	μ_+, cm$^2 \cdot$ V$^{-1} \cdot$ sec^{-1}					
	$\Delta\varepsilon = 0$ eV		$\Delta\varepsilon = 0.05$ eV		$\Delta\varepsilon \geqslant 0.1$ eV	
	$\Delta K = 0$	$\Delta K = 0,3$	$\Delta K = 0$	$\Delta K = 0,3$	$\Delta K = 0$	$\Delta K = 0,3$
0.5	9.6	10	9.0	9.1	9.0	9.0
0.7	14.0	15	13.0	13.0	12.5	12.5
0.8	16.0	18	14.5	15.0	14.0	14.0
1.0	20.0	24	18.0	19.0	18.0	18.0

TABLE 6

Substance	μ_-, cm$^2 \cdot$ V$^{-1} \cdot$ sec^{-1}	Reference	μ_+, cm$^2 \cdot$ V$^{-1} \cdot$ sec^{-1}	Reference	μ_-/μ_+
ZnS	140	[25]	15—10	Our results	10—15
ZnSe	530	[30]	28	[30]	19
ZnTe	340	[30]	110	[31]	3
CdS	300	[32]	50	[32]	6
CdS	350	[33]	25	[33]	14
CdSe	650	[30]	—	—	—
CdTe	1050	[30]	80	[30]	13

The great advantage of our method [34] is that it does not require the deposition of electrical contacts on a sample. Therefore, in this estimate of the mobility and (as shown later) of other parameters of semiconductors, we avoid basic difficulties which are associated with the contact phenomena that usually distort the results of measurements. It should be noted that it is very convenient to employ our method in estimating the transport properties of complex layered structures, such as various p–n–p and similar devices. In this case, a section of a structure can be excited in the transverse direction by a narrow strip of light. Then, photographic or even visual estimates of the relative positions of the boundaries of the luminescence region in each of the layers of the structure can give information on the diffusion characteristics. If the transport properties of one of the layers are known, they can be used to calibrate the method and the absolute values can be obtained in this way.

Estimates of the Quadratic Recombination Coefficient and Recombination Cross Sections

Equation (24) $D^2 = I_0/\alpha_0\gamma_0^2\nu$, which gives the relationship between D and ν, can be used to estimate the quadratic recombination parameter ν of ZnS:Cl crystals. A simple calculation, carried out for 0.5 cm^2/sec $\le$ D$\dagger \approx$ D $\le$ 1 cm^2/sec, gives values of $\nu^{b.i.}$ and ν (Table 7), which represent the recombination in the presence of background illumination ($\alpha_0^{b.i.} = 0.6$, $\gamma_0 = 0.54 \times 10^8$ cm^2) and in the absence of such illumination ($\alpha_0 = 0.75$, $\gamma_0 = 0.97 \times 10^8$ cm^2) on the assumption that the ionization rate is $I_0 = 6 \times 10^{18}$ cm$^{-3} \cdot$ sec^{-1}. We are assuming that the quantum efficiency of the photocreation of electron–hole pairs is 1, i.e., $I_0 = E$.

Using the values of ν found in this way, we can estimate from Eq. (9) the recombination cross section σ_{rec} of ZnS crystals. We shall use $\beta = \sigma_{rec} u$, where u is the thermal velocity of carriers participating in the recombination process, which is of the order of 10^7 cm/sec. In estimating σ_{rec}, we shall assume that the recombination of electrons and holes occurs at centers with identical physical properties, i.e., that, in the first approximation, we have $\beta \approx \beta_1$. In this case, we can use Eqs. (9), (81), and (82) to obtain

$$\sigma_{rec} = \frac{\nu}{10^{\kappa}a^{-12}e^{\varepsilon/kT}(10^{-\Delta\kappa}e^{\Delta\varepsilon/\kappa T} + 1)}. \tag{85}$$

Calculations based on this expression give the values of σ_{rec} listed in Table 8 for selected values of ΔK, $\Delta\varepsilon$, ε, and D.

TABLE 7

D, cm^2/sec	0.5	0.7	0.8	1.0
$\nu^{b.i.}$, cm^3/sec	$7.4 \cdot 10^{-5}$	$1.4 \cdot 10^{-4}$	$1.9 \cdot 10^{-4}$	$2.9 \cdot 10^{-4}$
ν, cm^3/sec	$3 \cdot 10^{-4}$	$5.8 \cdot 10^{-4}$	$7.6 \cdot 10^{-4}$	$1.2 \cdot 10^{-3}$

TABLE 8

$\Delta\varepsilon$; ε, eV	D, cm²/sec	σ_{rec}, cm²			
		$K_a = K_d = 18$ $\Delta K = 0$	$K_d = 18$ $\Delta K = 0.3$	$K_a = K_d = 17$ $\Delta K = 0$	$K_d = 17$ $\Delta K = 0.3$
$\Delta\varepsilon = 0.05$ $\varepsilon_1 = 0.30$ $\varepsilon = 0.25$	0.8	$9.4 \cdot 10^{-16}$	$3.4 \cdot 10^{-15}$	$9.4 \cdot 10^{-15}$	$3.4 \cdot 10^{-14}$
	0.7	$7.3 \cdot 10^{-16}$	$2.6 \cdot 10^{-15}$	$7.3 \cdot 10^{-15}$	$2.6 \cdot 10^{-14}$
$\Delta\varepsilon = 0.10$ $\varepsilon_1 = 0.35$ $\varepsilon = 0.25$	0.8	$2.1 \cdot 10^{-17}$	$8.5 \cdot 10^{-17}$	$2.1 \cdot 10^{-16}$	$8.5 \cdot 10^{-16}$
	0.7	$1.7 \cdot 10^{-17}$	$6.6 \cdot 10^{-17}$	$1.7 \cdot 10^{-16}$	$6.6 \cdot 10^{-16}$

The recombination cross section for the most probable values of the relevant parameters ($\Delta K = 0$, $K_a = K_d = 17$-18, $\varepsilon_1 = 0.25$ eV, and $\varepsilon = 0.30$ eV) is $\sigma_{rec} = 10^{-15}$ cm². This cross section is intermediate between the values of 10^{-13}-10^{-14} cm², which are typical of the centers that have an effective charge relative to the lattice [35-37], and the values $\sigma_{rec} < 10^{-16}$ cm², reported for neutral recombination centers [36, 37]. Thus, we may assume that, in the case of zinc sulfide, the recombination occurs at centers of dipole nature.

Estimates of the Densities of Free and Localized Carriers

The density of free holes at the center of the excitation region in a ZnS:Cl crystal can be estimated from Eq. (19), which gives $N_+^2(0) = I_0\alpha_0/\nu$. This estimate can be obtained quite simply in the presence of long-wavelength background illumination ($\alpha_0^{b.i.} = 0.6$) and without such illumination ($\alpha_0 = 0.75$). In these cases, we find that, if 0.5 cm²/sec $\leq D \leq 1$ cm²/sec and $I_0 = 6 \times 10^{18}$ cm⁻³·sec⁻¹, the hole densities are 2.2×10^{11} cm⁻³ $\geq N_+^{b.i.}(0) \geq 1.1 \times 10^{11}$ cm⁻³ and 1.2×10^{11} cm⁻³ $\geq N_+(0) \leq 6.2 \times 10^{10}$ cm⁻³. It should be noted that in estimates of this kind we do not need to know the depths of the levels ε_1 and ε.

The density of free electrons can be estimated using Eq. (5). Then, applying Eqs. (81) and (82), we obtain

$$N_-(0) = 10^{-\Delta K} e^{\Delta\varepsilon/kT} N_+(0). \tag{86}$$

If $\Delta K = 0$-0.3 and $\Delta\varepsilon = 0.05$ eV, we find that $N_-(0) = (6.8$-$3.4)N_+(0)$, and if $\Delta\varepsilon = 0.1$ eV we have $N_-(0) = (47$-$23)N_+(0)$.

These values of N_+ and N_- are sufficient to observe photoconductivity even in such a high-resistivity compound as ZnS.

The density of localized carriers can be found from Eq. (81). If $\delta/w = n/N_+$, which is valid if we allow for Eqs. (2) and (1d), we obtain

$$n(0) = N_+(0) \cdot 10^{K_d - 19} e^{\varepsilon/kT}. \tag{87}$$

Calculations show that if $K = 17$-18, $I_0 = 6 \times 10^{18}$ cm⁻³·sec⁻¹, $\varepsilon = 0.3$-0.25 eV, and 0.7 cm²/sec $\leq D \leq 0.8$ cm²/sec, we obtain 6.4×10^{15} cm⁻³ $\geq n(0) \geq 7.7 \times 10^{13}$ cm⁻³. This means that levels of this depth are far from saturation ($K_d = 17$), in agreement with the conditions assumed in the theoretical discussion of ambipolar diffusion.

We can use the inequalities (2), which underlie the theory, in estimating the lower limit of the frequency factor (w_0 and w_{01}) of the levels located at $\varepsilon = 0.3$ eV and $\varepsilon_1 = 0.25$ eV. If $\sigma_{rec} \approx 10^{-15}$ cm² and the values of N_+ are those estimated above, we find that $w_{01} \geq 3.3 \times 10^7$ sec⁻¹ and $w_0 \geq 1.5 \times 10^9$ sec⁻¹. These values are in good agreement with estimates of the frequency factor given in [38]. On the other hand, since $w_0 = \sigma_{cap} N_{c,v} u \approx \sigma_{cap} \times 10^{26}$ sec^{-1}, where σ_{cap} is the cross section for the capture of a free carrier by the localization level under investigation, it follows from the inequalities (2) that $\sigma_{cap} \gg 10^{-17}$ cm^2 for the capture of a hole by an acceptor center and $\sigma_{cap} \gg 4 \times 10^{-19}$ cm^2 for the capture of an electron by a donor center.

The estimates given above apply to ZnS:Cl crystals. In the case of ZnS:I crystals, these estimates can be obtained much more simply without recourse to experiments in alternating electric fields. The values of D'_+ and μ_+ can be calculated from Eqs. (33) and (34). Since the photoluminescence spectra of ZnS:Cl and ZnS:I are practically identical and the investigated crystals differ in respect of donor (and not acceptor) impurities, we can assume that the depths of the acceptor luminescence centers in both crystals are the same, i.e., $\varepsilon_I \approx \varepsilon_{Cl} = 0.30$ eV. On the other hand, the depth of the donor centers in these crystals, which are formed by the I and Cl ions, are different and such that $\varepsilon_{1Cl} > \varepsilon_{1I}$. In fact, since the electronegativities of I and S differ by less than those of Cl and S [39], we can assume that iodine, which replaces sulfur in the ZnS lattice, produces fewer distortions than chlorine. On the other hand, the experimental data indicate that the light sum stored in ZnS:I is smaller than that stored in ZnS:Cl and that the decay of the luminescence after the end of excitation at room temperature is faster for ZnS:I than for ZnS:Cl. These results and the absence of the influence of the long-wavelength background illumination on the photoluminescence of ZnS:I also indicate that $\varepsilon_{1Cl} > \varepsilon_{1I}$. Therefore, we shall assume that $\varepsilon_{1I} \approx 0.25\text{-}0.20$ eV, i.e., that 0.05 eV $\leq \Delta\varepsilon \leq 0.1$ eV. The value of $\Delta K = K_d - K_a$, representing the degree of compensation of a crystal, can be considered on the basis of the results reported in [15] and on the basis of the observation that the electrical conductivity of iodine-doped crystals is considerably higher than that of chlorine-doped samples. Consequently, we may conclude that the difference between the donor and acceptor concentrations is greater for the iodine-doped crystals. Therefore, we shall assume that $0.3 \leq \Delta K \leq 1.0$.

In this case, we find that, if $\sigma_{rec} = 10^{-15}$ cm^2, the hole mobility is of the order of 10 cm$^2 \cdot$ V$^{-1} \cdot$ sec^{-1}, which is close to the value of μ_+ that we obtained for ZnS:Cl crystals. Estimates of the values of $N_+(0)$, $N_-(0)$, and $n(0)$ show that, as in the case of ZnS:Cl crystals, the donor and acceptor centers are far from saturation ($K_d = 17$) and all the excitation rates used in our experiments and the inequalities (2) are again well satisfied.

Thus, estimates of various transport properties, which give values that are quite reasonable for ZnS-type crystals, confirm the validity of the theoretical assumptions underlying these estimates.

Conclusions

We must stress once again the advantages of the luminescence contour method developed in the present paper. This method not only makes it possible to show that at room temperature the excitation energy in ZnS crystals is transferred by the ambipolar diffusion, but also to estimate saveral properties of such crystals. In particular, it is possible to estimate the minority-carrier (hole) mobility, free-carrier recombination, and capture cross sections, as well as densities of free and localized electrons and holes. There is no need to deposit ohmic contacts on a crystal. The method is particularly useful because, under the quadratic recombination conditions, the traditional methods for the determination of these parameters are unsuitable and even some widely used concepts (for example, the average lifetime of minority carriers, diffusion length, etc.) are invalid.

The method may also be found useful in studies of other luminescent semiconductors in which energy is transferred under quadratic recombination conditions.

LITERATURE CITED

1. V. E. Lashkarev, Tr. Inst. Fiz. Akad. Nauk Ukr. SSR, No. 3, 3 (1952).
2. G. M. Guro, Usp. Fiz. Nauk, 72:711 (1960).
3. J. S. Blakemore and K. C. Nomura, J. Appl. Phys. 31:753 (1960).
4. S. M. Ryvkin, Photoelectric Effects in Semiconductors, Consultants Bureau, New York (1964).
5. M. V. Fok, Introduction to the Kinetics of the Luminescence of Crystal Phosphors [in Russian], Nauka, Moscow (1974).
6. W. E. Spear, J. Non-Cryst. Solids, 1:197 (1969).
7. N. N. Grigor'ev and M. V. Fok, Kratk. Soobshch. Fiz., No. 2, 19 (1972).
8. C. E. Bleil and J. Broser, Proc. Seventh Intern. Conf. on Physics of Semiconductors, Paris, 1964, Vol. 1, Physics of Semiconductors, publ. by Dunod, Paris; Academic Press, New York (1964), p. 897.
9. N. N. Grigor'ev and M. V. Fok, Fiz. Tekh. Poluprovodn., 3:874 (1969).
10. J. S. Blakemore, Semiconductor Statistics, Pergamon Press, Oxford (1962).
11. E. Kamke, Differentialgleichungen, Lösungsmethoden und Lösungen, Vol. 1, Gewöhnliche Differentialgleichungen, Chelsea Publ. Co., New York (1971) [reprint of earlier German edition].
12. É. N. Adirovich, Dokl. Akad. Nauk SSSR, 157:313 (1964).
13. G. S. Berlin, Prib. Tekh. Eksp., No. 5, 152 (1961).
14. V. V. Osiko and E. I. Panasyuk, Opt. Spektrosk., Sb. 1 Luminestsentsiya (Suppl. 1 Luminescence), 239 (1963).
15. A. V. Lavrov and G. E. Arkhangel'skii, Zh. Fiz. Khim., 44:297 (1970).
16. M. N. Alentsev and E. I. Panasyuk, Opt. Spektrosk., 5:207 (1958).
17. N. N. Grigor'ev and M. V. Fok, Proc. Nineteenth Conf. on Luminescence [in Russian], Part II, Riga State University (1970), p. 74.
18. Z. L. Morgenshtern, V. B. Neustruev, and M. I. Épshtein, Zh. Prikl. Spektrosk., 3:49 (1965).
19. I. K. Vereshchagin, V. G. Khavrunyak, and I. V. Khomyak, in: Electroluminescence of Solids [in Russian], Naukova Dumka, Kiev (1971), p. 148.
20. N. N. Grigor'ev and M. V. Fok, Kratk. Soobshch. Fiz., No. 8, 42 (1971).
21. D. Berlincourt, H. Jaffe, and L. R. Shiozawa, Phys. Rev., 129:1009 (1963).
22. Yu. V. Bochkov, A. N. Georgobiani, and G. S. Chilaya, Tr. Fiz. Inst. Akad. Nauk SSSR, 50:60 (1969).
23. T. S. Reshetina and V. F. Tunitskaya, Zh. Prikl. Spektrosk., 12:295 (1970).
24. V. V. Antonov-Romanovskii and L. A. Vinokurov, Opt. Spektrosk., 1:71 (1956).
25. M. Aven and R. E. Halsted, Phys. Rev. 137:A228 (1965).
26. M. Aven, Extended Abstracts, Electronics Division Meeting, Electrochemical Society, 11:46 (1962).
27. K. Era, H. Katayama, and S. Shionoya, J. Phys. Soc. Jap., 24:1180 (1968).
28. Ya. A. Oksman and É. E. Zablovskii, Fiz. Tverd. Tela, 6:1930 (1964).
29. D. Curie, Luminescence in Crystals, Methuen, London (1963).
30. M. Aven and J. S. Prener, Physics and Chemistry of II-VI Compounds, North-Holland, Amsterdam (1967).
31. M. Aven, J. Appl. Phys., 38:4421 (1967).
32. M. N. Islam and J. Woods, Solid State Commun., 7:1457 (1969).
33. P. G. Le Comber, W. E. Spear, and A. Weinmann, Brit. J. Appl. Phys., 17:467 (1966).
34. N. N. Grigor'ev and M. V. Fok, Abstracts of Papers presented at Second Conference on Physics and Chemistry of II-VI Compounds [in Russian], Institute of Physics, Academy of Sciences of the Ukrainian SSR, Kiev (1969), p. 65.

35. V. V. Antonov-Romanovskii, Kinetics of the Photoluminescence of Crystal Phosphors
     [in Russian], Nauka, Moscow (1966).
36. M. K. Sheinkman, in: Electroluminescence of Solids [in Russian], Naukova Dumka,
     Kiev (1971), p. 77.
37. N. V. Mitrofanova, Yu. P. Timofeev, S. A. Fridman, and V. V. Shchaenko, Izv. Akad.
     Nauk SSSR, Ser. Fiz., 35:1446 (1971).
38. H. Gobrecht and D. Hofmann, J. Chem. Phys. Solids, 27:509 (1966).
39. S. S. Batsanov, Electronegativity of Elements and Chemical Bonds [in Russian], Novosibirsk
     (1962).

RADIATIVE RECOMBINATION IN CADMIUM TELLURIDE CRYSTALS

Zh. R. Panosyan

Investigations were made of the radiative recombination and reflection spectra of CdTe crystals with different compositions. The measurements were carried out in the temperature range 4.2-300°K at wavelengths of 0.7-4.0 μ. The reflection spectra were used to calculate the free-exciton absorption and luminescence spectra. A comparison of the calculated results with the measured luminescence spectra revealed self-reversal of the free-exciton luminescence line. An analysis of the photoluminescence spectra as a function of the impurity — defect composition of the samples made it possible to interpret the majority of the observed luminescence lines and bands. A study was made of the strength of the electron — phonon interaction in optical transitions as a function of the nature of the centers involved. The profiles of the luminescence bands resulting from the recombination at deep levels were explained.

INTRODUCTION

Investigations of the optical properties of semiconductor crystals yield extensive information on the energy band structure. This structure is of great importance in various technical applications (for example, coherent and noncoherent light sources, quantum radiophysics and optoelectronic devices) and in the establishment of the fundamental properties of solids. The energy structure of defects in real crystals is of particular importance. Theoretical calculations meet with various difficulties of a basic nature and, therefore, experimental investigation methods are acquiring considerable importance.

A crystal may contain phonons, electrons and holes, excitons, polarons, vacancies and interstitial atoms or ions, foreign atoms, and dislocations.

Recent experience shows that investigations of the radiative recombination in semiconductors gives very useful information on the energy structure of defects and imperfections. This approach is particularly important in the case of wide-gap semiconductors such as II-VI compounds, and these include promising but little known cadmium telluride.

The present paper reports information on some defects in cadmium telluride crystals obtained by the radiative recombination method.

CHAPTER I

REVIEW OF PUBLISHED LITERATURE

§ 1. Principal Radiative Recombination Mechanisms and Energy Band Structure of II-VI Compounds

In contrast to metals, the carrier density in semiconductors and insulators is not constant. The disappearance or annihilation of an electron and a hole (equilibrium or non-equilibrium) in

a crystal is known as recombination. Recombination of electrons and holes may be nonradiative (in this case, all the energy liberated in the recombination is dissipated in the generation
of elementary excitations in a crystal) or radiative (in this case, all or part of the liberated
energy is emitted in the form of photons). The radiative recombination in semiconductors may
occur in the following ways: a) recombination of a free electron with a free hole (band—band
or interband transitions); b) recombination of a free carrier with one bound to a local center
(band—local center transitions); c) recombination of an electron with a hole, each bound to different local centers (interimpurity transitions); d) recombination of an electron with a hole
inside a local center (intracenter transitions); e) annihilation of a free exciton with one bound
to a local center.

Investigations of the radiative recombination spectra of semiconductors can yield information on the recombination mechanism and the energy spectrum of electrons. Radiative recombination of optically generated carriers is known as the photoluminescence and we shall
use the terms "radiative recombination" and "photoluminescence" as synonyms.

In this section, we shall discuss just the radiative annihilation of free excitons with those
bound to local centers.

Cadmium telluride is a II-VI semiconducting compound. Since the outer electrons of the
atoms participate in the formation of valence bonds in solids, it is assumed (and has been confirmed experimentally) that, in the case of the II-VI compounds, the conduction band has the
s-type symmetry of the wave functions (this band is formed from electron levels of the group
II atoms), whereas the valence band has the p-type symmetry (it is formed from the group VI
atoms) and, consequently, it should be triply degenerate. However, as a result of the spin—
orbit interaction, the degeneracy of the valence band is partly lifted and a crystal has the energy band structure shown in Fig. 1b. This structure is exhibited by cubic II-VI crystals with
the zinc-blende structure (including cadmium telluride). An external perturbation (for example, a pressure) can lift the remaining degeneracy of the valence band and this can give rise to
polarization phenomena not otherwise observed.

In hexagonal crystals, this perturbation is the internal crystal field which appears because
of a definite structure symmetry. The energy band structure of such crystals (which have the
wurtzite crystal structure) is shown in Fig. 1a. The more complex band structure of hexagonal
crystals causes difficulties in the observations and interpretation of the results, but yields more
extensive information on the nature and characteristics of electron processes occurring in the
II-VI crystals.

In the hexagonal subgroup of the II-VI compounds, the best known material is cadmium
sulfide and detailed studies of its optical properties have been going on for a long time. It has
been found that some of the recombination processes can be associated with the annihilation of
free excitons. A summary of the initial investigations intended to determine the main features
of the energy band structure is given in [1]. It has been found that three reflection peaks near
the interband absorption edge correspond to the formation of free excitons. An analysis of the

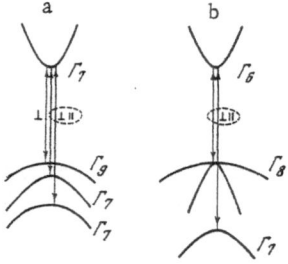

Fig. 1. Energy band structure of the II-VI
compounds: a) crystals with the wurtzite
structure; b) crystals with zinc-blende structure.

polarization of the reflected light has made it possible to refine the current theoretical ideas on the energy band structure of cadmium sulfide (Fig. 1a). An exact quantitative picture has been obtained using the reflection results to calculate the absorption (imaginary component of the permittivity) employing the Kramers–Kronig relationships (discussed later). Such calculations have yielded the binding energy of excitons (deduced from the reflection corresponding to excited states), forbidden band width, and energy gaps between the split valence bands. The values of these quantities have been combined with the quasicubic model Hamiltonian for a hexagonal structure in estimating the splitting due to the spin–orbit interaction in the absence of the internal crystal field. Luminescence studies have yielded results in agreement with the reflection (absorption) data for these crystals. At low temperatures (4.2°K) the relative intensity of the luminescence corresponding to the annihilation of free excitons is found to be small. This is explained in [1] assuming that the annihilation of free excitons is hindered by the large difference between the exciton and photon quasimomenta. The conservation of momentum is easily achieved by the simultaneous interaction of an exciton with lattice defects.

A method based on the Kramers–Kronig relationships was used successfully in [1] in studies of the main features of the energy band structure. It is interesting to consider this method in detail, particularly as its use in the interpretation of the experimental results represents a major point in the present paper.

The idea behind the method is as follows. If the absorption coefficient of a crystal is high (especially in the exciton absorption region), it is not possible to measure the transmission accurately and, therefore, the absorption has to be deduced from the experimental data on the reflection.

In general, the reflection coefficient is a complex function of the refractive index and the extinction coefficient (or of the real and imaginary components of the permittivity). This function is simplest for the normal incidence of light:

$$R = \frac{(n-1)^2 + k^2}{(n+1)^2 + k^2} .$$

(I.1)

It is convenient to introduce the functions r and φ, known as the complex reflection coefficient and phase angle, so that $r = (N-1)/(N+1)$, where $N = n + ik$ is the complex refractive index and $r = \sqrt{R}e^{i\varphi}$. Then, n and k can be represented by functions of $|r| = \sqrt{R}$ and φ

$$k = \frac{2\sqrt{R}\sin\varphi}{1 + R - 2\sqrt{R}\cos\varphi} , \qquad n = \frac{1 - R}{1 + R - 2\sqrt{R}\cos\varphi} ,$$

(I.2)

and the Kramers–Kronig relationship for the phase angle becomes

$$\varphi(E) = -\frac{E}{\pi} \int_0^\infty \frac{\ln R(E')}{E'^2 - E^2} dE' ,$$

(I.3)

where only the principal value of the integral is used [2, 3]. The absorption coefficient is $\alpha = (4\pi/\lambda)k$.

In calculating the phase angle, the upper limit of the integral is taken to be infinity. Thus, in practice, the application of the Kramers–Kronig integral method gives approximate results because the real upper limit is naturally finite. However, the parts of the spectrum located far from the region investigated directly make only a small contribution to the integral (this follows from the expression for the integrand). This explains why the Kramers–Kronig relationships can be used in practice. Moreover, the part of the spectrum where R(E) = const also makes no contribution to the integral. This can be demonstrated by evaluating the integral of Eq. (I.3) by parts:

$$\varphi(E) = -\frac{E}{\pi}\int_0^\infty \frac{\ln R(E')}{E'^2 - E^2}\, dE' = \frac{1}{2\pi}\int_0^\infty \ln\frac{E' + E}{|E' - E|}\cdot \frac{d}{dE'}\,[\ln R(E')]\, dE'. \qquad (I.4)$$

The Kramers—Kronig relationships have been used successfully in various studies. These have been mainly the investigations of the interband optical absorption (see, for example, [4, 5]), absorption in the exciton part of the spectrum [1], and strong lattice absorption [3]. Initially, the expression (I.3) for the phase angle was integrated graphically using the results given in [6]. Since the arrival of computers, this task is much easier to perform and the precision of the calculations is much higher than that of the graphical method.

The basic procedures in the use of the Kramers—Kronig relationships are as follows.

1. The reflection coefficient is measured carefully in the investigated part of the spectrum. It is desirable to determine the dependence R(λ) in the widest possible spectral interval.

2. The effect of the truncation of the integral in the calculation of the phase angle (I.3) is compensated by including various corrections. In this way, one can ensure that the calculated coefficient is equal to a specified values at some reference point (for example, it is assumed that the coefficient is zero for energies smaller than that corresponding to the forbidden band width).

By way of example, we shall consider the work of Thomas and Hopfield [1], whose ideas have been used in many subsequent investigations. The phase angle is calculated from the integral between finite limits and a trinomial $A\nu^2 + B\nu + C$ (ν is the frequency) introduced by way of correction. The parameters A, B, and C are selected to ensure that the absorption vanishes at three points selected in advance. Spitzer and Kleinman [7] compared the results of computer calculations based on the Kramers—Kronig relationships with the exact values of the refractive index and the extinction coefficient for quartz in the infrared part of the spectrum. They found that the error committed in the calculations did not exceed 10% in the strong-absorption region (k > 0.1). However, in the weak-absorption region (k < 0.1), the results of calculations were less accurate because the phase angle was of the order of the compensating term. Thomas and Hopfield [1] estimated the error in their calculations also to be 10%.

Apart from electron processes involving free excitons, some of the recombination mechanisms in this part of the spectrum can be interpreted and described semiquantitatively also by bound exciton complexes, i.e., electron—hole pairs localized near ionized or neutral point defects in the lattice. The use of the term "bound exciton complexes" is justified only if the binding energy of an exciton is higher than its binding energy to a center. Lampert [8] developed the concept of excitons in semiconductors and demonstrated the possibility of the existence of such complexes. An experimental confirmation of this concept in the case of Si was published by Haynes [9]; Thomas and Hopfield [10] later confirmed the same concept for CdS.

The luminescence lines due to the recombination of bound excitons were found to be much narrower than kT. The spectra of bound excitons in the II-VI compounds were identified successfully by investigations in magnetic field because the ground and excited states of the various complexes were split differently by these fields [8, 10, 11]. The results of an investigation of several groups of lines emitted by CdTe were reported in [12], particularly the results for the group with the shortest wavelengths adjoining the interband absorption edge. This group consisted of a series of partly resolved luminescence lines and the half-width of some of these lines was smaller than kT. These lines were attributed in [12] to the impurity recombination involving a shallow donor level with an ionization energy of 0.01 eV (according to [13]) and to the recombination involving many-particle exciton complexes (bound excitons). However, attempts to relate the luminescence spectra to the impurity content of the crystals were not successful.

An investigation of the absorption and luminescence of cadmium telluride crystals in the exciton part of the spectrum at 20°K [14] revealed that the exciton absorption maximum was located at 1.594 ± 0.001 eV. At this energy, the absorption reached 8×10^4 cm^{-1} and the line width was 3 meV.

An investigation of the energy band structure of cadmium telluride, based on a study of the influence of the pressure on the reflection spectra of normally incident light, was reported in [15].

The luminescence of CdTe crystals was also studied under bombardment with 150 keV electrons [16]. It was assumed that the shortest-wavelength lines at 1.59 eV (T = 10°K) and 1.573 eV (T = 80°K) were due to the exciton annihilation. It was reported in [16] that the coherent emission of CdTe appeared in this spectral region.

On the whole, the interpretations of the results obtained in studies of the luminescence and reflection spectra of CdTe crystals in the exciton absorption region were not in agreement. This was particularly true of the identification of the low-temperature (10°K) luminescence lines, including some very narrow unresolved lines [12, 16]. Moreover, a change in the luminescence mechanism responsible for this group of lines was observed when the temperature was raised to ~ 80°K.

Therefore, the present author undertook a detailed investigation of the radiative recombination spectra of CdTe in the exciton absorption region. This investigation was carried out in a wide range of temperatures and was combined with simultaneous measurements of the reflection spectra followed by calculations of the absorption spectra.

§ 2. Edge Luminescence

The edge luminescence of the II–VI compounds is understood to represent groups of equidistant lines corresponding to photon energies differing by a few hundredths to a few tenths of an electron-volt from the forbidden band width [17, 18]. The energy separation between the lines forming these groups is equal to the energy of longitudinal optical (LO) phonons in CdS, ZnS, and ZnO crystals [19]. According to the model proposed in [19], free electrons and holes recombine via exciton states emitting simultaneously one photon and several LO phonons. Therefore, this type of luminescence, emitted by all the II–VI compounds, has been termed the edge luminescence. However, the model was fairly soon found to be incorrect. The mechanism and nature of the edge luminescence emitted by the II–VI semiconducting compounds have been investigated in the greatest detail for CdS and the evolution of our understanding of this luminescence can best be followed by considering this compound.

Cadmium sulfide crystals emit two series of equidistant luminescence lines in the 5100–5600 Å range. The separation between the lines in each series is equal to the LO phonon energy, which is $\hbar\omega_0 = 0.038$ eV. A series which predominates at low temperatures and has a zero-phonon line at $\lambda_{max} = 5175$ Å is known as the long-wavelength series. At higher temperatures (> 25°K), the spectrum is dominated by a series with a zero-phonon line at $\lambda = 5135$ Å and called the short-wavelength series. Krivoglaz and Pekar [20] and, later, Hopfield [21] showed theoretically that a strong interaction of electrons with phonons could be explained only by a model postulating the localization (before recombination) of both carriers [20] or at least one [21]. It is shown in [20, 21] that the relative intensities of the lines in these two series can be described by the Poisson distribution.

Collins [22] suggested that at 77°K the short-wavelength series was due to the recombination of a free electron with a bound hole [23] because the density of free holes fell more rapidly (the hole lifetime was $\sim 10^{-8}$ sec at 77°K) than the intensity of the green edge luminescence band of CdS ($\sim 10^{-6}$ sec).

Halsted and Segall [24] studied the doublet structure of the edge luminescence of CdTe at 20°K or higher and suggested that the short-wavelength series of CdTe and other II–VI compounds was due to the recombination of a free hole with an electron localized in a doubly ionized acceptor level A^{2-}. In their opinion, the long-wavelength series appeared because of the annihilation of an exciton bound to the same acceptor. The existence of such an acceptor level was established in [25] from measurements of the electrical conductivity and it was found that in CdTe the level was separated by 0.056 eV from the bottom of the conduction band (at 100°K), whereas in CdS it was separated by 0.09 eV (at 250°K). This mechanism was extended in [24] to five II–VI compounds: CdTe, CdS, ZnSe, ZnS, and ZnO. The same mechanism explained the slower thermal shift of the position of the long-wavelength band compared with the temperature dependence of the forbidden band width $E_g(T)$; it also accounted for the temperature dependence of the relative intensities of the doublet lines. The slow thermal shift of the band, compared with $E_g(T)$, was attributed to the fixed position of the doubly charged A^{2-} acceptor relative to the bottom of the valence band.

The long-wavelength series observed at 4°K was explained in [26] by a donor–acceptor recombination and the short-wavelength series at 77°K was attributed to the recombination of a free with a bound carrier.

Investigations of the dependences of the intensity of the green edge luminescence of CdS on the excitation rate and of the temperature dependence of the afterglow spectrum (4.2-77°K) were used as the basis of a model proposed in [27]. According to this model, the long-wavelength and short-wavelength series were due to radiative transitions between bound states, which could be levels of donor–acceptor pairs. The short-wavelength luminescence series appeared due to electron transitions from shallower donor levels, whereas the long-wavelength series was due to transitions from deeper donor levels to the same acceptor level. However, to explain the temperature dependences of the intensities of the edge luminescence bands in accordance with this model, it was necessary to assume that the probability of the thermal liberation of electrons to the conduction band was greater for the deeper donor levels than for the shallower. This model was adopted also in [28], where it was reported that the luminescence shifted in the direction of higher energies when the excitation intensity was raised (this was observed for the short-wavelength series at 77°K and for the long-wavelength series at 4°K). The shift of the short-wavelength series was considerably smaller and a much greater increase in the excitation intensity (compared with the long-wavelength series) was needed to observe this shift at all.

A theory of the energy levels of donor–acceptor pairs, employing the effective mass and hydrogenic molecule approximations, was proposed in [29] and supported by the experimental results in [30], where this theory was used to explain the nature of the edge luminescence emitted from GaP.

The energy dependence of a luminescence line of a donor–acceptor pair on the distance r between the partners in such a pair can be described by

$$E\,(r) = E_g - (E_a + E_d) + \frac{e^2}{\varepsilon_0 r}, \qquad\qquad (I.5)$$

where E_g is the forbidden band width; E_a and E_d are the activation energies of acceptors and donors, respectively; ε_0 is the low-frequency permittivity. The value of E(r) shifts toward shorter wavelengths with decreasing distance r and, at the same time, the radiative recombination probability rises, i.e., the spontaneous emission time of pairs becomes longer with increasing r. Therefore, the shift of the edge luminescence in the direction of higher energies, observed on increase in the excitation intensity, is an important argument in support of the donor—acceptor recombination mechanism. However, as pointed out in [31], the shift of the short-wavelength series in the direction of higher energies with rising excitation intensity can also be explained by assuming that a free electron recombines with a hole bound to an acceptor. Two explanations are possible.

1. Since the Boltzmann distribution of free electrons has a maximum at $E = kT_{eff}$, the energy of the emitted photons is given by

$$\hbar\omega = E_g - E_a + kT_{eff} \tag{I.6}$$

At high excitation levels, the effective electron temperature T_{eff} may exceed the lattice temperature at 77°K and the line shift for $T_{eff} \sim 130°K$ should be ~ 4 meV.

2. The line shift is possible because of a reduction in the ionization energy of acceptors as a result of screening by free carriers.

Following the experiments on GaP, a study was made of the edge luminescence kinetics of CdS [32] and it was found that at 1.6°K the long-wavelength series was due to recombination of an electron bound to a donor with a hole bound to an acceptor.

Colbow and Nyberg later [33–35] carried out a more detailed study of the kinetics and temperature dependences of the intensities of the short- and long-wavelength series of the green edge luminescence of CdS and of the photoconductivity of this compound. They concluded that the long-wavelength series was due to the donor—acceptor or bound—bound (bb) recombination [26]. The short-wavelength series was attributed to the radiative recombination of a free electron with a hole bound to an acceptor, i.e., a free—bound (fb) transition.

On the other hand, Goede and Gutsche [36] analyzed the profiles of the luminescence lines and the temperature dependences of their intensities in the 18-150°K range and concluded that the short-wavelength series of CdS was due to radiative transitions in donor—acceptor pairs with the shortest internal distances r. They demonstrated that the experimentally observed profiles could be represented by the sum of equidistant Gaussian curves with the same half-width H. Hence, we assumed that the short-wavelength series appeared due to radiative transitions between donors and acceptors. The donor—acceptor mechanism was extended by Goede and Gutsche to the long-wavelength series, which predominated at temperatures below the range which they investigated. However, this did not resolve the question of the temperature dependences of the lines in both series [27].

Thus, it follows from this review that, in spite of the large amount of work already done on the edge luminescence of CdS, there is as yet no agreement on the mechanism responsible for the edge luminescence. According to one of the models proposed for CdS [26, 35, 37, 38], the long-wavelength series predominating at low temperatures is due to radiative transitions between donors and acceptors, whereas the short-wavelength series predominating at higher temperatures results from the radiative capture of free electrons by acceptors. According to the second model, both series are due to interimpurity radiative transitions within donor—acceptor pairs [27, 28, 36].

Studies of the optical and electrical properties of the II-VI semiconducting compounds have been hindered by the difficulties encountered in the growth of single crystals with controllable parameters. The problems of growth of the II-VI crystals, the energy structure of impurities and defects, and their electrical, optical, and other properties are considered in an extensive review [39] devoted to the II-VI semiconducting compounds.

CHAPTER II

EXPERIMENTAL METHOD

§1.   Apparatus Used in Determination of Luminescence Spectra

The apparatus used in the studies of luminescence spectra consisted of four main units, which were a light source, a cryostat, a monochromator, and a recording system.

A block diagram of the apparatus is shown in Fig. 2.

Light Source.   Electron guns and lasers can be used to achieve very high power densities under pulsed and continuous excitation conditions.   However, such sources are often fairly complex and difficult to use.   Therefore, conventional excitation sources, i.e., incandescent lamps or gas-discharge tubes, are still frequently used to provide a fairly high but not extreme excitation power.   This power can be increased by increasing the brightness of the source (this is limited by the construction of the lamp or tube) or by increasing the relative aperture (D/f) of the condenser lens.   If spherical optics is employed, the relative aperture of the condenser is limited by the spherical aberration of the lenses, which rises sharply with increasing D/f.   However, if parabolic optics is used, one can employ much larger relative apertures because of the freedom from sphericla aberrations.

Figure 3 shows the construction of a high-power light source with a parabolic optical system [40], which was used by the author and his colleagues in investigations of the photoluminescence of such semiconductors as CdTe, Si, and Ge.   The optical system consisted of a condenser (4a and 4b) and a mirror (11).   The condenser had two quartz lenses with parabolic surfaces (Fig. 4).   Each lens had a relative aperture D/f = 1:1.   The parabolic mirror had an aperture D/f = 2:1.   The light source was located 70 mm (2f of the mirror) from the mirror and 45–50

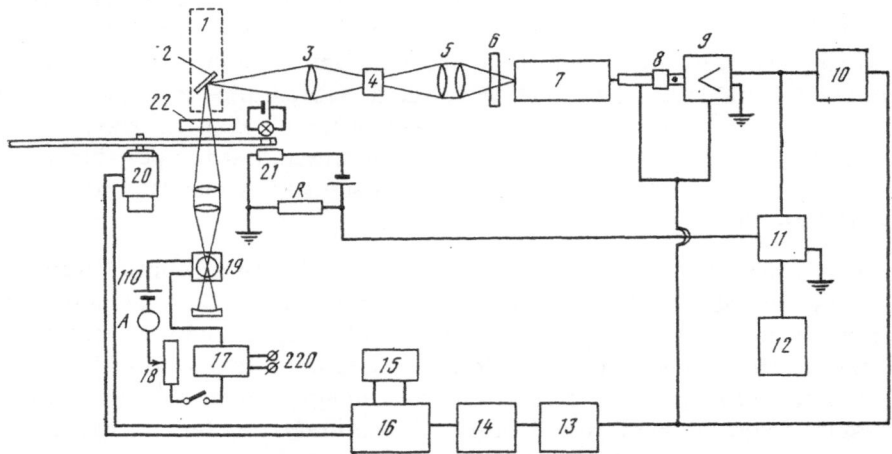

Fig. 2.   Block diagram of the apparatus used in studies of the recombination radiation spectra: 1) cryostat; 2) sample; 3) parabolic lens; 4) optical beam splitter; 5) condenser; 6) KS-19 filter; 7) DFS-12 spectrometer; 8) FÉU-22 photomultiplier; 9) resonance amplifier; 10) vacuum-tube voltmeter; 11) synchronous detector; 12) automatic plotter; 13) voltage stabilizer; 14) af oscillator; 15) power supply; 16) power amplifier; 17) high-frequency discharger; 18) water-cooled resistance; 19) high-luminosity light source; 20) synchronous motor; 21) photoresistor; 22) SZS-22 filter.

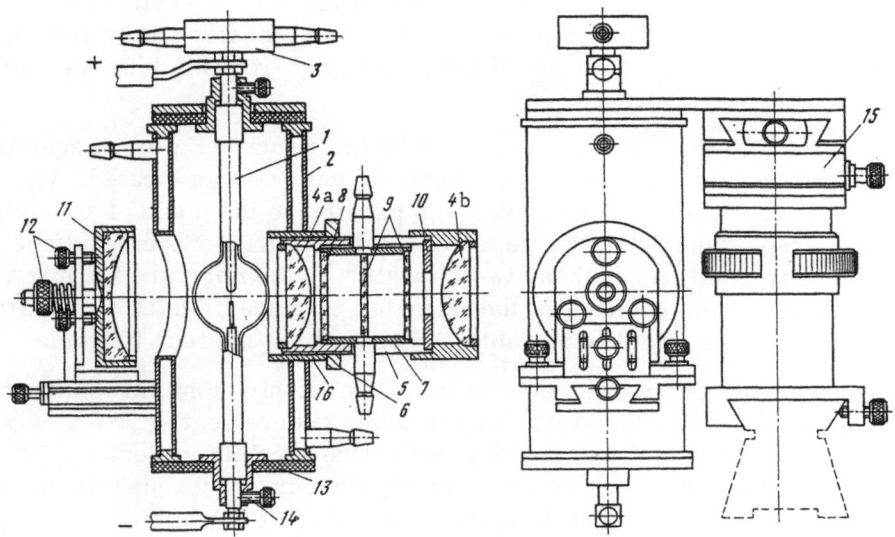

Fig. 3. Construction of high-luminosity light source: 1) DKsSh-1000 lamp; 2) water-cooled protective jacket; 3) cooling water for the anode; 4a-4b) two-lens condenser; 5) tube; 6) locknut; 7) cell wall; 8) optical quartz window; 9) filters; 10) diaphragm; 11) mirror; 12) mirror alignment device; 13) Teflon insulator; 14) lamp holder; 15) three-coordinate table; 16) casing.

Fig. 4. Parabolic quartz lens (a) and parabolic mirror (b). Here, curve 1 is the parabolic surface $Y = x^2/54$; curve 1' is a parabolic surface $Y = x^2/140$.

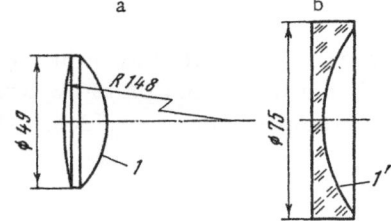

mm from the condenser (~2f from the condenser).* For this configuration of the optical elements, the light source produced, at a distance of 45 mm from the condenser, two superimposed images of the source of natural size. The total solid angle from which light was collected was about $\pi/2$.

The actual source of light was a very-high-pressure xenon lamp of the DKsSh-1000 type (its output was 1000 W). This lamp was supplied from a dc source via a water-cooled resistance made of Nichrome wire, 1.5 mm thick. The lamp (1) was placed inside a water-cooled protective jacket (2). Experience showed that the service life of the lamp could be increased by cooling its anode (3) with water. The spectral composition of the exciting light was determined by filters (9) located between the two condenser lenses. The filters employed depended on the actual experiment to be performed. In particular, when the luminescence of cadmium telluride (forbidden band width ~1.6 eV) was excited, we used a water filter in combination with SZS-17 and SZS-22 colored glasses. This selection of filters passed light in the range 0.36-0.57 μ.

---

*Quartz was selected as the lens material not only because it was desirable to retain the ultraviolet component of the emission spectrum of the lamp but also because the first lens, closest to the lamp, was heated very strongly.

The construction of the light source was such that the optical system could easily be adjusted. The distance between the lamp and the first lens (4a) was set by rotating a tube (5) in a casing in such a way that the beam of light emerging from the lens was parallel; this position was locked with a nut (6).

The image of the DKsSh-1000 lamp produced by the condenser was an equilateral triangle with a high-brightness region near its vertex where the cathode was located. When the mirror was oriented in an appropriate way, the image was parallel to the plane of the source but was rotated by 180° (reversed triangle). The geometrical dimensions of the resultant spot were governed by the relative positions of these two triangles. In particular, the mirror could be oriented in such a way that the vertices of the triangles coincided. In this case, the brightness in a spot of ~1 mm diameter was considerably higher than the average brightness of the image.

The light source unit as a whole could be moved smoothly along three directions within ~30 mm on a three-coordinate table (15). This was very convenient, particularly in investigations of the luminescence of small samples located inside a helium cryostat. The intensity of the exciting light could be varied either by altering the current through the lamp or using an iris shutter, whihc could reduce the intensity by a factor of 50.

The parabolic optics of the system made it possible to employ lenses and mirrors with large relative apertures and thus achieve fairly high excitation intensities. According to our estimates, the excitation flux in the spectral range set by the filter combination could be 10^{20} photons$\cdot$cm$^{-2}\cdot$sec^{-1} (under continuous operating conditions). This flux density was estimated from the short-circuit current of a silicon photodiode,

$$i_{\text{sc}} = \frac{S}{6.25 \cdot 10^{12}} \int\limits_{0.36}^{0.57} N(\lambda)\, K(\lambda)\, d\lambda, \tag{II.1}$$

where S is the illuminated area of the photodiode; $K(\lambda)$ is the coefficient characteristic of a given diode but dependent on the wavelenght; $N(\lambda)$ is the number of photons of a given wavelength incident on the photodiode surface in 1 sec, which can be expressed by

$$N(\lambda) = A\varphi(\lambda), \tag{II.2}$$

where A is a constant and the function $\varphi(\lambda)$ is the spectrum of the excitation source determined experimentally using a photomultiplier corrected for its spectral sensitivity. Substituting Eq. (II.2) into Eq. (II.1), we find that

$$A = 6.25 \cdot 10^{12} \frac{i_{\text{sc}}}{S} \frac{1}{\int\limits_{0.36}^{0.57} K(\lambda)\, \varphi(\lambda)\, d\lambda}. \tag{II.3}$$

The total number of photons reaching the photodiode in 1 sec is

$$N = \int\limits_{\lambda_1}^{\lambda_2} N(\lambda)\, d\lambda = A \int\limits_{\lambda_1}^{\lambda_2} \varphi(\lambda)\, d\lambda = 6.25 \cdot 10^{12} \frac{i_{\text{sc}}}{S} \cdot \frac{\int\limits_{0.36}^{0.57} \varphi(\lambda)\, d\lambda}{\int\limits_{0.36}^{0.57} K(\lambda)\, \varphi(\lambda)\, d\lambda}. \tag{II.4}$$

A numerical estimate based on Eq. (II.4) gave $N \sim 10^{20}$ photons$\cdot$cm$^{-2}\cdot$sec^{-1}. This was the maximum excitation flux density, whereas we often used flux densities about an order of magnitude lower. The steady-state density of nonequilibrium carriers produced by such continuous illumination was

$$\Delta n = g\tau = \gamma N\alpha(1 - R)\tau, \tag{II.5}$$

where g is the carrier generation rate; τ is the carrier lifetime; γ is the quantum efficiency; α is the absorption coefficient; R is the reflection coefficient.

Substituting in Eq. (II.5) the appropriate constants for CdTe ($\gamma = 1$, $N = 10^{20}$ photons · $cm^{-2}\cdot sec^{-1}$, $\alpha = 10^4$ cm^{-1}, $R = 0.28$, $\tau = 10^{-8}$ sec [41–43]), we found that the strongest illumination corresponded to a steady-state nonequilibrium carrier (pair) density of 10^{16} cm^{-3} in CdTe.

Cryostat. The radiative recombination and reflection spectra of cadmium telluride crystals were determined in the temperature range 4.2–300°K. Three types of cryostats were used in this temperature range.

Between 4.2 and 100°K we employed two different variants of a KR-15 metal cryostat developed at the Lebedev Physics Institute. These variants differed in the construction of the lower part, which was soldered to the inner chamber of the cryostat filled with the cooling agent (liquid helium, hydrogen, or nitrogen).

In the first variant, a copper rod, acting as a heat sink, was soldered to the inner chamber. A sample was bonded by a silver paste to the polished surface of the rod.

The lower part of the cryostat had three plane-parallel quartz windows of 40 cm diameter. The exciting light was admitted through one of these windows and the other two were used to study the recombination radiation. This construction made it possible to study the luminescence emitted from the surface of the sample facing the lamp and from the opposite side of the sample. A thermocouple was attached to the sample with a silver paste. The sample was a plane-parallel plate thinner than 1 mm. Therefore, we could assume that the temperature of the sample was practically the same throughout its volume. (The thermal conductivity of CdTe was known to be 0.015 cal·$deg^{-1}\cdot cm^{-1}\cdot sec^{-1}$ at T = 300°K [44].) We used copper−constantan and gold−copper thermocouples. The latter thermocouple had a steeper characteristic at temperatures below 80°K. Therefore, at temperatures up to 80°K we used mainly the gold−copper thermocouple (the gold was alloyed with 3% Fe but the copper was pure). The emf developed by the thermocouple was measured with an R306 dc potentiometer. A galvanometer with a sensitivity of 0.3×10^{-6} V/div was used as a null detector. The temperature was measured to within ~1 deg K.

The first variant of the cryostat had a number of shortcomings. We were unable to obtain temperatures below 13°K and the temperature of the sample changed when high excitation intensities were employed. In this situation, the spectrum could change under the simultaneous action of two factors which were the excitation and temperature rise. It was sometimes impossible to distinguish the influence of these two factors. Moreover, the mounting of a sample in the cryostat took a fairly long time and it was not possible to interchange the samples during measurements.

In the second variant of the cryostat, we used an intermediate copper tube to connect the sample with the inner chamber. This tube had a quartz can at its end. The can contacted the copper tube via a stainless steel compensator and an Invar ring. The bottom of the can was also made of Invar. The Invar was bonded to the quartz with Araldite at 200°C. The quartz can was made of a high-quality (optical) tube which did not weaken significantly the transmitted radiation and did not scatter this radiation. The cooling liquid filled the whole can and the sample was immersed into the can through the tube used to fill the cryostat. The samples could be interchanged easily because the cooling liquid was at atmospheric pressure. Spectra could be recorded at 4.2, 20.4, and 77.4°K because the sample was in direct contact with the cooling liquid. Variation of the excitation intensity during an experiment by a factor of 50 did not alter the temperature of the sample because, even at the maximum flux density reaching the sample, a gas layer was not formed on the surface.

In the temperature range 77–300°K, the luminescence spectra were studied using a cryostat made of optical-quality quartz. A quartz Dewar flask had three plane-parallel windows and was very convenient to use. In this case, the samples was again in direct contact with the cooling liquid (liquid nitrogen, alcohol mixed with dry ice, or water).

Monochromator. The recombination radiation spectrum of a semiconductor frequently consists of fairly narrow and weak lines. Therefore, in order to achieve the maximum resolution, it is desirable to use high-luminosity monochromators. It is shown in [45] that the luminosity of monochromators with diffraction gratings is several times higher than the luminosity of the prism monochromators.

We used two diffraction-grating monochromators: DFS-12 in the spectral range 0.76–1.1 μ and SP-102 in the 1–3.5 μ range. The parameters of these monochromators are listed in Table 1.

The focal lengths of the entry and exit collimators of both monochromators were equal $f_1 = f_2 = f$ and, moreover, the angular magnification was unity, i.e., $S_1 = S_2 = S$, where S_1 and S_2 are the cross-sectional areas of the beams incident and reflected by the grating.

The light flux leaving the exit slit of a monochromator is

$$\Phi = B_\lambda \tau_\lambda (\Delta\lambda)^2 \frac{h}{f} S \frac{d\varphi}{d\lambda}, \tag{II.6}$$

where B_λ is the monochromatic brightness of the source; τ_λ is the transmission coefficient of the monochromator; $\Delta\lambda$ is the spectral width of the exit slit (in monochromators the width of the exit slit is usually equal to that of the entry slit).

The luminosity L of a monochromator is equal to the light flux at the exit from the monochromator for $B_\lambda = 1$ and $\Delta\lambda = 1$, i.e.,

$$L = \tau_\lambda \frac{h}{f} S \frac{d\varphi}{d\lambda}. \tag{II.7}$$

The relationship between the dimensions of the diffraction grating and the size of the effective aperture is ignored in Eqs. (II.6) and (II.7) because the coefficients representing this relationship are close to unity.

A comparison of the luminosities of the DFS-12 and SP-102 monochromators indicated that the value of τ_λ could be assumed to be the same in both cases. Substituting the parameters from Table 1 into Eq. (II.7), we found that the luminosity of the DFS-12 monochromator was 20 times higher than the luminosity of the SP-02 instrument. Because of the higher luminosity of the DFS-12 monochromator, the spectral width of the slit used in recording the spectra of

TABLE 1

Type of monochromator	h, mm	f, mm	h/f	$d\varphi/d\lambda$, μ^{-1}	S, cm^2
DFS-12	40	822	0.049	1.216	210
SP-102	6	270	0.022	0.218	132

Note. Here, h is the slit height; f is the collimator focal length; $d\varphi/d\lambda$ is the angular dispersion of the grating; S is the grating area.

cadmium telluride at low temperatures usually amounted to ~1 Å (0.2 meV) and sometimes even less. This ensured a good resolution of the spectral lines (the linear dispersion of the DFS-12 monochromator was 10 Å/mm at 9000 Å). Naturally, the resolution depended also on the sensitivity of the radiation detector.

The recombination radiation of a sample was focused (by an optical system composed of a parabolic lens, optical beam splitter, and a condenser – Fig. 2) onto the entry slit of the monochromator in such a way that the image of the sample completely covered the slit and the angular aperture of the monochromator was completely filled.

Since the dispersive element was a diffraction grating, the overlap of the various orders of the spectral lines (orders I and II the in case of the DFS-12 monochromator) was avoided by placing a KS-19 filter in front of the entry slit of the monochromator. This filter stopped the radiation of wavelengths shorter than 6500 Å. Moreover, the KS-19 filter absorbed the exciting light from the lamp so that only the luminescence of CdTe entered the monochromator.

In a different variant, when we used the SP-102 monochromator, the overlap of the spectral lines of different orders was even greater in the 1-3.5 μ wavelength range. Therefore, to separate the various orders, we had to use such filters as germanium, water, and colored glasses (IKS-1, IKS-3, KS-19, and others). The water filters were placed in front and after the condenser to absorb the thermal radiation emitted by the hot quartz lens (4b in Fig. 3), which was of ~3 μ wavelength and could reach the detector.

In this variant, the optical system for focusing the radiation on the entry slit of the monochromator differed somewhat from that described. The recombination radiation of a sample was focused in the plane of the entry slit of the monochromator by a condenser consisting of NaCl lenses. This lens material was selected to reduce chromatic aberrations because the dispersion of NaCl in the investigated part of the spectrum was weak. The parameters of the optical system were selected so as to ensure that the image of the sample covered completely the slit and the angular aperture of the monochromator was filled. This could be done without an optical beam splitter because of the relatively small height of the slit of the SP-102 monochromator.

An emission line of a sodium lamp (λ = 5896 Å) was used to calibrate the SP-102 monochromator and check the calibration of the DFS-12 instrument.

Recording System. The resolving power of the apparatus as a whole was governed by the luminosity of the monochromator and the sensitivity of the radiation detector. When the DFS-12 monochromator was used, the luminescence was detected with an FÉU-22 photomultiplier whose spectral sensitivity set the long-wavelength limit of the investigated spectral interval (1.1 μ). The FÉU-22 photomultiplier was selected because of the low value of the thermal current (~10^{-9} A) compared with other photomultipliers suitable for this spectral range. The signal/noise ratio was improved still further by cooling the photomultiplier with liquid nitrogen in an evacuated chamber. The construction of this special chamber made it possible to focus the image of the exit slit of the monochromator on the cathodes of photomultipliers with side (FÉU-22) and end (FÉU-28) windows. Several photomultipliers of the FÉU-22 and FÉU-28 type were subjected to this cooling treatment. In this way, the signal/noise ratio was improved by two orders of magnitude but only for the FÉU-22 photomultiplier. Therefore, we used a cooled FÉU-22 photomultiplier as the radiation detector. This increase in the signal/noise ratio by two orders of magnitude made it possible to improve the resolution of the system as a whole (the monochromator and the detector) by one order of magnitude.

An electric signal picked up from a 1.5 MΩ load resistance in the anode circuit of the FÉU-22 photomultiplier was applied to the input of a selective narrow-band amplifier (f = 580 Hz, Δf = 10 Hz) with a maximum amplification factor of ~10^6. The noise at the amplifier output was practically entirely due to the noise of the FÉU-22 photomultiplier. A selective amplifier could be used because the light flux was interrupted mechanically at a frequency of 580 Hz. The signal produced by the amplifier was applied either to an SD-1 synchronous detector or to a vacuum-tube voltmeter to measure the amplitude of the signal and the signal/ noise ratio (Fig. 2). A reference voltage was applied to the SD-1 detector from a photoresistor illuminated with an incandescent lamp modulated at the same frequency as the light flux produced by the excitation source. A static voltage produced by the SD-1 detector drove an ÉPP-09 automatic plotter.

The spectral width of the monochromator slit was made several times smaller than the width of the investigated luminescence line to avoid distortion of the true line profile. Accurate reproduction of the line profile by the recording system was ensured by satisfying the condition $S > 5V_\tau$, where S is the spectral width of the monochromator slit and V_τ is the spectral range scanned during the time constant of the recording system.

A lead sulfide photoresistor, cooled with dry ice, was used as a detector in conjunction with the SP-102 monochromator. The range of sensitivity of the PbS photoresistor extended from the visible wavelengths to 3.5 μ.

§2. Determination of the Reflection Spectra

The reflection spectra of cadmium telluride were investigated in the exciton absorption region using the same crystals as those employed in the study of the radiative recombination spectra. The measurements were carried out at temperatures from 4.2 to 77.4°K using the cryostats described above.

The reflection spectra were recorded employing an incandescent lamp (17 V, 170 W) placed inside the jacket of the light source (Fig. 3) in place of the DKsSh-1000 lamp. All the elements of the optical system were removed with the exception of the second parabolic lens (4b in Fig. 3). This lens produced an image of the lamp filament on the surface of the sample. The light from the lamp passed through a KS-19 filter fixed to the cryostat window. The reflection spectra of CdTe were recorded mainly for normal incidence of light on the sample. The rest of the apparatus used in recording the reflection spectra was the same as that employed in the determination of the radiative recombination spectra. The reflection coefficient $R(\lambda)$ was determined for angles of incidence not exceeding 7°. The aperture of the light beam used in these measurements was ~3°. The value of $R(\lambda)$ was calculated from the ratio of the intensities of light reflected from the investigated crystal and from a silver mirror, whose reflection coefficient was assumed to be 96% in the investigated range of wavelengths.

The experimentally obtained dependences $R(\lambda)$ were used in calculations of the phase angle and then in the evaluation of the dependences $n(\lambda)$, $\alpha(\lambda)$, and $k(\lambda)$ (see §1 in Chap. I), performed on a BÉSM-4 computer.

§3. Preparation of Cadmium Telluride Samples

Crystals of CdTe were prepared in the Laboratory of Semiconductor Physics at the Lebedev Physics Institute by S. A. Medvedev, S. N. Maksimovskii, and Yu. V. Klevkov.

The photoluminescence of CdTe was investigated in parallel with the development of the technology of its synthesis and helped to improve this technology. Moreover, a correct interpretation of the luminescence spectra could only be provided when maximum information was available on the composition and method of preparation of the investigated crystals. We determined the luminescence spectra of a large number of samples prepared by different methods.

Some of our CdTe crystals were grown by a variant of the Bridgman method (directional crystallization under a controlled vapor pressure) and the crystals obtained in this way were denoted by BM. The crystal properties were controlled by varying the Cd vapor pressure. This pressure could be altered by varying the temperature of a reservoir of pure cadmium between 720 and 850°C. When the Cd vapor pressure corresponded to the temperature of pure Cd in the 790–800°C range, the crystals obtained were coarse-grained and semi-insulating with a resistivity of 10^4–10^8 $\Omega \cdot$ cm (the composition was close to stoichiometric) [46]. We obtained n-type crystals at higher pressures and p-type crystals at lower pressures because of deviations from stoichiometry. The compound CdTe could have considerable deviations from stoichiometry: the maximum deviation corresponded to $\sim 10^{17}$ cm^{-3} of excess atoms [48].

During growth, the crystals were sometimes doped with In, Al, Na, As, and other impurities.

Moreover, we employed the method of horizontal zone refinement under a controlled Cd vapor pressure (crystals of this type were denoted by ZR); this was done to produce a uniform distribution of impurities and defects along a crystal [48].

Finally, we also used the method of vacuum sublimation from the vapor phase at relatively low crystallization temperatures ($\sim$900°C); such crystals were denoted by SL. These crystals were coarse-grained and only of p-type with a resistivity from 5 to 10^8 $\Omega \cdot$ cm [47].

The methods and conditions during heat treatment of CdTe were described in detail in [48–50].

Heat treatments of CdTe crystals in Cd or Te vapors could be used to control deviations from the stoichiometric composition in a wide range. The heat treatments in the Cd vapor were carried out at T = 900°C at vapor pressures from 0.2 to 5.8 atm and in the Te vapor at T = 700°C. The duration of a heat treatment ranged from a few hours to 530 h.

These treatments were also applied to CdTe + Cd solutions at T = 800°C and to CdTe + Te solutions at T = 900°C. The duration of a heat treatment ranged from 24 to 120 h.

Moreover, we annealed CdTe samples in vacuum at 700°C. The volume of an evacuated ampoule in which the annealing was carried out exceeded that of the sample by a factor of 2–3.

Before the heat and annealing treatments, each sample (up to 1 mm thick) was etched in an HCl:HNO$_3$:H$_2$0 = 1:1:1 mixture and washed in deionized water. The heat treatments were carried out in quartz ampoules first heated in vacuum.

The photoluminescence spectra were investigated using 6 × 4 × 1 mm plates. After mechanical polishing, the damaged surface layer ($\sim$50–150 μ) was removed by etching in a polishing solution of the composition HNO$_3$:K$_2$Cr$_2$O$_7$:H$_2$O = 10 ml:4 g:20 ml. In spite of the identical etching conditions, the quality of the surfaces obtained in this way was not always constant and this altered the luminescence intensity. Therefore, control measurements were carried out on samples with freshly cleaved surfaces.

CHAPTER III

EXCITON LUMINESCENCE AND ABSORPTION IN CADMIUM TELLURIDE CRYSTALS

§ 1. General Description of the Photoluminescence Spectra of Cadmium Telluride

The present chapter is concerned with the luminescence spectra resulting from the annihilation of free and bound excitons. We shall also compare these spectra with the absorp-

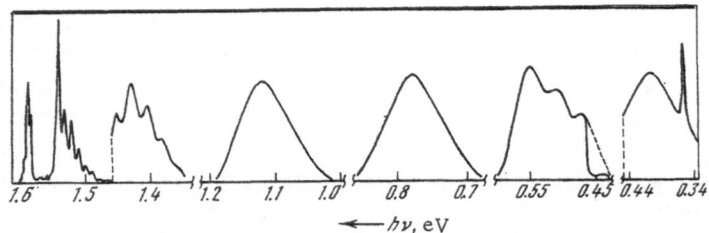

Fig. 5. Photoluminescence spectrum of CdTe crystals
at 20.4°K.

tion by free excitons. However, since the discussion of the results will touch upon other radiative recombination mechanisms and the corresponding luminescence bands, it is convenient
to begin this chapter with some general information on the luminescence spectra of the investigated CdTe crystals.

Figures 5 and 6 show the photoluminescence spectra of cadmium telluride crystals recorded at 20.4 and 77.4°K. The spectra were not obtained for a particular crystal but include
all the bands observed in n- and p-type CdTe crystals.

In all the figures which will be given later, the amplitudes of the curves obtained for different crystals and under different experimental conditions are made approximately the same
for convenience of comparison. The actual intensities may differ very considerably from one
figure to another.

The bands observed in the luminescence spectra can be divided arbitrarily into the
following groups, in accordance with their energy: a) exciton absorption; b) edge luminescence;
c) 1.43 eV; d) 1.1 eV; e) 0.78 eV; f) 0.5 eV; g) 0.41 eV.

We shall not consider here the nature of electron transitions and centers responsible for
the groups of lines (this will be done elsewhere), but we shall note the following characteristic
features of the observed spectra: 1) when the temperature is lowered, the luminescence band
corresponding to the exciton absorption region splits into narrow lines, whereas the edge luminescence and 1.43 eV band acquire an equidistant structure due to the interaction with LO
phonons (phonon replicas); 2) the relative intensities of the various groups of lines in the luminescence spectra depend on the excitation rate, temperature, and impurity composition of
a crystal; 3) although the division into groups, given above, corresponds to more or less different radiative recombination mechanisms, the phonon replicas belonging to different groups
frequently overlap. Therefore, to observe and analyze bands corresponding to a specific recombination mechanism, it is usually necessary to have a crystal with a specific impurity composition and it is necessary to carry out measurements at a particular temperature and excitation rate.

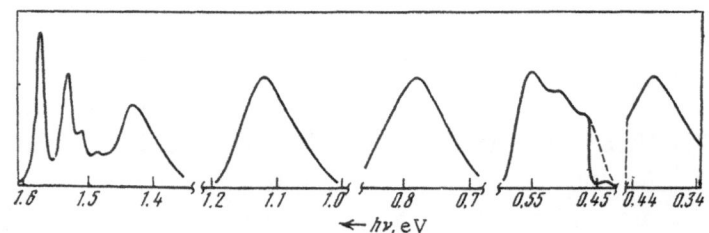

Fig. 6. Photoluminescence spectrum of CdTe crystals
at 77.3°K.

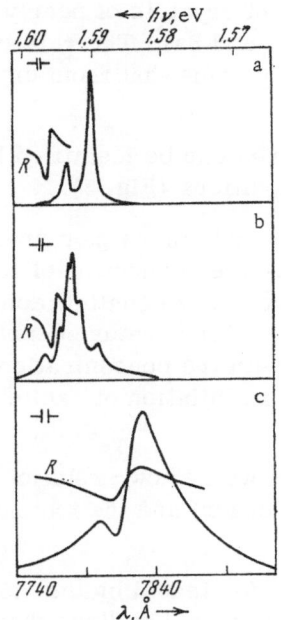

Fig. 7. Photoluminescence spectra of a CdTe crystal with a nearly stoichiometric composition (n-type, n = 1 × 10^14 cm^{-3} at 77.3°K) recorded at different temperatures (°K): a) 4.2; b) 20.4; c) 77.3.

§ 2. Photoluminescence of Cadmium Telluride Crystals in the Exciton Reflection Region

The group of the CdTe photoluminescence lines with the shortest wavelengths is located in the exciton reflection region at 1.602-1.585 eV (20.4°K). At low temperatures (4.2-20.4°K), up to ten fairly narrow (~1 meV) and only partly resolved lines can be observed for the crystals with fewest defects. A group of lines reported in [12] is similar to the blue edge luminescence of cadmium sulfide and is probably of the same origin, i.e., it is due to the radiative annihilation of free excitons and exciton–impurity complexes and to the radiative recombination involving impurities.

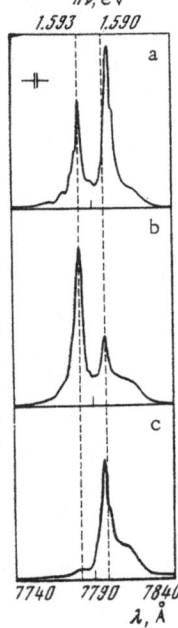

Fig. 8. Influence of the impurity – defect composition of a crystal on the photoluminescence spectrum of CdTe at 4.2°K: a) semi-insulating crystal with a nearly stoichiometric composition; b) n-type crystal prepared at a high Cd vapor pressure; c) p-type crystal prepared by annealing for 530 h in Te vapor at T = 750°C.

The greatest number of lines is observed in the spectra of crystals of nearly stoichiometric composition in which none of the defects predominates (Figs. 7 and 8a). The relative intensities and the energy positions of the various luminescence lines vary somewhat from crystal to crystal at a fixed temperature.

The photoluminescence lines recorded at low temperatures can be identified by comparing the results obtained for crystals with different impurity compositions (Fig. 8).

The photoluminescence spectra of crystals grown at high cadmium vapor pressures (Figs. 8b and 9) are dominated, at 4.2°K, by the 1.593 eV line. The n-type conduction of these crystals is due to the presence of interstitial cadmium atoms which give rise to shallow donor levels with ionization energies of 0.01-0.02 eV [13, 51, 52]. The strong temperature dependence of the intensity of this line and the increase in its relative intensity with the concentration of interstitial cadmium atoms allow us to attribute it to the radiative annihilation of excitons bound to neutral donors (interstitial cadmium atoms).

The ionization energy of these donors, calculated using a well-known relationship between the binding energy of a complex (an exciton bound to a neutral donor) and the ionization energy of a donor [53], is ~0.015 eV.

The spectra of the p-type crystals recorded at 4.2°K are dominated by the 1.590 eV line. The relative intensity of the 1.593 eV line at this temperature is very low (less than in crystals with nearly stoichiometric composition, as shown in Fig. 8c).

These observations and the characteristic width of the line as well as the very strong temperature dependence of its intensity suggest that the 1.590 eV line is due to the radiative annihilation of an exciton bound to an acceptor. Electrical measurements carried out on the investigated crystals reveal the presence of an acceptor level with an ionization energy of 0.05 eV [50, 54]. This energy is in agreement with the value calculated on the basis of the results given in [53] assuming that the 1.590 eV line represents excitons bound to acceptors.

The attribution of the 1.590 and 1.593 eV lines to the annihilation of bound exciton rather than to the radiative annihilation involving band—impurity transitions is based on an examina-

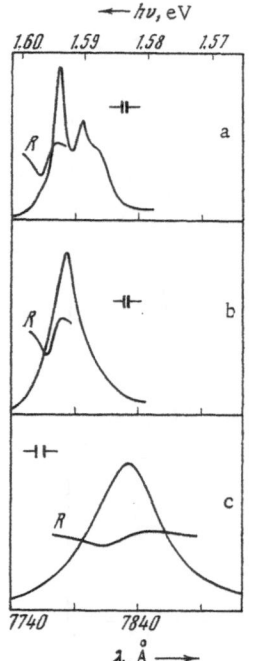

Fig. 9. Photoluminescence spectra of a CdTe crystal with a considerable deviation from stoichiometry (n-type, $n = 7.8 \times 10^{16}$ cm^{-3} at 77°K) recorded at different temperatures (°K): a) 4.2; b) 20.4; c) 77.3.

tion of the line width and temperature dependence of the intensity. The nature of these lines can be finally determined on the basis of measurements in magnetic fields, in the same way as has been done in the case of cadmium sulfide [10].

Crystals with moderate deviations from stoichiometry (up to $n \sim 10^{16}$ cm^{-3}) exhibit weak luminescence lines in the 1.577-1.568 eV range at 4.2-20.4°K. The published information on the energy of longitudinal optical phonons in CdTe (see, for example, [12]) allow us to regard these lines as the phonon replicas of the 1.593 and 1.590 eV lines. The phonon replica of the 1.590 eV line at T = 4.2°K is strong even in the spectra of those n-type crystals ($n = 2 \times 10^{16}$ cm^{-3} at 77.3°K) which are dominated by the 1.593 eV line. This shows that the interaction between phonons and acceptors is stronger than between phonons and donors (see §3 in Chap. IV). In those cases when the width of the phonon replicas can be determined reliably, its value is found to be of the same order as that of the main (zero-phonon) lines.

It is interesting to note that the phonon replicas (1.577-1.568 eV) include a line which can be regarded as a LO phonon replica of the free-exciton line. We must bear in mind that the profile of the main exciton luminescence line and the profile of its phonon replica should be, in general, different. It is shown in [55] that, in the case of direct excitons, the profile of a phonon replica line should reproduce approximately the energy distribution of excitons in the same way as in the case of indirect excitons [56], whereas the profile of the main line exhibits its resonance nature more strongly. Although, in our case, the exact profile of the phonon replica line of the exciton luminescence line is difficult to determine because of its overlap by other lines, nevertheless, we can say that at 20.4°K it has an asymmetric profile which corresponds approximately to the Boltzmann distribution. This shows that excitons are in thermal equilibrium with the lattice vibrations.

One of the main tasks of the present investigation is to identify the line which results from the annihilation of free excitons and to determine the relationship between the luminescence and absorption spectra. This is tackled, first, by an analysis of the influence of temperature on the photoluminescence spectra. The intensities of the lines corresponding to the annihilation of bound excitons should decrease more rapidly with rising temperature than the intensity of the free-exciton line.

In the case of crystals with nearly stoichiometric composition, we can follow the characteristic temperature dependences of the relative intensities of these lines (Fig. 7). At 4.2°K, the line at 1.590 eV predominates. At higher temperatures, its relative intensity decreases rapidly so that at ~ 32°K the line is no longer observed. The relative intensity of the line with the shortest wavelength rises with temperature up to 40°K. This provides some grounds for attributing this line to the annihilation of free excitons. However, when the temperature is raised, the intensities of the other lines grouped around the 1.593 eV line also change (20.4°K). The individual temperature dependences of the intensities of these lines cannot be determined because of the strong overlap.

Moreover, the lines in this group become broader with rising temperature and, beginning from about 40°K, they merge into a single fairly wide band and it is not possible to determine the role of each of the lines distinguishable at lower temperatures. Consequently, at 77.3°K we obtain a spectrum shown in Fig. 7c. The profile of the luminescence spectrum also recorded at lower temperatures (from about 40°K) is similar. Only a broadening of both lines is observed in the 40-77°K range and a detailed analysis of the profile shows that the outer wings are broadened preferentially, whereas the central part changes less rapidly with rising temperature. This suggests self-absorption, whose specific features will be discussed later. Changes in the relative intensities and energy separations between the lines are not very great in this range of temperatures.

In the case of crystals of compositions quite far from stoichiometry (n-type crystals), the recombination radiation spectra differ only slightly from the spectra of crystals of nearly stoichiometric composition. The concentration quenching prevents the observation of many of the lines that can be resolved in the case of crystals with nearly stoichiometric composition (Fig. 9). In particular, at 77.3°K it is not clear whether we are dealing with a single broad line or whether the resultant luminescence is due to the broadening of the 1.589 and 1.583 eV lines (one or both), observed at this temperature in the spectra of nearly stoichiometric samples (Fig. 7c).

In the case of p-type crystals, the spectra change with temperature in approximately the same way as those of n-type crystals. It is only worth pointing out that at 77.3°K the luminescence maximum is shifted somewhat (by 1.5-2.0 meV) toward shorter wavelengths. Thus, an analysis of the temperature dependences of the intensities of the photoluminescence lines of CdTe single crystals fails to establish reliably which of the observed lines is due to the annihilation of free excitons.

A study of the reflection spectra in the exciton part of the spectrum can provide useful information on the free-exciton luminescence line of the investigated crystals. With this point in mind, we carried out a detailed study of the reflection spectra of cadmium telluride crystals with different deviations of stoichiometry. This study was carried out in the 4.2-77.3°K range under normal incidence conditions. Some of the results are given in Figs. 7, 9, and 10. The characteristic profile, changes in the absolute value of the reflection coefficient, energy position of the recorded reflection line, as well as comparison of the results obtained with those reported by other workers for cadmium telluride and other thoroughly investigated materials [15] leave no doubt that the $n = 1$ exciton reflection was recorded.

A comparison of the reflection and luminescence spectra (Fig. 7) shows that the exciton reflection region coincides with the range of energies where there are two luminescence lines with the shortest wavelengths. This suggests that one or both of these lines are due to the

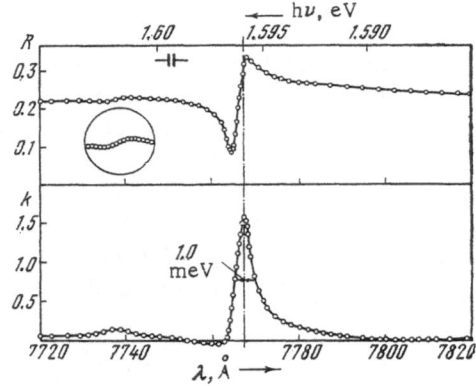

Fig. 10. Experimentally obtained spectrum of the reflection coefficient R in the exciton region of a crystal with a nearly stoichiometric composition ($n = 4 \times 10^{15}$ cm^{-3} at 77°K) and the calculated, on the basis of the reflection coefficient, spectral dependence of the extinction coefficient k at 4.2°K (the inset shows the reflection spectrum with a feature corresponding to the absorption associated with the formation of the excited $n = 2$ state of free excitons).

radiative recombination of free excitons with n = 1. It is not possible to draw more definite conclusions from the luminescence and reflection data, especially as the recorded luminescence spectra are distorted by self-absorption.

Additional information on the position of the free-exciton resonance line can be obtained by considering the position of its phonon replica. However, the fairly strong overlap of the phonon replicas with other short-wavelength lines makes it impossible to determine the position of the main line with the necessary degree of precision. Calculations of this kind simply confirm that the luminescence lines with the shortest wavelengths are associated with the recombination of free excitons.

The final identification can be made by comparing the observed recombination radiation with the data on the absorption of the same crystals in the investigated part of the spectrum. This was the next stage of our investigation.

§3. Free-Exciton Absorption in Cadmium Telluride Crystals

In studies of the optical properties of solids, it is desirable to carry out comprehensive studies of the radiative recombination and absorption or transmission of crystals. However, it is difficult to determine the absorption spectra directly because even in the case of fairly low absorption coefficients ($\alpha \sim 10^3$ cm^{-1}) it is necessary to prepare samples whose thickness does not exceed 10μ. It is quite difficult to prepare such samples without disturbing the surface too seriously. Obviously, in the case of even higher absorption coefficients ($\alpha \sim 10^5$ cm^{-1}), typical — in particular — of the exciton absorption region of interest to us, this approach to the study of the absorption spectra is entirely inappropriate. The only solution is an indirect approach using the Kramers–Kronig relationships which allow us to obtain information on the absorption from the experimental results on reflection. The basic way of using the Kramers–Kronig relationships and the final mathematical formulation of the problem are given in §1, Chap. I.

We calculated the absorption spectra in the following way. The experimentally determined reflection spectra $R(\lambda)$ were determined in the wavelength range 7700-7900 Å (see §2 in Chap. II). It is clear from the results in Fig. 10 that at the ends of this spectral interval the changes in the reflection coefficient are very small compared with the exciton reflection region. Moreover, it follows from investigations reported by several workers [15, 57-59] that this variation is also small in a wider spectral range extending in both directions away from the exciton reflection region. Hence, the range of integration in the calculation of the phase angle [Eq. (I.3)] was extended so that it was initially about 50-60 times as wide as the range in which the experimental observations were made. However, it was found later that the quality of the results obtained was not greatly affected when this range was shortened by a factor of 6 and thus exceeded the range of direct experimental measurements just by a factor of 10. It was assumed that the reflection coefficient was constant throughout this extended region and its value was taken to be equal to the value at the corresponding edges of the spectral interval where the observations were made.

This resulted in some error in the final results. However, it was clear that the error was small because, in accordance with Eq. (I.3), the structure of the absorption spectrum, i.e., of the function $\varphi(E)$, in the spectral range of interest to us was governed primarily by the structure of the function $R(E)$ in the same part of the spectrum. The regions included in the integration and located far from the spectral interval of interest to us made a contribution to $\varphi(E)$ which varied weakly within this interval. The contributions to $\varphi(E)$ of the regions located symmetrically with respect to the interval under investigation had different signs and the absolute values of these contributions were comparable. Therefore, an extension of the limits of integration to a wider spectral region should shift the calculated absorption spectrum along the ordinate without altering its profile significantly.

On this basis, the final limits of integration were selected so that the calculated value of the extinction coefficient in the forbidden band region on the low-wavelength side of the exciton absorption edge was zero.

The principle of this calculation method did not differ from the principles of the methods described in the literature. A comparison with [4, 5] indicated that, in our case, the results of observations were extrapolated subject to the necessary conditions (zero absorption in the forbidden band).

The phase angle $\varphi(E)$ and next the dependences $k(E)$ and $\alpha(E)$ were calculated using a BÉSM-4 computer at the Lebedev Physics Institute. We used a program with a selection of steps and with a linear interpolation of $R(E)$. A singularity of the integral (I.3) was bypassed in the usual way: the integral was calculated in two steps, first from the lower limit to a point $E-\varepsilon$ and then from $E + \varepsilon$ to the upper limit. In our calculations, we used a value of ε equal to 10^{-6} eV. This gave the necessary precision, as demonstrated by a direct calculation of the integrand at the point $E \pm \varepsilon$ and a comparison of the results of calculations obtained for different values of ε. It was discovered, as reported in the literature (see, for example, [1, 3]), that the greatest error in the final results was due to the inaccuracy of the experimentally determined reflection coefficient. As demonstrated clearly by the expression for the phase angle $\varphi(E)$, this inaccuracy contributed an error to the absolute value of the absorption coefficient and distorted somewhat the calculated absorption spectrum. However, the distortion was, in practice, very small because a change in the absolute value of the absorption coefficient simply resulted in the multiplication of the absorption curve by a constant. Thus, we estimated the computation error as 15-20% and found that the error was mainly in the absolute intensity and was less in the profile of the calculated absorption (Fig. 11).

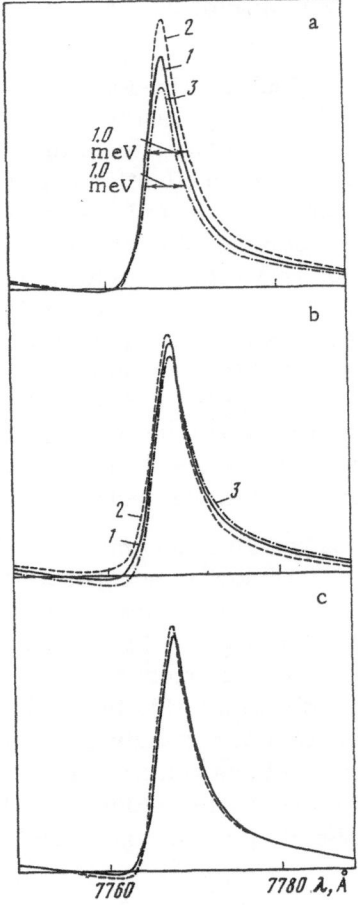

Fig. 11. Absorption spectra calculated: a) for different values of the reflection coefficient on the long-wavelength side of the exciton peak (curves 1-3 correspond to R = 0.238, 0.265, and 0.215 in this region); b) by varying the lower limit of the integral (I.3) in steps of 0.25 eV (curves 1-3); c) by varying ε ($\varepsilon = 10^{-4}$ eV for the continuous curve and $\varepsilon = 10^{-6}$ eV for the dashed curve).

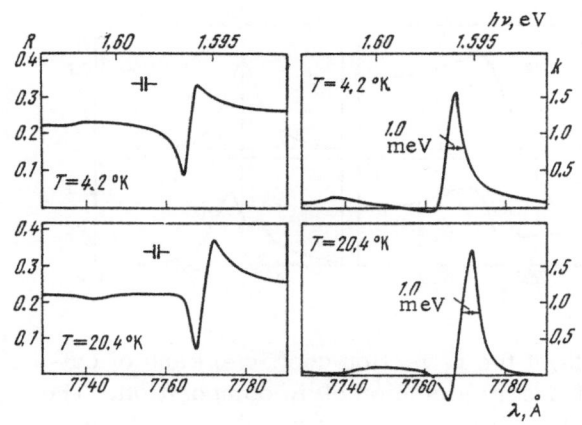

Fig. 12. Reflection and absorption (extinction) spectra of a crystal with a nearly stoichiometric composition ($n = 4 \times 10^{15}$ cm^{-3} at 77.3°K) at 4.2 and 20.4°K.

The results of calculations of the reflection and absorption spectra for one of the CdTe crystals are plotted in Figs. 10 and 12 for 4.2 and 20.4°K. The absorption lines obtained at both temperatures have approximately the same width (1.0 meV) and the maximum value of the extinction coefficient ($k = 1.5$, $\alpha = 2.5 \times 10^5$ cm^{-1}). The absorption maximum at 4.2°K is located at 1.5958 eV and the maximum at 20.4°K is located at 1.5951 eV.

The reflection spectra of some of the crystals with the fewest defects have a weak maximum at 1.6011 eV in the spectra recorded at 4.2°K (Figs. 10 and 12). Calculations of the absorption spectra reveal a weak maximum at 1.6020 eV. The absence of any other features in the reflection spectrum suggests that the maximum corresponds to the first excited ($n = 2$) state of excitons. In this case, the energy separation between the excited and ground states of excitons is 6.2 ± 0.3 meV, the binding energy of excitons is $G = 4/3[E(n = 2) - E(n = 1)] = 8.3 \pm 0.4$ meV, and the forbidden band width is 1.6041 ± 0.0006 eV at 4.2°K. The effective Bohr radius of the ground states of excitons is ~ 60 Å ($m_e = 0.1$ m, $m_h = 0.4$ m, $\varepsilon_0 = 10$).

It should be pointed out that the ionization energies of excitons differ slightly from the values reported by other workers. An estimate of this ionization energy for cadmium telluride crystals was first obtained by the extrapolation of similar results for the II-VI compounds [15]. The value quoted in [15] was 11 meV. Calculations of the absorption spectra from the experimental data on the reflection based on the Kramers–Kronig relationships [60] yielded the 2°K absorption maxima of the ground and first excited states of excitons located at 1.5960 and 1.6030 eV. The binding energy of excitons given in that paper[*] was 10 meV. The difference $E(n = 2) - E(n = 1)$ obtained in [61] was 9 meV.

It was reported in [60] that the first excited state of excitons could be observed in the reflection spectra at temperatures up to 22°K. Our study of these spectra at 20.4°K failed to reveal any structure which could be identified with $n = 2$ exciton states.

It is interesting to note that when an excited exciton state is observed in the absorption (reflection) spectrum, the luminescence spectrum of the same crystal has a very weak line at an energy corresponding approximately to the exciton absorption maximum.

The results of our study of the reflection spectra at 77.3°K and of the corresponding calculated absorption spectra are plotted in Figs. 13a and 13b. Since the first problem is to identify the nature of the luminescence lines, we shall discuss some of the reflection results and particularly the details in Figs. 12 and 13 in the next section.

---

[*]It follows from the values of the energy $E(n = 2) - E(n = 1) = 7$ meV given in [60] that the binding energy is 9.3 and not 10 meV.

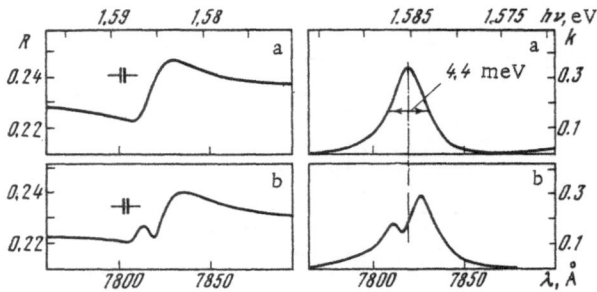

Fig. 13. Experimental reflection spectra (R) and calculated absorption spectra (k) of CdTe crystals with nearly stoichiometric compositions (T = 77°K).

Figure 13a shows the results of measurements of the reflection coefficient and of calculations of the extinction coefficient of a crystal of nearly stoichiometric composition. The electrical properties and the observed recombination radiation of this crystal were close to the properties of a crystal whose low-temperature (4.2-20.4°K) properties were described earlier. It is clear from Fig. 13 that the absorption maximum was located at 1.5853 eV and the widths of the absorption line was 4.4 eV (0.7 kT). The maximum value of the absorption coefficient was 5×10^4 cm^{-1}. The shift of the absorption line due to the increase in temperature from 4.2 to 77.3°K was approximately 11 meV, in agreement with the results of other workers [62, 63].

The obtained information on the temperature dependences of the width of the exciton absorption line is in general agreement with the conclusions reached in [64], where this dependence is derived theoretically allowing for the interaction of excitons with longitudinal optical phonons. However, it should be pointed out that the investigated temperature range (4.2-20.4-77.3°K) is not wide enough to test the dependence obtained in [64].

§ 4. Comparison of the Free-Exciton Absorption and Luminescence Spectra

We calculated the luminescence lines to obtain further information on the energy position and profile of the free-exciton line. We used considerations based on the principle of detailed equilibrium. Similar calculations were carried out for germanium [65, 66] (see also review [67]).

According to the formulation of the principle of detailed equilibrium for the interband radiative recombination [65], under thermal equilibrium conditions the rates of generation and recombination of equilibrium carriers balance out at each frequency, i.e., the balance is detailed. This makes it possible to express the radiative recombination rate in terms of quantities which describe the absorption in the system:

$$R(\nu)\,d\nu = \frac{8\pi}{c^2}\,n^2(\nu)\,\frac{\alpha(\nu)}{e^{\frac{h\nu}{kT}}-1}\,\nu^2 d\nu \approx \frac{8\pi}{c^2}\,n^2(\nu)\,\alpha(\nu)\,e^{-\frac{h\nu}{kT}}\,\nu^2 d\nu \quad \text{for } h\nu \gg kT. \qquad \text{(III.1)}$$

Under steady-state conditions, we obtain

$$R_{\text{st}} = \frac{np}{n_i^2}\,R_{\text{equil.}} \qquad \text{(III.2)}$$

where n and p are the electron and hole densities, respectively; n_i is the density of electrons in an intrinsic semiconductor. It should be pointed out that the use of the above relationship implies that the energy distributions of nonequilibrium electrons and holes are the same as those of equilibrium carriers. It is usual to assume that the carrier distribution remains of the quasi-Boltzmann type, i.e., it has the form of an equilibrium (Boltzmann) distribution with the equilibrium value of the Fermi level replaced by the quasi-Fermi level representing the deviation from equilibrium. This assumption is based on a comparison of the lifetime of non-equilibrium carriers and of the characteristic time of their interaction with thermal lattice vibrations (see, for example, [68]).

A reasoning similar to that given above for the interband recombination case applies naturally to an exciton system [66] provided that thermodynamic equilibrium between excitons and phonons is established during the exciton lifetime. According to theoretical estimates, the exciton lifetime in crystals relatively free of defects is $\sim 10^{-9}$ sec and the time between the collisions of an exciton with phonons is $\sim 10^{-11}$ sec [69]).

An estimate based on the excitation levels used in our experiments shows (§ 1, Chap. II) that, at the nonequilibrium carrier densities established in our samples, the carrier gas is not yet degenerate and this makes it possible to calculate the luminescence line profile from

$$I(\nu) \propto \nu^2 n(\nu) \alpha(\nu) e^{-\frac{h\nu}{kT}} \tag{III.3}$$

Our calculations were made allowing for the dispersion of the refractive index $n(\nu)$, calculated in the same way as the extinction coefficient from the reflection data using the Kramers–Kronig relationships. However, control calculations carried out without allowance for this dependence gave results not too different from the first set of results.

Figure 14 compares the exciton absorption (k) and luminescence (I_{calc}) spectra calculated using the principle of detailed equilibrium with the experimentally observed luminescence spectrum (I_{exp}). We can see that the absorption and the calculated luminescence maxima are located at 1.5852 and 1.5846 eV, respectively. However, the two maxima of the experimental luminescence spectrum are located at 1.5894 and 1.5829 eV, i.e., the calculated maxima are located between the experimental maxima. This result suggests that the free-exciton luminescence line is partly self-absorbed at 77.3°K. However, the self-absorption not only reduces the intensity of the luminescence line. We can see that there are additional special features which have not yet been taken into account. Since the exciton absorption and luminescence line profiles are almost identical, the exciton luminescence lines exhibit a self-reversal at high optical densities of the excited layer; this is analogous to the well-known self-reversal of the spectral lines observed in atomic spectroscopy. This self-reversal produces a dip at the center of the atomic emission line, which is observed for sources of finite optical density in the presence of a gradient

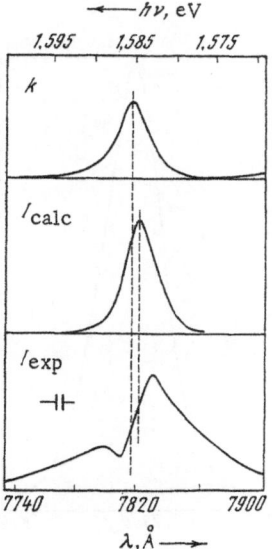

Fig. 14. Comparison of the calculated absorption spectra (k) with the luminescence spectra calculated using the principle of detailed equilibrium (I_{calc}) and with the experimental luminescence spectrum (I_{exp}) of a crystal of a nearly stoichiometric composition (T = 77.3°K).

of the concentration of the emitting atoms directed into the source [70–71]. The presence of a similar exciton concentration gradient may result from the high velocity of the nonradiative recombination on the surface. Moreover, the shift of the absorption line relative to the luminescence maximum results in the location of the dip outside the center of the luminescence line and this produces two asymmetric maxima.

A comparison of the calculated and experimental profiles can be made by selecting the same relative intensity scale. We can do this if we know the attenuation of the recombination radiation as a result of self-absorption. This can be estimated, only very approximately, by postulating a simple spatial distribution of the luminescence-emitting excitons. If we assume that

$$g(x)\,dx = \begin{cases} ax\,dx & \text{for } x \leqslant L, \\ Ae^{-\alpha_1 x}\,dx & \text{for } x \geqslant L, \end{cases} \tag{III.4}$$

where $g(x)dx$ is the number of emitting excitons in an interval dx at a distance x from the surface; α_1 is the absorption coefficient at the luminescence wavelength; L is the value of x at which the exciton concentration reaches its maximum; a and A are constants, and if we select reasonable values of L, we can easily show that the attenuation of the luminescence at the absorption maximum is by a factor of 6–8. A comparison of the calculated and experimental luminescence line profiles is made in Fig. 15 bearing this point in mind. It is clear from this figure that the attenuation of the luminescence line corresponding to the ground exciton state is very considerable.

The temperature dependences of the relative intensities of the individual luminescence lines exhibited by crystals relatively free of defects (n-type crystals with $n \leq 10^{15}$ cm^{-3} and high-resistivity crystals with $\rho \sim 10^6$ $\Omega \cdot$cm nearly stoichiometric compositions) suggest that the long-wavelength edge of the recorded luminescence in these crystals includes a contribution of some of the lines of the 1.593 eV group (impurity radiative recombination). However, if we consider all the temperature dependences of the photoluminescence spectra and the results of our comparison of the luminescence and absorption spectra, we may conclude that this contribution is slight and that the luminescence is basically of exciton nature.

The self-reversal of the exciton luminescence lines, observed at T = 77°K (Fig. 14), may also occur at lower temperatures but then a detailed comparison of the spectra is difficult, partly because the intensities of the lines corresponding to the annihilation of bound excitons rise

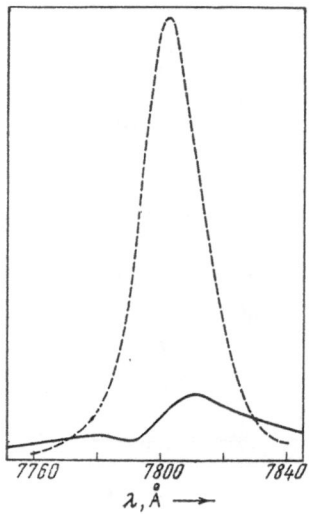

Fig. 15. Reconstruction of the true luminescence spectrum (dashed curve) from the experimentally observed spectrum (continuous curve) of a sample with nearly stoichiometric composition (T = 77.3°K).

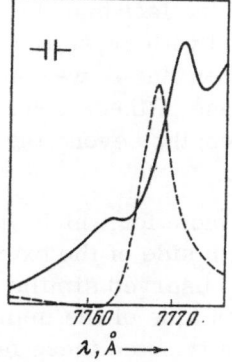

Fig. 16. Comparison of the recorded photoluminescence spectrum (continuous curve) with the calculated absorption spectrum (dashed curve) of a crystal with a nearly stoichiometric composition (T = 4.2°K).

strongly so that the luminescence which may be attributed to free excitons is found in the short-wavelength wing of a group of bound-exciton lines. A comparison of the exciton (n = 1) absorption line with the luminescence spectrum at 4.2°K shows that the absorption peak is located between the two maxima of the luminescence lines with the shortest wavelengths (Fig. 16). This indicates that both these lines represent the free-exciton luminescence. The minimum at the center is due to self-reversal. The strong absorption ($\alpha \approx 2 \times 10^5$ cm^{-1}) is responsible for the fact that only a small part of the line appears in the luminescence, whereas the major part is absorbed.

The low-temperature reflection spectra of crystals with considerable deviations from stoichiometry (n ~ 10^{17} cm^{-3}) also have a weak structure in the absorption spectrum due to the formation of free excitons (Fig. 17a). Calculations based on the Kramers–Kronig relationships indicate that the absorption coefficient at 4.2°K reaches ~6 × 10^4 cm^{-1} (k = 0.35) and the half-width of the absorption line is about 3 meV (Fig. 17b). The changes in the absorption spectrum between 4.2°K and 20.4°K are slight. At 77.4°K the exciton region exhibits a weak absorption with a maximum value of the absorption coefficient α ~ 2 × 10^4 cm^{-1}. At all three temperatures, the maximum of the absorption (extinction) coefficient is located at energies which agree approximately with the absorption maxima of crystals with nearly stoichiometric compositions.

The luminescence spectra of crystals with considerable deviations from stoichiometry exhibit, beginning from 20°K, a single wide luminescence band (Fig. 9). We mentioned earlier the difficulties encountered in the determination of the nature of the recombination radiation emitted by these crystals in the exciton part of the spectrum [72]. It follows from the reflection and absorption data that this luminescence band includes a contribution of the free-exciton recombination. Unfortunately, this contribution cannot be estimated quantitatively. An analysis

Fig. 17. Experimental reflection (a) and luminescence (c) spectra and calculated absorption (extinction) spectrum (b) of a crystal with a considerable deviation from stoichiometry (n-type, n = 7.8 × 10^{16} cm^{-3} at 77.3°K), recorded at 4.2°K.

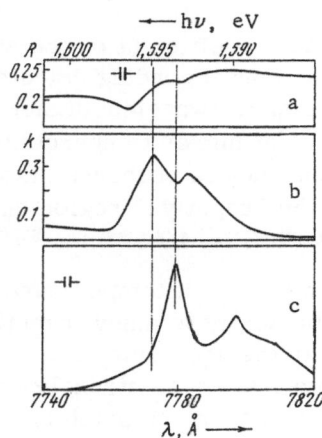

of the profile of the experimentally observed band must allow for the fact that, because of the reduction in the absorption coefficient by a factor of 3-4 compared with crystals with nearly stoichiometric compositions, the self-absorption of the free-exciton line is weaker (the dip at the center of the line disappears). We may assume that, in the case of these nonstoichiometric crystals, a considerable proportion of the luminescence at wavelengths beyond the exciton absorption maximum is due to the impurity radiative recombination.

It should be pointed out that the reflection spectrum of a sample for which the results are plotted in Fig. 17a has an additional feature on the long-wavelength side of the exciton peak. The origin of this feature is not clear. Thomas and Hopfield [1] observed similar features in the reflection spectrum of CdS (weak peaks on the long-wavelength side of the main reflection maximum associated with the formation of free excitons). They attributed these features to some spectral lines excited by the light of the source used in recording the reflection spectra. This was deduced from the observation that the reflected light had a pronounced angular directionality, whereas the spontaneous luminescence did not. Control experiments, in which a crystal was illuminated in such a way that the light reflected from its surface did not reach the spectrograph, made it possible to distinguish clearly these luminescence lines.

When a similar procedure was applied in our experiments, no luminescence lines were observed. Moreover, a comparison of the luminescence and reflection spectra indicated that the strongest luminescence line at 1.593 eV did not correspond to an additional long-wavelength reflection maximum (Fig. 17). This additional maximum could not be attributed either to the strong luminescence line at 1.590 eV superimposed on the reflection spectrum, because samples with nearly stoichiometric composition (whose spectra were dominated by this line at 4.2°K) did not have an additional absorption maximum.

The calculated absorption spectra have an additional maximum on the long-wavelength side of the exciton absorption peak. This additional maximum disappears at high temperature.

Thus, our comparison of the luminescence and absorption spectra and an analysis of the influence of temperature and impurity composition on these spectra has enabled us: a) to identify the lines corresponding to the free-exciton annihilation and to determine the relative contribution of the exciton recombination to the luminescence spectra of crystals with different amounts of impurity; b) to confirm the existence of the self-reversal of the exciton luminescence lines and to explain it by the special nature of the exciton luminescence spectra.

However, some of the details of the absorption (reflection) and luminescence spectra observed for the first time cannot yet be explained. They include, apart from the features in the reflection spectra of heavily doped crystals (Fig. 17), some features in the reflection spectra of nearly stoichiometric crystals (Figs. 12 and 13). Comparing the reflection spectra of one of the crystals obtained at 4.2 and 20.4°K (Fig. 12), we can see that, in the latter case, the rise of the reflection coefficient on the short-wavelength side of the minimum is much faster than at 4.2°K. The corresponding calculated absorption spectrum has a characteristic "negative" region on the short-wavelength side of the maximum (Fig. 12). An additional reflection maximum appears at higher temperatures on the short-wavelength side of the main reflection peak. The same tendency is observed also in the calculated absorption spectra. At the same time, the form of the "negative" region changes. The reflection and absorption spectra of this crystal are shown in Fig. 13b.

This influence of temperature on the reflection and absorption spectra differs in detail but applies basically to many investigated crystals. The differences is manifested mainly by the changes in the spectrum between 20.4 and 77.3°K: the additional maximum in the reflection spectrum rises less strongly with increasing temperature and at 77.3°K it is partly suppressed or disappears completely (Fig. 13a).

This new fine structure in the reflection spectra is observed for cleaved and etched surfaces and there is no correlation between the type of surface and recorded reflection spectra.

Going back to the luminescence spectra of the investigated crystals, we have to make two main points. Firstly, in all the cases when the reflection spectrum has the features just reported, the maximum of the calculated absorption does not coincide exactly with the luminescence minimum located between the two luminescence maxima with the shortest wavelengths. Secondly, the luminescence line with the shortest wavelength is stronger at 4.2 and 20.4°K in the spectra of those crystals which exhibit the aforementioned features most clearly.

The information available is insufficient to draw final conclusions about these additional features. Splitting of the exciton absorption lines has been mentioned in the literature. This splitting has been explained as follows.

1. The splitting of the exciton peak in the reflection spectrum of CdTe crystals under pressure is reported in [15] and it is found to lift partly the valence-band degeneracy.

2. The reflection spectra of CdS and ZnTe crystals recorded at 1.6-4.2°K [73] have a fine structure. This structure is explained in a well-reasoned manner by allowance for the spatial dispersion in the absorbing medium. Very special effects, which arise near crystal surfaces [73], disappear at higher temperatures.

3. Fine effects of the spin—orbit interaction modify the energy expansion W(k) at the energy band extrema by the appearance of terms linear in k. When this is taken into account together with the spatial dispersion of the absorbing medium [74], the foregoing considerations indicate that a fine structure should appear in the reflection spectra of crystals in the exciton absorption region and this structure is due to the lifting of the degeneracy at the band extrema. A structure which can be interpreted in this way has indeed been observed for CdS [75] and CdTe [60].

It is not clear to what extent the mechanisms just discussed can explain the features observed in our spectra. A preliminary analysis shows that, in order to explain our results on the basis of the model mentioned in point 3 above, we have to assume that the band splitting is approximately an order of magnitude greater than that found in [60]. A possible role of random stresses in the investigated crystals (point 1) is in conflict with the relative constancy of the splitting in different crystals and with the specific temperature dependence of the effect. This dependence (in general, fairly weak) also militates against the full acceptance of the explanation put forward in [73]. There is no doubt, however, that, in some measure, the observed features (particularly the "negative absorption region" in Fig. 12) are due to the fact that the Kramers—Kronig relationships employed by us ignore the spatial dispersion of the optical constants and the special nature of the boundary conditions on the surfaces of various crystals.

We may thus conclude that we are dealing here with some new effects whose nature has not yet been determined.

CHAPTER IV

EDGE LUMINESCENCE AND PHONON EFFECTS IN THE PHOTOLUMINESCENCE OF CADMIUM TELLURIDE

§1. Dependence of the Edge Luminescence Spectrum on the Impurity Composition of Crystals

To determine the nature of the edge luminescence of cadmium telluride, we paid special attention to the correlation between the form of the photoluminescence spectrum and the im-

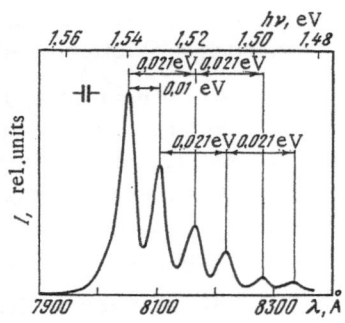

Fig. 18. Typical edge luminescence spectrum of a high-resistivity ($\rho \sim 10^7$ $\Omega \cdot$ cm) nearly stoichiometric sample of CdTe. T = 20.4°K.

purity—defect composition of a crystal in the temperature range 4.2-150°K. With this in mind, we carried out a systematic study of the luminescence spectra of samples cut from approximately 100 ingots differing in respect of the growth method and degree of doping with various impurities. Some of these samples were subjected to a heat treatment (annealing) in vacuum or under controlled Cd and Te vapor pressures.

We shall first consider the results obtained for high-resistivity self-compensated undoped crystals. Most of such samples exhibit a luminescence spectrum of the kind shown in Fig. 18. The six lines present in the spectrum can be divided into two series, each of three lines; we shall call them the long-wavelength (I) and the short-wavelength series (II). The two series appear as a result of simultaneous emission of a photon and n longitudinal optical (LO) phonons (in the present case, n = 0, 1, 2). The energy of an LO phonon in CdTe is 0.021 eV. The energies of the zero-phonon lines in the series I and II are 1.528 and 1.539 eV, respectively. As in the case of the green edge luminescence of CdS (§ 2, Chap. I), the temperature dependences of the intensities in these two series are quite different: at 4.2°K the series I predominates, but at higher temperatures its relative intensity falls so that at T > 25°K the spectrum is dominated by the series II. The same temperature dependence is reported by Halsted and Segall [24] and it is explained on the basis of a model mentioned earlier: according to this model, the series I is due to transitions of bound holes and the series II to transitions of free holes to the same acceptor. The energy separation between the lines in the series corresponding to the same value of n is equal to the binding energy of a hole to a doubly charged acceptor.

Our results raised doubts about this model. It is found that the relative intensities of the series I and II and the separation between the lines corresponding to the same value of n vary from sample to sample. This is observed when the excitation rate is kept strictly constant and samples are immersed in liquid nitrogen (Fig. 19). This is in conflict with the Halsted—Segall model and is evidently due to the fact that the impurity—defect compositions of high-resistivity self-compensated samples are not constant (it is not possible to control this composition accurately even when special measures are taken to stabilize the growth conditions). This conclusion is supported by the observation that some of these crystals exhibit not only the series I and II, but also an additional series III with the zero-phonon line located at 1.548 eV (curve 2 in Fig. 19).

Fig. 19. Parts of the edge luminescence spectra obtained for two different high-resistivity undoped CdTe samples at 20.4°K. The excitation intensity was the same for both samples but their resistivities were different: 1) $\rho \sim 10^8$ $\Omega \cdot$cm; 2) $\rho \sim 10^6$ $\Omega \cdot$cm.

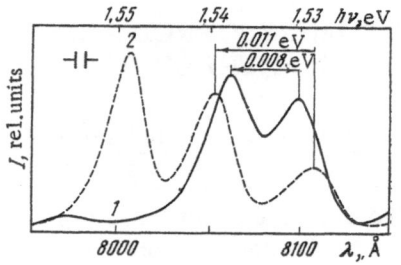

Thus, investigations of the photoluminescence of high-resistivity self-compensated samples of CdTe have shown that the two series of the edge luminescence lines reported in [12, 24] and found in the present study cannot be explained by the presence of centers of just one type in CdTe crystals.

Our investigation of the photoluminescence of samples with different impurity compositions has yielded information on the changes in the spectra with the type and concentration of impurities or defects. Figure 20 shows spectra of three types which correspond to three different ratios of the donor and acceptor concentrations. Figure 20a gives the photoluminescence spectrum of high-resistivity self-compensated samples ($\rho \sim 10^6$-10^8 $\Omega \cdot$cm), in which the donor and acceptor concentrations are approximately equal. The spectrum shown in Fig. 2b is typical of low-resistivity n-type samples ($\rho \sim 0.1$ $\Omega \cdot$cm), prepared either by increasing the deviation from stoichiometry (by raising the Cd vapor pressure during growth) or by doping with a donor impurity (for example, In). We can see that, in this case, the intensity of the series I is considerably higher than the intensity of the series II, so that the latter is practically invisible at low temperatures and can be seen only at T > 40°K.

The spectrum shown in Fig. 2c is typical of p-type samples ($\rho \sim 10$ $\Omega \cdot$cm) prepared at low Cd vapor pressure during growth [76]. A similar spectrum is also exhibited by samples

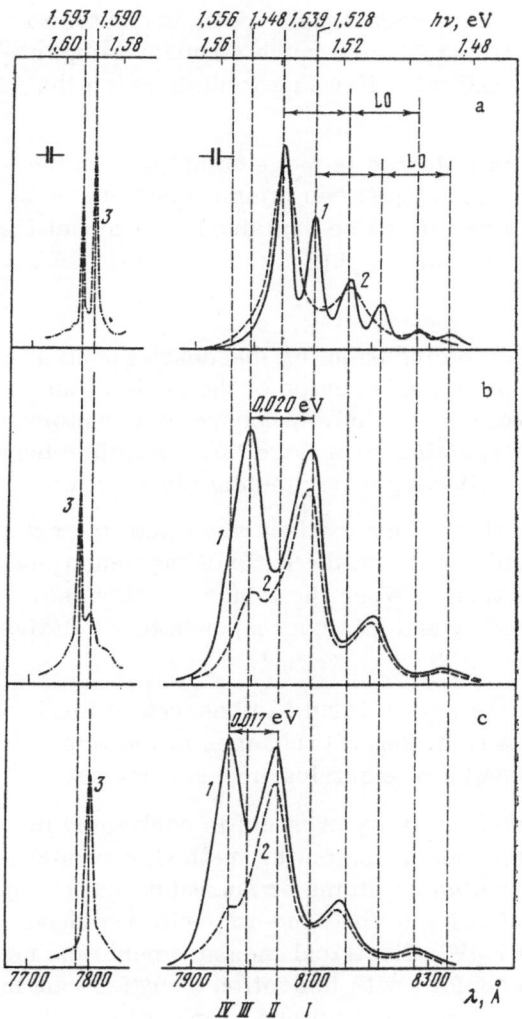

Fig. 20. Influence of the impurity composition of a sample on the photoluminescence spectra of CdTe at various temperatures: a) p-type, $\rho \sim 10^8$ $\Omega \cdot$cm; b) n-type, $\rho \sim 0.1$ $\Omega \cdot$cm; c) p-type, $\rho \sim 10$ $\Omega \cdot$cm; 1) 20.4°K; 2) 35°K; 3) 4.2°K.

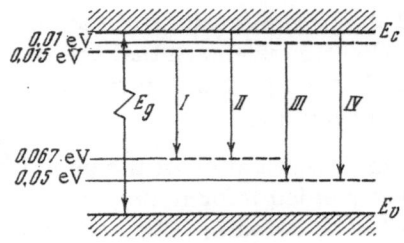

Fig. 21. Electron transitions responsible for
the edge luminescence of CdTe at T = 20.4°K;
E_g = 1.605 eV; E_I = 1.528 eV; E_{II} = 1.539 eV;
E_{III} = 1.548 eV; E_{IV} = 1.556 eV.

annealed in vacuum for 90-120 h at 700°C. The spectrum of CdTe crystals in which acceptor impurities predominate exhibits only a strong series II; the I series is not observed even at the lowest temperatures.

It is clear from Fig. 20 that the differences between the photoluminescence spectra of samples with different impurity compositions are not restricted only to changes in the relative intensities of the series I and II. An increase in the donor or acceptor concentration not only enhances the intensity of the series I and II but also produces luminescence lines at R = 1.548 eV (Fig. 20b) and E = 1.556 eV (Fig. 20c). We shall show later that there are grounds for assuming that these are the head lines of additional series which we shall call III and IV. The energy separation between the series I and II and between the series II and IV is, respectively, 0.020 and 0.017 eV, which is fairly close to the LO phonon energy (0.021 eV). This may be why the lines belonging to two series (for example, I and III) have occasionally been attributed to one series. We can see from Fig. 20a that the series III and IV have much stronger temperature dependences than the series I and II, respectively, and this allows us to distinguish the lines belonging to the various series.

Thus, the edge luminescence spectrum of cadmium telluride is more complex than the corresponding spectrum of cadmium sulfide, which consists of just two luminescence series. We shall show in the following sections that the series I and II can be explained by a model that has been invoked to interpret the luminescence of CdS [26, 35, 37, 38].

§ 2. Nature of Radiative Transitions

The observed influence of the impurity composition on the photoluminescence spectra allows us to suggest the following model (Fig. 21) to explain the behavior of the series I and II: the series I is due to radiative transitions of electrons from shallow donors to acceptors, whereas the series II is due to radiative transitions of free electrons from the conduction band to the same acceptors. This model is supported by the following experimental observations.

According to this model, when the donor and acceptor concentrations are fixed, the ratio of the intensities of the lines in the series I and II should depend on the ratio of the density of the electrons bound to donors to the density of free electrons. When the temperature rises, the population of the shallow donor levels decreases rapidly and this corresponds to a relative reduction of the intensity of the I lines.

Changes in the relative intensities of the lines in the series I and II, observed in the spectra recorded at fixed temperatures and excitation levels (Fig. 20), with changes in the impurity composition can be explained by this model using the results of electrical measurements.

The presence of shallow donors ($\sim$ 0.01 eV) in low-resistivity crystals is confirmed by the electrical and optical measurements [51, 52, 54, 71]. Depending on the method of preparation of crystals, these donors can be either interstitial cadmium atoms or substitutional group III impurities (In, Al). The depth of the acceptor participating in the band-impurity radiative transitions (series II) is, according to our results, 0.067 eV. Electrical measurements carried out on p-type crystals have revealed the presence of acceptors with ionization energies ranging from 0.05 to 0.3 eV [50, 77], depending on the conditions during growth and preparation.

The concentrations of donors and acceptors in high resistivity self-compensated crystals are approximately equal and the ratio of the probabilities of the donor—acceptor and band—acceptor radiative transitions is such that at 20.4°K the intensities of the lines in the I and II series are of the same order of magnitude.* Electrical measurements show that low-resistivity n-type crystals contain not only large amounts of donors but also a considerable amount of acceptors (the degree of compensation of crystals with electron densities $n = 1.6 \times 10^{16}$ cm^{-3} at $T = 300°$K is 25% [51]). At 20.4°K, the thermal ionization of donors can be ignored so that the ratio of the intensities of the series I and II depends on the ratio of the donor concentration to the density of nonequilibrium electrons. This ratio increases with increasing donor concentration and it explains the observed relative increase of the series I (Fig. 20b). The relatively low intensity of the series I in p-type crystals may be explained if we bear in mind that the conditions during growth or heat treatment of such crystals favor a reduction in the donor concentration. Changes in the profile of the zero-phonon line of the series I with the donor concentration and the excitation rate may be explained by the model of interimpurity transitions described by Eq. (I.7) in § 2, Chap. I.

When the donor concentration is increased, the line maxima shift in the direction of shorter wavelengths. This is due to the fact that the Coulomb term $e^2/\varepsilon_0 r$ occurring in the expression (I.7) for the energy of the photons emitted as a result of the interimpurity recombination becomes greater when the average distance between the donors and acceptors in pairs decreases [29, 30]. A similar shift of the maxima is also observed when the excitation rate is increased. According to the theory of the interimpurity recombination mechanism, this shift is due to the fact that when the excitation rate is increased, the role of the pairs with small values of r and, consequently, with large photon energies in Eq. (I.7) increases because the recombination in pairs with largers values of r become saturated.

A detailed analysis of the line profiles is difficult because of the fairly strong overlap of the lines belonging to different series. In the case of high-resistivity self-compensated crystals, this analysis can be carried out for the zero-phonon lines of the series II. It is found that the width of this line is ~2.5 kT. The short-wavelength wing of the line is described by an exponential function. These data on the line profile are in agreement with the proposed model which attributes the series II to radiative transitions of free electrons to acceptors. It should be pointed out that when the excitation rate is increased, the series II and I lines shift in the direction of shorter wavelengths. The origin of this shift is not quite clear. It is possibly due to an increase in the effective temperature of the electrons or to a reduction in the ionization energy of acceptors as a result of screening by free carriers [31] (it is estimated that the nonequilibrium carrier density is 10^{15}-10^{16} cm^{-3}; see § 1 in Chap. II).

It is interesting to note that the changes in the relative intensities of the I and II lines with the impurity composition are correlated with the changes in the relative intensities of the lines which result from the annihilation of bound excitons (§ 2 in Chap. III). It is clear from Fig. 20 that an increase in the donor concentration is accompanied by an increase in the relative intensity of the series I lines and a simultaneous increase in the intensity of the 1.593 eV line due to the annihilation of excitons bound to donors. Acceptor centers predominate in p-type crystals and the relatively high intensity of the series II is associated with the strong line at 1.590 eV, which is due to the annihilation of excitons bound to acceptors. At 20.4°K the intensities of the I and II series lines are comparable in the spectra of high-resistivity self-compensated crystals; in this case, the 1.593 and 1.590 eV lines are also of approximately equal intensity. This correlation suggests that the 1.593 and 1.590 eV lines are due to the radiative

---

*We must bear in mind that the equality of the total donor and acceptor concentrations of different types does not imply that the concentrations of the specific donors and acceptors between which radiative transitions take place are also equal.

annihilation of excitons bound to the same donors and acceptors which are responsible for the series I and II. The ionization energies of donors and acceptors, calculated from the binding energy of excitons and the energy of transitions corresponding to the annihilation of bound excitons, are in good agreement with the ionization energies of donors and acceptors participating in the I and II series. (The ionization energy of an acceptor is deduced from the photon energy of the zero-phonon lines in the series II, whereas the ionization energy of a donor is deduced from the difference between the energies of the zero-phonon lines in the I and II series.)

Thus, an analysis of the available experimental results shows that the behavior of the I and II series can be explained by a model which allows for the band—acceptor and donor—acceptor transitions.

As mentioned earlier, suitable donors are interstitial Cd atoms (E_d = 0.015 eV, see § 2, Chap. III) and substitutional In (E_d = 0.01 eV [52]) or Al (E_d = 0.014 eV [54]) impurities which have approximately equal ionization energies. However, the results of heat treatments of CdTe in the Cd vapor suggest that the interstitial Cd atoms are most likely to be responsible for the series I. The nature of the corresponding acceptors with an ionization energy of 0.067 eV is not quite clear. The presence of these acceptor centers in undoped crystals suggests that they may be either residual impurities or large complexes which include intrinsic lattice defects and residual impurities. Prolonged heat treatments of semi-insulating samples of CdTe in the saturated Te vapor (T = 750°C, t = 530 h) destroy completely all the edge luminescence series. Such heat treatment obviously destroys the complexes corresponding to the acceptor level at E_v + 0.067 eV, although new defects with a higher nonradiative recombination probability may form in a crystal because the intensity of the other luminescence bands also decreases strongly.

The III series (1.548 eV) is strongest in the spectra of crystals doped with In (E_d = 0.01 eV [52]). The spectra of such crystals with electron densities n $\sim 10^{17}$ cm^{-3} are dominated by the series III and its phonon replicas. This series shifts, like the series I, toward shorter wavelengths when the excitation rate or the concentration of In atoms is increased. Consequently, the series III may be attributed to optical transitions between indium (shallow donor) and acceptor levels (in this case, the acceptor is somewhat shallower than the acceptor responsible for the series I and II).

The IV series (1.556 eV) is of considerable intensity in the spectrum of crystals prepared at low Cd vapor pressures and is strongest in crystals annealed in vacuum (§ 1, Chap. IV). The appearance of the series IV after such annealing gives some grounds [71, 72] for attributing this series to electron transitions from the conduction band to the lower level of the double acceptor at E_v + 0.05 eV observed by Lorenz and Segall [78] in similar crystals. It is possible that this acceptor is also the final state of an electron in radiative transitions from donors, which give rise to the series III (Fig. 22). This hypothesis is supported by the transition energies in the III and IV series and by the fact that the nature of the acceptors responsible for the appearance of both series is the same (this will be demonstrated later).

The conditions during annealing of CdTe crystals in vacuum are such that the cadmium vacancy concentration is likely to increase. The temperature dependences of the Hall effect in such samples reveal only the E_v + 0.05 eV acceptor [50, 54]. This acceptor is identified in [50] with cadmium vacancies and in [54] with complexes which include intrinsic lattice defects.

Lorenz and Segall [78] report that the E_v + 0.05 eV level in similar crystals is the lower state of a double acceptor discovered by them (this acceptor is attributed either to cadmium vacancies or to complexes which include such vacancies).

The concentration of similar acceptors in CdTe probably rises as a result of doping with In because of self-compensation. This self-compensation has been established for CdTe crystals on several occasions [49, 79, 80] and its essence as is follows: when an impurity is in-

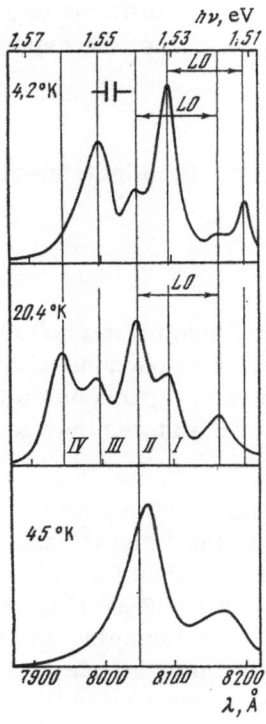

Fig. 22. Influence of temperature on the photoluminescence spectrum of an n-type CdTe crystal prepared by V. A. Chapnin by a special heat treatment (n = 1.15×10^{14} cm^{-3}, $\rho = 8.4$ $\Omega \cdot$ cm, $\mu = 7100$ cm$^2 \cdot$ V$^{-1} \cdot$ sec^{-1} at 90°K).

troduced into CdTe, the maintenance of the electrical neutrality demands the appearance either of an ionized interstitial atom or a vacancy in one of the sublattices, depending on the nature of the impurity. A determination of the degree of compensation of donor centers by acceptors on the introduction of In into CdTe led S. N. Maksimov [48] to the conclusion that these acceptors were mainly cadmium vacancies.

The existence of the simplest defects in CdTe is unlikely because of the high concentration of residual impurities. It follows from the cited investigations [48, 50, 54, 78, 80] that the $E_v + 0.05$ eV level may be attributed to cadmium vacancies but it is not clear whether the centers responsible for this level simply consist of such vacancies or are complexes which include these vacancies.

It is interesting to note that investigations of the threshold of formation of defects by electron bombardment at ~15°K [81] have shown that the simplest defect is a cadmium vacancy. (The concentrations of the residual impurities in the CdTe crystals employed in [81] were as follows: 3×10^{16} cm^{-3} Zn, 10^{17} cm^{-3} S, 5×10^{16} cm^{-3} Se, 4×10^{17} cm^{-3} Si.) An investigation of the luminescence spectra at the same temperature [81] has revealed lines similar to the III and IV series described above and these lines have been attributed to the same mechanism as that invoked by us; however, the shortest-wavelength line generated by electron irradiation [81] is shifted in the direction of shorter wavelengths relative to our IV series and it is wider, probably because of the overlap with lines of other origin. It is therefore likely that the ionization energy of the acceptors (assumed to be cadmium vacancies) obtained in [81] is slightly underestimated. However, it is also possible that the line in question is of different origin from that of our IV series.

The similarity of the mechanism responsible for the III and IV series to the mechanism of the I and II series (Fig. 21) is also supported by the observation that the 4.2°K spectrum is dominated by the I and III series (Fig. 22) whereas at 20.4°K the II and IV series are dominant.

However, it is worth pointing out an unexpected difference between the temperature dependences of the series II and IV at higher temperatures (Fig. 22). The intensity of the series IV rises more rapidly with temperature than that of the series II and it is difficult to explain this difference simply by the difference between the ionization energies of the corresponding acceptors.

§ 3. Electron—Phonon Interaction

The intensity distribution in the series of lines resulting from the electron—phonon interaction in optical transitions is given by the formula [20]

$$I_n = I_0 \frac{\overline{N}^n}{n!} \qquad (n = 0, 1, 2, \ldots,), \qquad (IV.1)$$

where I_n is the intensity of a line resulting from simultaneous emission of a photon and n phonons, whereas the parameter N represents the intensity of the interaction of phonons with electrons in a given transition (this parameter represents the average number of phonons emitted in an electron transition). The above formula is valid if at least one of the states between which transitions take place is localized [21].

If the spectra contain only the I and II series and there is no overlap with the III and IV series, the distribution of the intensities of the phonon replicas in the I and II series is in good agreement with the above formula. The value of N for both series is then found to be the same and equal to 0.34. If the spectra also include the lines at 1.548 and 1.556 eV (Fig. 21), the intensity distributions in the I and II series are no longer given by this formula. Deviations from the formula increase with increasing relative intensities of these lines. Hence, we may conclude that the 1.548 and 1.556 eV lines are the zero-phonon lines of the series III and IV. Since, as mentioned earlier, the separation between these lines and between the zero-phonon lines in the I and II series are similar to the LO phonon energy, it follows that the lines belonging to the III and IV series should overlap the I and II lines and this will be manifested as an apparent change in the relative intensities of the latter lines. The low-temperature (4.2-20.4°K) spectra of p-type crystals prepared by vacuum annealing are dominated by the IV series (the intensity of the II series is so low that the profiles of the zero-phonon line of the IV series and of its phonon replicas are identical). In this case, we can determine the value of N for the IV series. It is found that $\overline{N} = 0.19$.

The parameter $\overline{N}$ is a measure of the intensity of the interaction between the LO phonons and the centers involved in optical transitions. In this connection, it is worth mentioning that the parameter $\overline{N}$ is the same for the series I and II, in spite of the fact that these series are due to electron transitions of different types. This is evidently due to the fact that, in both cases, the dominant influence is the interaction between phonons and the acceptor, which is the final state in the I and II series. This is also supported by the following experimental observation. It is found that the 1.590 eV line, attributed to the annihilation of excitons bound to acceptors, has a phonon replica, whereas the 1.593 eV line, attributed to the annihilation of excitons bound to donors, has no visible phonon replica (§ 2, Chap. III).

A comparison of the intensity distributions in the phonon replica series associated with different centers can be used to determine $\overline{N}$ as a function of the parameters of the centers involved. The depth of the acceptor level responsible for the I and II series is $E_a = 0.067$ eV. This corresponds to N = 0.34. The photoluminescence spectra of CdTe exhibit an additional group of luminescence lines near 1.43 eV and this group may be attributed to the interaction between LO phonons and electrons in donor—acceptor transitions [71, 72] (see § 1, Chap. V).[*]

---

[*]This group of lines is called the "1.42 eV band" in [71, 72], in accordance with the energy of the maximum of the envelope of this group of lines at T = 80°K. The energy of the zero-phonon line is 1.453 eV.

The distribution of the intensity between the phonon replicas in this group of lines is also described by Eq. (IV.1) with N = 1.5 (Fig. 23c). Thus, this group of lines is characterized by an interaction of electron transitions with LO phonons stronger than that in the series I. It should be noted that the series I and the group of lines in the region of 1.43 eV are due to the same shallow donor because the acceptor energy is, respectively, E_a = 0.067 and 0.14 eV. The parameter $\bar{N}$ is quite different for the two lines which means that LO phonons interact mainly with acceptors. The value of N increases with increasing depth of the level.

This result is in qualitative agreement with Hopfield's theory [21]. He obtained the following expression for the interaction of phonons with electrons in transitions between an allowed band and local centers:

$$\bar{N} = \left(\frac{e^2}{a}\right)\left(\frac{1}{\hbar\omega_0}\right)\frac{1}{\sqrt{2\pi}}\left(\frac{1}{n^2} - \frac{1}{\varepsilon_0}\right),\tag{IV.2}$$

where $\hbar\omega_0$ is the phonon energy; n is the refractive index; ε_0 is the static permittivity; a is a constant representing the degree of localization of a carrier in a center. The charge distribution is given by

$$\rho(r) = (\pi^{\frac{1}{2}} a)^{-3} \exp\left(-\frac{r^2}{a^2}\right).\tag{IV.3}$$

It is clear from Eq. (IV.2) that $\bar{N}$ increases with the degree of localization in the final state. Hopfield's theory is, strictly speaking, inapplicable to the two groups of luminescence lines discussed above because they are due to interimpurity transitions. However, since in both cases the initial state is a shallow donor, which can be described by a combination of electron wave functions with wave vectors concentrated in a small region near the bottom of the conduction band [82], we may expect that, even in this case, there should be a tendency for $\bar{N}$ to increase with decreasing a. It should be pointed out that the relatively small value of this constant ($\bar{N}$ = 0.19) for the series IV is also in qualitative agreement with Hopfield's theory because

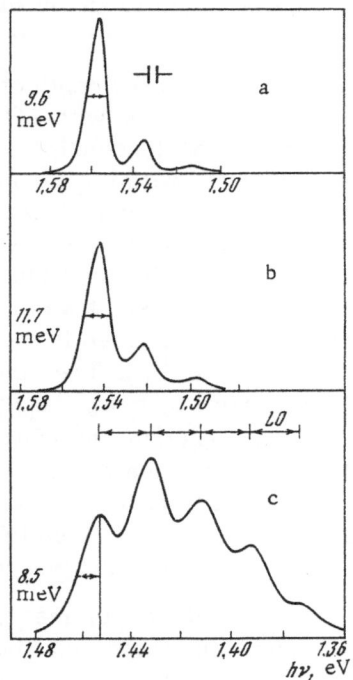

Fig. 23. Dependence of $\bar{N}$ and the line width on the depth of acceptors in CdTe at 20.4°K: a) series IV, $\bar{N}$ = 0.19, E_a = 0.05 eV; b) series II, $\bar{N}$ = 0.34, E_a = 0.067 eV; c) 1.43 eV band, E_a = 0.14 eV.

this series is associated with acceptors which have a lower ionization energy, i.e., a higher value of a than the acceptors responsible for the series II.

It is difficult to compare quantitatively Hopfield's theory and the experimental values of $\bar{N}$ deduced for different series of lines. First, the value of a which occurs in Eq. (IV.2) is not the Bohr radius of the ground state of a donor or acceptor. It follows that this quantity cannot be calculated from the ionization energies of hydrogen-like impurities. Secondly, at least one of the acceptors responsible for the group of lines at 1.43 eV is known to be definitely not hydrogen-like. Moreover, as pointed out earlier, the formula (IV.2) is derived only for the band–impurity transitions.

In the case of interimpurity optical transitions, the parameter $\bar{N}$ depends generally on the properties of the acceptor and donor. The predominance of the interaction of phonons with acceptors in the case of the series I and the band at 1.43 eV can also be explained by the fact that, in both cases, shallow donors ($E_d \sim 0.015$ eV) participate in transitions and these donors interact relatively weakly with phonons because of the large electron orbits. Since the parameter $\bar{N}$ is the same for the series I (shallow donor–acceptor transitions) and for the series II (transitions from the conduction band to the same acceptor), we can ignore (at least in the first approximation) the donor contribution and find the change in $\bar{N}$ as a function of the ionization energy of the acceptor, irrespective of whether the radiative transitions occur from the conduction band or from the shallow donor to the acceptor in question. An analysis of our results shows that $\bar{N}$ is described satisfactorily by a quadratic function of E_a:

$$\bar{N} = 76\,(\text{eV}^{-2})\,E_a^2. \tag{IV.4}$$

It is assumed that $\bar{N} = 0$ when $E_a = 0$ [21]. The deviation of the experimental data for CdTe from the dependence (IV.4) is about 1% for $E_a = 0.05$–0.14 eV.

Changes in the depth of level at which radiative recombinations terminate are manifested not only by this strong variation of the parameter $\bar{N}$, which represents the interaction between the radiative centers and LO phonons, but also by a change in the line width (Fig. 23). This broadening is probably due to an enhancement of the interaction with acoustic phonons when the degree of localization of carriers increases.

Thus, an increase in the depth of the level is accompanied by a simultaneous increase in the strength of the interaction with LO phonons (an increase in $\bar{N}$) and with acoustic phonons (line broadening). If both tendencies are maintained with increasing level depth, we may expect the structure associated with the LO phonons in the recombination via deep levels (~ 0.3–0.4 eV) not to be resolved except at very low temperatures (because of the line broadening); moreover, the line profile should be described by the Poisson distribution (IV.1). This hypothesis was checked by investigating the profiles of the lines with maxima at 1.11 and 1.02 eV, which corresponded to deeper centers. It was found that at $\sim 15°$K the LO phonon structures were not observed. The profiles obtained in the range 15–150°K were described satisfactorily by the Poisson distribution with $\bar{N}$ deduced from Eq. (IV.4). The ionization energy of the acceptor E_a was calculated from $E_a = E_d - E_0$, where E_0 is the energy of a zero-phonon optical transition found by extrapolation of the short-wavelength wing of the luminescence line on the basis of Eq. (IV.1). This operation was analogous to that employed for other lines in the case when the phonon replicas were resolved. The error in the determination of E_a did not exceed the LO phonon energy. It was pointless to extend the investigated temperature range because the formula (IV.1) was valid only subject to the condition $kT \ll \hbar\omega_0$ [21].

A comparison of the experimentally determined profile of the 1.11 eV line (continuous curve) with the envelope of the distribution $N^n/n!$ (dashed and chain curves) is made in Fig. 24. The profile of the line at 1.11 eV can be described satisfactorily by a Poisson distribution with

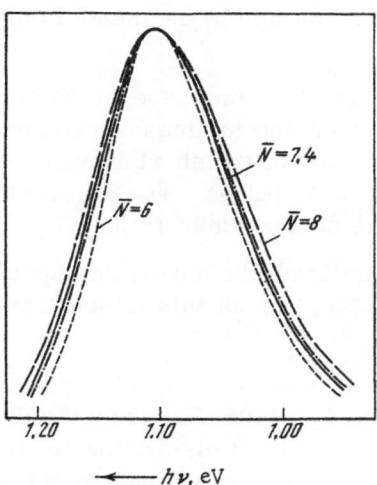

Fig. 24. Comparison of the experimental (continuous curve) and calculated (dashed and chain curves) profiles of the 1.11 eV line. The calculated profiles are based on the Poisson distribution with different values of $\bar{N}$.

$\bar{N}$ = 7.4. This value of $\bar{N}$ is in agreement with $\bar{N}$ = 7.78, obtained from Eq. (IV.4) for the ionization energy of the acceptor 0.32 eV deduced as above. The agreement indicates that the line appears because of radiative transitions from the conduction band to deep acceptors (§ 2 in Chap. V).

It is interesting to note that acceptors with a thermal ionization energy of 0.35 eV appear in CdTe crystals after prolonged annealing (180 h); this information is deduced from electrical measurements in [83]. The luminescence spectra of the same crystals are dominated by a line with a maximum at 1.11 eV.

The line at 1.0 eV is also described satisfactorily by a Poisson distribution with $\bar{N}$ = 11, which agrees with $\bar{N}$ = 10.97, deduced from Eq. (IV.4) for an ionization energy of acceptors of 0.38 eV. When the $\bar{N}$ increases, the line profile corresponding to the Poisson distribution of Eq. (IV.1) becomes more symmetric (Fig. 25) and at still higher values of $\bar{N}$, it corresponds to the Gaussian distribution. This is why the profiles of deep luminescence bands are frequently described incorrectly by the Gaussian distribution [38, 81] ignoring the value of $\bar{N}$.

It follows from the above analysis that the band width is governed by the value of $\bar{N}$ and the temperature dependence of this width becomes weaker with increasing $\bar{N}$ because the contribution of the thermal broadening of the individual phonon replica lines in the combined band decreases. Even the width of the band at 1.43 eV exhibits a very weak temperature dependence, which can hardly be measured (Fig. 27).

Two points are worth stressing in connection with the analysis of the profiles of wide luminescence bands.

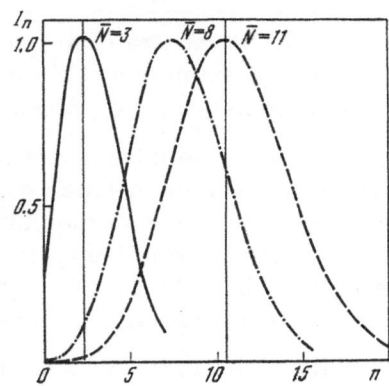

Fig. 25. Influence of the parameter $\bar{N}$ on the envelope of the Poisson distribution [$I_n = I_0$ ($\bar{N}^n/n!$), n = 0, 1, 2, . . .].

1. A comparison of the band profile with the Poisson distribution is meaningful only at moderately high temperatures (the condition $kT \ll \hbar\omega_0$).

2. According to the proposed model, the ionization energy of a luminescence center should be deduced not from the position of the maximum of the corresponding luminescence band, as is frequently done (see, for example, [12, 84]), but by extrapolation of the short-wavelength edge of the band to the zero-phonon line with the aid of the Poisson distribution. The difference between the energies obtained by these two methods may be quite considerable ($\sim \overline{N}\hbar\omega_0$).

It should be noted that in calculating the ionization energy from the absorption spectra of ionic insulator crystals, it is usual to include the Stokes shift [20, 85] but this is not done for semiconducting crystals.

The empirical dependence of $\overline{N}$ on the ionization energy of acceptors in CdTe, given by Eq. (IV.4), is satisfied both by the hydrogen-like acceptor with $E_a = 0.05$ eV and by the deep acceptors which are very far from the hydrogen-like state. Theoretical discussions of this dependence are made difficult by the fact that, in the case of deep centers which are not hydrogen-like, we must include the wave functions of both energy band [86], i.e., we can no longer use the one-band approximation for the energy structure.

We have so far considered the profiles of the luminescence bands associated with the interaction between electrons and phonons of the host lattice. It is interesting to note that the structure of the 0.5 eV luminescence band (Figs. 5 and 6) is due to the interaction of electrons with local phonons. Evidently, in this case, there are three partly overlapping lines separated by an interval of 0.045 eV. The fact that the relative intensities of these lines are independent of the temperature and impurity composition of the sample (§ 3, Chap. V) means that the observed luminescence structure cannot be explained by radiative transitions to the levels of various centers or to the system of levels of one center. Moreover, the energy of 0.045 eV is twice as high as the LO phonon energy. If we assume that the observed luminescence structure is due to local phonons, it follows that this band is associated with impurity atoms of a mass considerably smaller than the mass of the atoms in the host lattice [87].

Thus, we may conclude that all electrons undergoing radiative transitions in CdTe crystals interact with longitudinal optical phonons. This interaction is described well by the empirical expression (IV.4), which is deduced from an analysis of the experimental data.

CHAPTER V

RADIATIVE RECOMBINATION IN DEEP CENTERS IN CdTe CRYSTALS

§ 1.   Radiative Recombination between Donors and Deep Acceptors

The luminescence spectra of n- and p-type CdTe crystals recorded at 77.3°K always include a band with a maximum at 1.43 eV. The intensity of this band rises with change of conduction from p- to n-type, i.e., when the concentration of excess cadmium is increased.

The dependence of the intensity of the 1.43 eV band on the concentration of excess cadmium is best studied by considering n-type CdTe. Table 2 gives the results of electrical measurements as well as the energy positions of the maxima of luminescence lines, their widths at mid-amplitude, and intensities (in relative units). It is clear from Table 2 and Fig. 26 that the intensity of the 1.43 eV band increases with rising cadmium vapor pressure (or with increasing temperature of the reservoir with pure cadmium) during crystal growth.

The intensity of the 1.43 eV band is highest (last four samples in Table 2) for those crystals which are doped with Cd during a heat treatment either from a high-pressure Cd vapor

TABLE 2

Sample*	T_{Cd}, °C†	n, cm^{-3}	μ, cm$^2 \cdot$V^{-1} $\cdot$sec^{-1}	ρ, $\Omega \cdot$cm	Band at 1.43 eV			Band at 1.1 eV		
					position, eV	width, eV	intensity, rel. units	position, eV	width, eV	intensity, rel. units
BM-47	825	4,4·10^{15}	5680	0,4	1.42	0,11	80	—	—	—
BM-32	830	5,2·10^{15}	2000	0,6	1.42	0,11	30	—	—	—
BM-34	835	8,8·10^{15}	2700	0,3	1.43	0,11	440	1.08	0,20	50
BM-51	840	2·10^{16}	2500	0,1	1.43	0,11	650	1.04	0,13	50
BM-41	845	7,8·10^{16}	1800	0,02	1.44	0,10	570	1.07	0,16	15

Samples annealed in Cd vapor	T_{CdTe} °C‡	n, cm^{-3}	μ, cm$^2 \cdot$V^{-1} $\cdot$sec^{-1}	ρ, $\Omega \cdot$cm	Band at 1.43 eV			Band at 1.1 eV		
					position, eV	width, eV	intensity, rel. units	position, eV	width, eV	intensity, rel. units
SL–30 (220 h)	900	2,4·10^{17}	570	0,05	1.43	0,11	9 800	1,0	0,12	25
SL–30 (530 h)	700	1,5·10^{17}	770	0,05	1,43	0,14	12 000	—	—	—
In CdTe + Cd solution										
BM–52 (90 h)	800	1,2·10^{15}	4560	1,10	1.43	0,1	440	—	—	—
BM–52 (120 h)	800	2,9·10^{16} (300° K)	910 (300° K)	0,35 (300° K)	1.43	0,12	11 200	—	—	—

*The designations of the samples include an indication of the preparation method: BM denotes crystals prepared by the Bridgman method; ZR denotes zone-refined crystals; SL denotes sublimated crystals.

†T_{Cd} (°C) is the temperature corresponding to a selected Cd vapor pressure during growth of a crystal.

‡T_{CdTe} (°C) is the temperature during annealing.

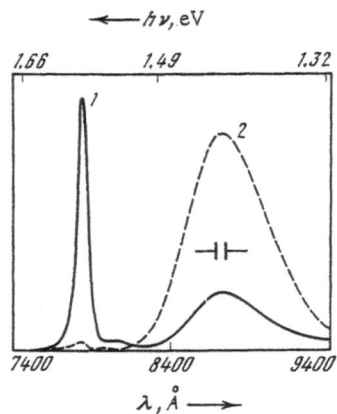

Fig. 26. Photoluminescence spectra recorded at 77.3°K for two n-type CdTe samples with different concentrations of excess cadmium: 1) sample BM-47, $n = 4.4 \times 10^{15}$ cm^{-3}; 2) sample BM-51, $n = 2 \times 10^{16}$ cm^{-3}. The intensity of the 1.58 eV line is almost the same for these two samples so that it can be used to compare the relative intensities of the 1.43 eV band.

or from a saturated solution of cadmium telluride in cadmium. The intensity of this band rises with the duration of diffusion.

Some of the cadmium telluride plates prepared from a Cd + CdTe solution had the highest concentration of excess Cd corresponding to the limit of existence of the compound at the crystallization temperature of 840°C.

The intensity of the 1.43 eV band increases equally strongly when In and Al donor impurities are introduced into a crystal. In this case the band shifts in the direction of longer wavelengths (the maximum of the band corresponds to 1.418 eV when the indium concentration is ~10^{18} cm^{-3} and in the case of a crystal with the same concentration of Al the band is located at ~1.406 eV). An increase in the intensity of this band as a result of doping with In (5×10^{17} cm^{-3}) was also observed by de Nobel at 77°K [13] but he attributed this band to the interband transitions.

Annealing of semi-insulating crystals in vacuum and in Te vapor, which reduces the concentration of excess cadmium, lowers strongly the intensity of the 1.43 eV band (the reduction depends on the annealing temperature and duration). Annealing for 120 h at 900°C destroys the 1.43 eV band. It follows from these results that the shallow donor states associated with In, Al, and excess Cd play a definite role in the appearance of the 1.43 eV band. This band may be due to interimpurity electron transitions between shallow donors and relatively deep acceptors.

The luminescence spectrum recorded in the region of 1.43 eV below 77°K splits into separate lines separated by the LO phonon energy (Fig. 27). As mentioned earlier (§ 3, Chap. IV), the distribution of the intensities between the lines resulting from the emission of a photon and $n = 0, 1, 2, \ldots$ LO phonons corresponds to $\bar{N} = 1.5$ and it is obtained because of the high value of the ionization energy of the acceptor. The profile of the 1.43 eV band obtained at

Fig. 27. Photoluminescence spectra recorded in the 1.43 eV range at two temperatures (°K): 1) 77.3; 2) 25.

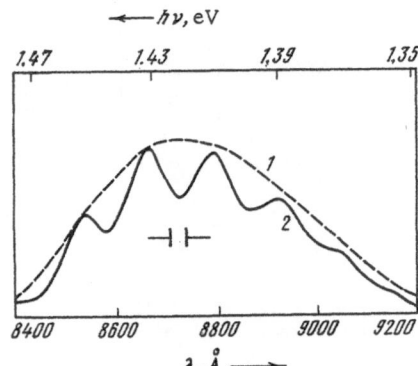

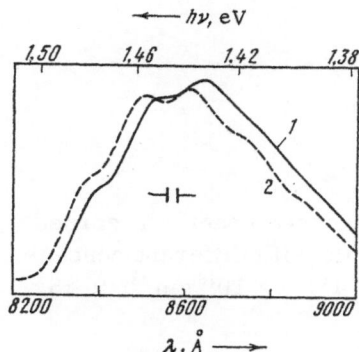

Fig. 28. Shift of the 1.43 eV band with the rate of excitation i_0 recorded at 20.4°K for a sample with the maximum concentration of excess Cd ($\approx 5 \times 10^{17}$ cm^{-3}): 1) i_0; 2) $10i_0$.

77.3°K is an envelope of the phonon replicas which can be observed separately at 25°K. At 77.3°K the thermal broadening results in merging of the separate phonon replicas so that a structure-free profile is observed. For the sake of brevity we shall denote this band by 1.43 eV, which is the energy of its maximum at 77.3°K, and we shall do this irrespective of the temperature under consideration.

The width of the envelope of the 1.43 eV band is practically independent of temperature because it is governed by the parameter $\bar{N}$ (§ 3, Chap. IV). It has been pointed out in [88] that the width of this band is approximately the same in the temperature range 300–100°K.

The donor–acceptor mechanism of the radiative recombination responsible for the 1.43 eV band is supported also by the observation that the band shifts toward shorter wavelengths when the excitation rate is increased. Figure 28 illustrates the shift of this band as a function of the exciton rate of a crystal with the maximum possible concentration of the excess Cd atoms. When the excitation rate is increased by a factor of 10, the band shifts by 5 meV in the direction of shorter wavelengths. The phonon structure of the 1.43 eV band of this crystal is barely noticeable because of the concentration quenching of the individual phonon replicas. The shift of the band with the excitation rate is observed also for low-resistivity p-type crystals (with hole densities of $\sim 10^{16}$ cm^{-3}), but when the excitation rate is increased by a factor of 20, the shift is only 2 meV (Fig. 29).

The 1.43 eV band also shifts toward shorter wavelengths when the concentration of excess Cd is increased. Figure 30 shows the spectra of two n-type CdTe samples in which the excess Cd concentration is 1×10^{16} and 5×10^{17} cm^{-3}, respectively. The photoluminescence

Fig. 29. Shift of the 1.43 eV band with the rate of excitation i_0 recorded at 20.4°K for a p-type sample with a hole density of $\sim 10^{16}$ cm^{-3}: 1) i_0; 2) $20i_0$.

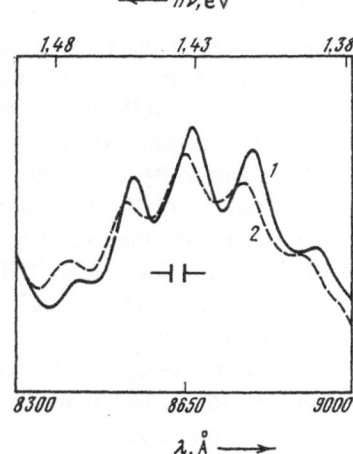

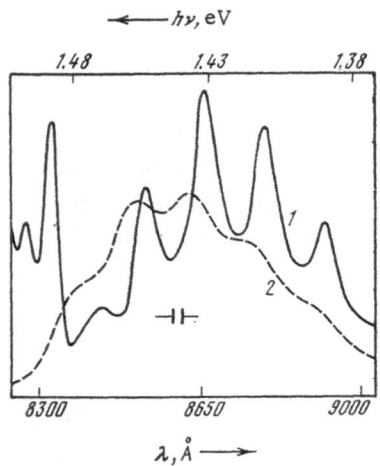

Fig. 30. Photoluminescence spectra recorded at 4.2°K for two crystals with different concentrations of excess Cd: 1) $\approx 1 \times 10^{16}\,cm^{-3}$; 2) $\approx 5 \times 10^{17}\,cm^{-3}$.

spectra in Fig. 30 correspond to the same excitation rate and the same temperature. When the concentration of excess Cd is increased by two orders of magnitude, the 1.43 eV band shifts by 5 meV in the direction of shorter wavelengths.

Thus, we may conclude that the 1.43 eV band of CdTe crystals with excess Cd appears because of radiative transitions of electrons from a shallow Cd donor ($E_c - 0.15$ eV) to an acceptor. It follows from Eq. (I.7) that if we know the ionization energy of the donor, we can determine the optical ionization energy of the corresponding acceptor to within the Coulomb interaction term. This can be done by investigating the luminescence spectra at low temperatures when the optical transitions without and with the participation of n = 1, 2, 3, . . . LO phonons can be distinguished in the 1.43 eV band. This operation is permissible if only the 1.43 eV band is present and there are no lines due to other radiative transitions, as is the case in the edge luminescence region (see Chap. IV).

It is clear from Fig. 30 that on the short-wavelength side of the 1.43 eV band there is a line at ~1.47 eV and the relative intensity of this line increases with rising concentration of the excess Cd. The appearance of the ~1.47 eV line in the spectra of two crystals shown in Fig. 30 is due to different causes. At low concentrations of the excess Cd the intensities of the edge luminescence series I and II in the luminescence spectrum of CdTe at 4.2°K are only slightly higher than the intensity of the 1.43 eV band. The phonon replicas of the I and II series (Fig. 20a) with n = 2 and 3 are located approximately at 1.49 and 1.47 eV, respectively (curve 1 in Fig. 30); the intensities of these replicas decrease strongly with increasing n and, therefore, the distortion of the 1.43 eV band because of the overlap by the edge luminescence can be ignored. As mentioned in the preceding chapter, the intensity of the edge luminescence series I rises strongly with increasing excess concentration of Cd. When this concentration of Cd reaches ~$10^{17}\,cm^{-3}$, the edge luminescence masks the 1.43 eV band at low temperatures and this band can be distinguished only if the spectrum is recorded in great detail. It should be noted that this happens in spite of the fact that the intensity of the 1.43 eV band also rises with increasing concentration of the Cd donors. Obviously, this is why the luminescence in the region of 1.43 eV band is not reported in an investigation of the 20°K photoluminescence of undoped n-type CdTe with excess Cd [12]. The appearance of this band is attributed to Cd vacancies because this band appears clearly in the spectra of p-type crystals. As pointed out above, the luminescence spectrum of CdTe is dominated by the 1.43 eV band at 77.3°K and its intensity rises strongly with increasing Cd concentration. It is possible that Halstead, Lorenz, and Segall [12] failed to notice that at low temperatures (4-20°K) the intensity in the region of 1.43 eV is much greater for n-type crystals that for p-type crystals and that a detailed record of the spectrum makes it possible to separate the 1.43 eV band from the "tail" of the edge luminescence.

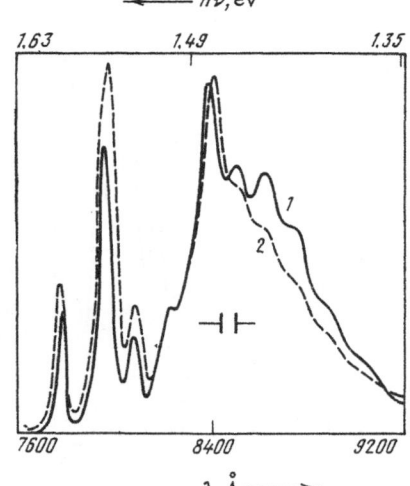

Fig. 31. Photoluminescence spectra of a sample annealed in vacuum at 700°C for 90 h, recorded at different temperatures (°K): 1) 22; 2) 33.

Moreover, at still higher concentrations of excess cadmium ($\sim 5 \times 10^{17}$ cm^{-3}) the photoluminescence spectrum is dominated by the 1.43 eV band even at 4.2°K (curve 2 in Fig. 30), whereas the intensity of the edge luminescence is then low. However, even in this case there is a line of considerable intensity in the region of 1.47 eV and this line distorts strongly the 1.43 eV band. The short-wavelength edge of this band is lifted somewhat and the distribution of the intensity in the 1.43 eV band is no longer described by the Poisson distribution of Eq. (IV.1). The deviation from this distribution increases with increasing intensity of the 1.47 eV line, which also has its phonon replicas probably with a different parameter N. The ~ 1.47 eV line cannot be associated with the 1.43 eV band and it cannot be regarded as the zero-phonon component of the latter band because the experimentally determined relationship between the intensities is not in agreement with the Poisson distribution. The 1.47 eV line is clearly separate from the 1.43 eV band in the spectrum of a crystal annealed at 700°C in vacuum (Fig. 31). We can see from Fig. 31 that the intensity of the 1.43 eV band decreases with rising temperature much more rapidly than the intensity of the 1.47 eV line so that at 33°K this line can be seen quite readily. It is not clear whether the structure of the spectrum is due to the phonon replicas of the 1.43 eV band or of the 1.47 eV line. In any case, it is clear that the value of $\overline{N}$ for the 1.47 eV line is considerably smaller than for the 1.43 eV band. An approximately similar band is also obtained when samples are annealed in vacuum at 1000°C. The temperature dependences of the Hall effect of such samples reveal a deep donor level at $E_c - 0.15 \pm 0.02$ eV [54]. We may assume that the 1.47 eV line is due to the radiative capture of a hole by a deep donor. This conclusion is supported by the observation that the intensity of the electron–phonon interaction is different for the centers which have similar ionization energies (these centers are associated with the 1.43 eV band and with the 1.47 eV line). We can explain this by the fact that the structure of these centers is quite different. Our example shows that in order to separate the lines belonging to different centers we can use not only the influence of the temperature, excitation rate, impurity composition, etc. on the spectrum but also the strong dependence of the value of $\overline{N}$, governed by the intensity of the electron–phonon interaction, on the properties of the centers. Thus, the value of $\overline{N}$ can be used as a characteristic of a given center. The usefulness of this characteristic in the separation of lines with similar energies but belonging to different centers is demonstrated also in the edge luminescence case (§ 3, Chap. IV). The same characteristic has been used earlier to predict the existence of the series III and IV, which were confirmed experimentally subsequently to this prediction.

When the intensity of the edge luminescence is low compared with the intensity of the 1.43 eV band and the 1.47 eV line or other lines near the 1.43 eV band are not observed, we

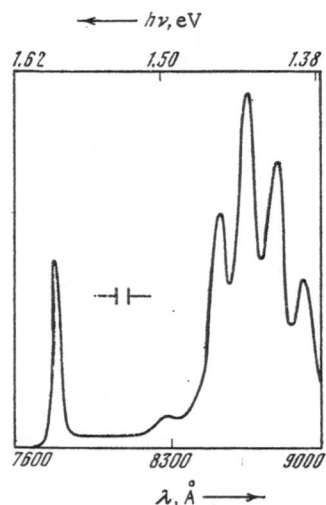

Fig. 32. Photoluminescence spectrum recorded at 4.2°K for a CdTe crystal subjected to a prolonged (530 h) annealing in saturated Te vapor at 750°C.

can determine the ionization energy of the acceptor associated with the 1.43 eV band, As pointed out in § 2, Chap. IV, the spectrum of a sample subjected to a prolonged (530 h) annealing in the saturated Te vapor shows no edge luminescence (Fig. 32) and only a weak band at 1.43 eV. Since at 4.2°K the emission of a photon is accompanied only by the emission of a phonon, the zero-phonon transition corresponds to the line with the shortest wavelength at 1.453 eV. Consequently, the ionization energy of the acceptor involved in the donor—acceptor luminescence band at 1.43 eV is 0.14 ± 0.0085 eV.

Chapnin [54] determined the temperature dependences of the Hall coefficient of the investigated p-type crystals and found an acceptor level with a thermal ionization energy E_v + 0.15 ± 0.02 eV, which he attributed either to accidental impurities or to complexes formed from intrinsic defects and impurity atoms. Thus, the thermal and optical ionization energies were found to be the same (within the limits of the experimental error).

This value of the ionization energy is also frequently found in the investigations of electrical properties [13, 89, 90], where it is attributed to centers formed from cadmium vacancies. For example, Halsted, Lorenz, and Segall [12] used the results of de Nobel [13] and attributed the 1.43 eV band to electron transitions from the conduction band to a cadmium vacancy level. However, the dependences of the intensity of this band on the concentration of excess Cd and on the annealing conditions in vacuum are in conflict with these conclusions. It is difficult to imagine a crystal in equilibrium which would have a high concentration of excess cadmium and a high concentration of cadmium vacancies. It is likely that the E_v + 0.15 eV level is associated with residual acceptor impurities which may be present in CdTe in sufficient numbers. Since the concentration of the residual acceptors varies from sample to sample, the intensity of the 1.43 eV band should be quite different for batches of samples with the same resistivity.

The nature of these acceptors has not yet been finally established. It may be assumed that introduction of excess cadmium atoms is accompanied by the contamination of CdTe with various acceptor impurities from the walls of the quartz ampoule [49]. This is probably why n-type CdTe cannot be grown in quartz boats, but such crystals can be prepared in vitreous carbon boats [48]. It is most likely that the contaminants captured from quartz ampoules are Si or O_2. We checked this hypothesis by introducing small amounts of Si, SiO_2, and CdO during growth of CdTe in vitreous carbon boats. In all cases we obtained either high-resistivity p-type crystals or strongly compensated n-type crystals [48]. The intensity of the 1.43 eV band in the spectra of these crystals was very high compared with the spectra of the samples prepared under the same conditions but without the deliberate addition of Si and O_2. Clearly, the acceptor with the energy $E_v + 0.15 \pm 0.02$ eV represents a large complex which includes silicon or oxygen impurity atoms. This hypothesis explains the results given in [90] where it is reported that the intensity of the 1.43 eV cathodoluminescence band increases if a chemically etched sample is kept not in vacuum but in an oxygen atmosphere. A scatter of the thermal ionization energy values, amounting to 0.02 eV, is reported in [50] for the $E_v + 0.15$ eV level of CdTe. A reduction in the luminescence energy of the 1.43 eV band within 0.02 eV is also observed for our samples when they are doped with In and Al. Obviously, this shift can be attributed to a change in the ionization energy of the acceptor of the type discovered in [50].

§ 2. Band at 1.1 eV

The radiative recombination spectra of cadmium telluride crystals exhibit a wide band at about 1.1 eV. The position of the maximum of this band varies from 1.12 to 1.0 eV at 77.3°K depending on the impurity and defect compositions of the investigated crystals. The reported positions of the maximum of this band vary considerably: 1.10 and 1.06 eV [13]; 1.02 eV [88]; 1.1 eV [81, 84, 91]. However, in all these investigations it is assumed that the band is the same in spite of the considerable differences in the positions of its maximum.

Changes in the position of the maximum of this 1.1 eV band with the impurity–defect composition are reported also in [84], where the shift is attributed to an overlap by the stronger band at 1.43 eV. However, it is then necessary to assume that with increasing overlap the 1.1 eV band shifts away from the 1.43 eV band, whereas in fact the 1.1 eV band should shift in the opposite direction.

A detailed investigation of the position and width of the 1.1 eV band in the spectra of crystals with different impurity–defect compositions has established that its width also varies from 0.11 to 0.21 eV, depending on the position of the band maximum at 77.3°K. In those cases when the band maximum is at one of the extreme positions, either at 1.11 eV or at 1.0 eV, the width of the band is least, whereas in the intermediate positions of the maximum the width increases to 0.21 eV. In the case of cadmium telluride crystals annealed in a vacuum at 700°C the dominant band in this part of the spectrum is located at 1.11 eV. The intensity of this band increases with increasing duration of annealing. A similar position of the band is observed for a sample doped with Al and annealed in a vacuum at 600°C. On the other hand, in the case of n-type crystals with large deviations from stoichiometry the band maximum is located near 1.0 eV.

When the temperature is lowered to 20°K the maximum of the band shifts anomalously strongly in the direction of shorter wavelengths. This shift exceeds the shift which might result from the temperature dependence of the forbidden band width. The forbidden band width changes by ~ 0.01 eV between 77 and 20°K, whereas the band maximum shifts from ~ 1.02 eV (77.3°K) to ~ 1.10 eV (20°K).

In the case of crystals with small deviations from stoichiometry (in which electron densities are $n < 5 \times 10^{15}$ cm^{-3}) this band is practically invisible at 77.3°K (Table 2), but at 20°K its intensity is quite high and the maximum is located at ~ 1.11 eV, which is a higher energy than that observed for other n-type samples at 77.3°K.

The profile of the 1.1 eV band also varies strongly with the excitation rate.

The observed dependences of the profile of the 1.1 eV band on the impurity–defect composition, temperature, and excitation rate suggest that it is not a simple band but it is composed of at least two strongly overlapping bands which are frequently observed at the same time. They can be isolated only in those cases when the maximum occupies the extreme positions of 1.11 and 1.0 eV and then the width does not exceed 0.11–0.12 eV.

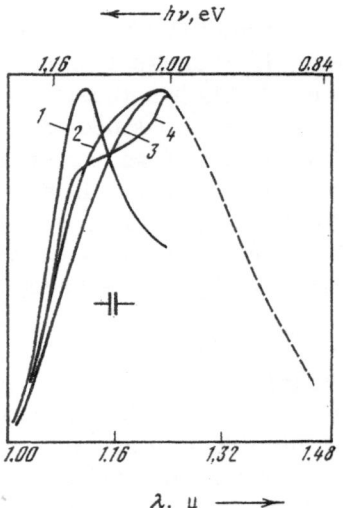

Fig. 33. Changes in the cathodoluminescence spectra recorded at 80°K using different rates of excitation and different electron energies: 1) 40 keV, $j = 5 \times 10^{20}$ pairs $\cdot$ cm^{-3} $\cdot$ sec^{-1}; 2) 33 keV, $j = 6 \times 10^{21}$ pairs $\cdot$ cm^{-3} $\cdot$ sec^{-1}; 3) 20 keV, $j = 6 \times 10^{21}$ pairs $\cdot$ cm^{-3} $\cdot$ sec^{-1}; 4) 33 keV, $j = 10^{22}$ pairs $\cdot$ cm^{-3} $\cdot$ sec^{-1}.

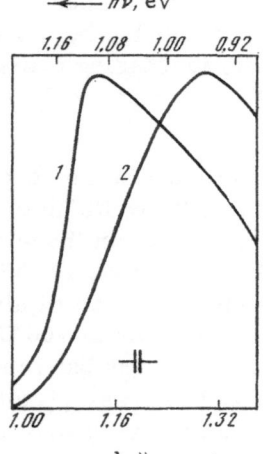

Fig. 34. Cathodoluminescence spectra recorded at two temperatures (°K): 1) 28; 2) 80.

However, the two bands cannot be distinguished clearly under photoluminescence conditions. This can be done only when the luminescence is excited by electron bombardment.*

Figure 33 shows the cathodoluminescence spectra of semi-insulating CdTe crystals obtained using different excitation rates at 80°K. At low excitation rates ($\sim 5 \times 10^{20}$ pairs $\cdot$ cm^{-3} $\cdot$ sec^{-1}) a band with a maximum at 1.1 eV is observed and this band is accompanied by a weak luminescence at longer wavelengths. When the rate of excitation is increased, the intensity of the long-wavelength luminescence increases more rapidly and a band with a maximum at 1.0 eV, accompanied by a slight kink in the region of 1.1 eV, is observed for an excitation rate of $\sim 10^{22}$ pairs $\cdot$ cm^{-3} $\cdot$ sec^{-1}. When the temperature is lowered to 28°K, the same excitation rate produces a band in the region of 1.1 eV (Fig. 34).

As pointed out earlier, the profile of the 1.11 eV band in the photoluminescence (Fig. 24) and cathodoluminescence spectra can be described satisfactorily by the Poisson distribution (IV.1) with the parameter $\overline{N} = 7.4$. The relationship (IV.4) gives 0.315 eV for the ionization energy of the acceptors responsible for this band. An approximately similar value (0.32 eV) is

---

*Measurements of the cathodoluminescence spectra were carried out by V. D. Kulakovskii and V. S. Konoplev.

obtained by extrapolating the short-wavelength edge of this band in accordance with Eq. (IV.1) to the zero-phonon line. In the case of vacuum-annealed crystals, which exhibit only the 1.11 eV band, electrical measurements yield the acceptor level $E_v + 0.35$ eV [83]. It is assumed in [83] that this level represents either the first charge state of a double acceptor A_2 or a copper impurity. Contamination of cadmium telluride crystals with copper impurities from the walls of quartz ampoules is very likely during such prolonged annealing.

The intensity of the 1.0 eV band increases when crystals are grown at high Cd vapor pressures. At the excitation rates quoted above this band does not saturate in the cathodoluminescence spectrum. The profile of the 1.0 eV band can also be described by a Poisson distribution with $\overline{N} = 11$. The ionization energy, deduced by extrapolation of the short-wavelength edge of the Poisson distribution to the zero-phonon line, is 0.38 eV. According to the empirical relationship (IV.4), the ionization energy of 0.38 eV corresponds to $\overline{N} = 10.97$, which is in good agreement with the experimentally obtained value $\overline{N} = 11$.

The forbidden band of such a material as CdTe may easily include an acceptor level of the required energy. Investigations of the temperature dependence of the Hall coefficient of p-type crystals have demonstrated the presence of energy levels from 0.05 to 0.5 eV in the lower half of the forbidden band [54] and the actual levels depend on the conditions during preparation of a crystal.

An increase in the intensity of the 1.1 eV band during electron bombardment at $T \sim 10°K$ is reported in [81]. It is also reported that the intensity increases after heat treatment at high Cd vapor pressures [84] and it is concluded that tellurium vacancies are the only defects which can explain these experimental results. However, the conclusions reached in [81] should be approached with caution because the 1.11 and 1.0 eV bands are not distinguished although they are of different origins. As pointed out in the beginning of this section, the observed shift of the 1.1 eV band with the impurity composition is attributed erroneously in [84] to the overlap with the 1.43 eV band. The ionization energy of tellurium vacancies (0.46 eV) given in the same paper is also unreliable because, first of all, the overlap of the separate bands is ignored, secondly, the ionization energy is deduced from the maximum of the 1.1 eV band, and thirdly, the forbidden band width at 77.3°K is erroneously taken to be 1.56 eV instead of 1.595 eV.

The nature of the acceptors responsible for the 1.0 eV band with an ionization energy of 0.38 eV is not quite clear. It is probable that these are large complexes including tellurium vacancies in spite of the fact that free vacancies have a donor level.

§ 3. Recombination Radiation at $\lambda > 1.2\,\mu$

This section is descriptive and its aim is to give a preliminary account of the luminescence bands of CdTe crystals observed in the $\lambda > 1.2\,\mu$ range. The nature of these bands cannot yet be established because of the insufficient information.

The recombination radiation in the wavelength range in excess of $1.2\,\mu$ was investigated using n- and p-type cadmium telluride crystals.

The spectra of n-type cadmium telluride crystals with large deviations from stoichiometry show no strong luminescence band in this region, in spite of the fact that the integrated intensity of the luminescence emitted from n-type CdTe crystals with $n \geq 10^{16}$ cm^{-3} is over an order of magnitude higher than the intensity of the luminescence emitted from p-type crystals with an approximately similar hole density. The spectra of the n-type crystals are dominated primarily by bands at shorter wavelengths, particularly by the 1.43 eV band whose intensity is sometimes more than three orders of magnitude higher than the intensity of the same band in the spectra of p-type crystals. Clearly, the luminescence with $\lambda > 1.2\,\mu$ is not observed in the case of low-resistivity n-type crystals because of the stronger short-wavelength recombination channels.

TABLE 3

Sample	ρ, $\Omega \cdot cm$ (290°K)	n, cm^{-3} (290 °K)	μ, $\frac{cm^2}{V \cdot sec}$ (290°K)	Line position, eV (80°K)	Line width, eV (80°K)	Line intensity, rel.units (80°K)
SL-2	800	$9.5 \cdot 10^{12}$	82	0.78 0.50	0.18 0.20	10 5
SL-4	80	$2.5 \cdot 10^{15}$	35	1.08	0.13	30
SL-6	10^7	—	—	0.86	0.18	10
SL-10	10^8	—	—	1.48 1.36	0.10 0.10	2 1
SL-20	10^7	—	—	1.42 1.12	0.10 0.21	60 5
SL-23	10^8	—	—	1.43 1.08	0.11 0.15	10 7
SL-30 SL-30 SL-30	10^7 10^7 10^7	— — —	— — —	1.43 1.07 0.50	0.10 0.14 0.20	4 10 10

The greatest number of lines is exhibited by p-type CdTe and the spectra of the majority of such crystals include long-wavelength bands. Particularly rich is the luminescence spectrum of the crystals grown from the vapor phase by the vacuum sublimation method. The low temperature during growth of these crystals reduces the concentration of thermal equilibrium point defects and the concentration of contaminants captured from the quartz ampoule walls [48]. Table 3 gives the results of electrical measurements carried out on such crystals and it includes the positions of the maxima of the luminescence lines, their widths at mid-amplitude, and their intensities in relative units. The intensities of these lines are relatively low (compare with Table 2). However, we may assume that these crystals contain no dominant impurities or defects.

The most interesting of the features observed in the spectra of these crystals is a band at 0.5 eV. This band is exhibited by the majority of p-type crystals prepared by other methods. Its profile is somewhat unusual for the deep bands exhibited by CdTe. It consists of three or four partly overlapping lines separated by 0.045 eV (Figs. 5 and 6). The profile of this band is strongly distorted in the ~2.7 μ range because of the absorption by water vapor contained in air. As pointed out in § 3, Chap. IV, the relative intensities of the lines composing the band are independent of temperature (between 20 and 80°K) and of the impurity composition. The impurity—defect compositions of the different samples were quite different, as manifested by differences between their spectra. In particularly, the spectrum of a p-type crystal with a relatively low resistivity (p ~5 × 10^{15} cm^{-3}) was found to be dominated mainly by the 0.5 eV band, whereas crystals annealed in vacuum exhibited also other strong bands (1.11 and 0.41 eV); this was true of crystals which were undoped or doped with Al before annealing. If the lines composing the 0.5 eV band were due to electron transitions to different local levels [84] or to a system of levels of the same center, their relative intensities should vary with the impurity—defect composition or with temperature. However, the profile of the combined band was unaffected. Moreover, the distance between the lines (0.045 eV) was more than twice as large as the LO phonon energy. Therefore, it seems justified to ascribe the observed structure of the 0.5 eV band to radiative transitions of electrons interacting with local phonons. The local phonons in CdTe are known from studies of the optical absorption [92, 93]. In the case of Li-doped CdTe the energy of a local phonon is 0.0341 eV. A line with an energy of 0.0213 eV is attributed to local vibrations associated with some defects in the crystal lattice. In the case of Be-doped CdTe there are absorption lines at 0.0485, 0.0965, 0.141, and 0.145 eV due to the excitation of various harmonics of local vibrations [93]. The vibration with the 0.0485 eV energy [93] agrees, within the experimental error, with the energy of the local phonons (0.045 eV) of the 0.5 eV luminescence band. In view of the proposed interpretation it would be interesting to determine the infrared absorption spectra of the crystals under investigation.

The nature of the centers responsible for the ~ 0.5 eV band is not clear. The appearance of local vibrations requires that an impurity atom should be approximately 1.5-2 times lighter than the host lattice atoms or the elastic constants should increase by a factor of 3-5 [87, 94]. Suitable residual impurity atoms in CdTe may be Cu or Zn.

The luminescence spectrum of vacuum-annealed CdTe crystals exhibits, at 20-30°K, a wide structure-free band with a maximum at 0.41 eV, which is independent of the composition of the crystal (it is observed for semi-insulating crystals and those doped with Al). The profile of this band is strongly distorted by the atmospheric absorption and will not be analyzed.

The 0.41 eV band is probably due to transitions to a level of the centers whose concentration increases as a result of annealing because this band is not observed in the investigated crystals before annealing.

A narrow line with its maximum at 0.364 ± 0.002 eV at $T = 20-30°K$ is observed for a crystal doped with Al and then annealed in a vacuum at 600°C. The line width is limited by the width of the monochromator slit and it does not exceed 1.6 meV. This line is likely to be due to an intracenter transition which interacts weakly with acoustic phonons because the energy position at 0.364 eV ($E_g = 1.605$ eV) and width of this line support this hypothesis.

The nature of the centers in which this intracenter transition takes place has not been considered. These centers may be accidental impurities with suitable valence electrons captured from the quartz ampoule walls during vacuum annealing of CdTe.

The luminescence bands with energies at 0.41 and 0.364 eV are reported here for the first time.

The author is deeply grateful to his scientific directors V. S. Vavilov and A. A. Gippius for their constant interest and invaluable help. He is also grateful to B. M. Vul for his interest. The author thanks S. A. Medvedev, S. N. Maksimovskii, and Yu. V. Klevkov for supplying cadmium telluride crystals, and V. A. Chapnin for supplying the results of his electrical measurements. Thanks are also due to V. A. Sedlov, V. V. Ushakov, and V. D. Kulakovskii for considerable help in the various stages of this investigation.

LITERATURE CITED

1. D. G. Thomas and J. J. Hopfield, Phys. Rev., 116:573 (1959).
2. T. S. Moss, Optical Properties of Semi-Conductors, Butterworths, London (1959).
3. F. C. Jahoda, Phys. Rev., 107:1261 (1957).
4. H. R. Philipp and E. A. Taft, Phys. Rev., 113:1002 (1959).
5. M. P. Rimmer and D. L. Dexter, J. Appl. Phys., 31:775 (1960).
6. H. Rode, Network Analysis and Feedback Amplifier Design, Van Nostrand, New York (1946), Chaps. XIV, XV.
7. W. G. Spitzer and D. A. Kleinman, Phys. Rev., 121:1324 (1961).
8. M. A. Lampert, Phys. Rev. Lett., 1:450 (1958).

9. J. R. Haynes, Phys. Rev. Lett., 4:361 (1960).
10. D. G. Thomas and J. J. Hopfield, Phys. Rev., 128:2135 (1962).
11. D. G. Thomas and J. J. Hopfield, Phys. Rev. Lett., 7:316 (1961).
12. R. E. Halsted, M. R. Lorenz, and B. Segall, J. Phys. Chem. Solids, 22:109 (1961).
13. D. de Nobel, Philips Res. Rep., 14:361 (1959).
14. R. E. Halsted, D. T. F. Marple, M. R. Lorenz, and B. Segall, Bull. Am. Phys. Soc., 6:148 (1961).
15. D. G. Thomas, J. Appl. Phys. Suppl., 32:2298 (1961).
16. V. S. Vavilov and É. L. Nolle, Fiz. Tverd. Tela, 8:532 (1966).
17. D. C. Reynolds, C. W. Litton, and T. C. Collins, Phys. Status Solidi, 12:3 (1965).
18. R. E. Halsted, "Radiative recombination in the band edge region," in: Physics and Chemistry of II-VI Compounds (ed. by M. Aven and J. S. Prener), North-Holland, Amsterdam (1967), p. 385.
19. F. A. Kroger and H. J. G. Meyer, Physica (Utr.), 20:1149 (1954).
20. M. A. Krivozlaz and S. I. Pekar, Tr. Inst. Fiz., Akad. Nauk Ukr. SSR, No. 4, 37 (1953).
21. J. J. Hopfield, J. Phys. Chem. Solids, 10:110 (1959).
22. R. J. Collins, J. Appl. Phys., 30:1135 (1959).
23. J. J. Lambe, C. C. Klick, nan D. L. Dexter, Phys. Rev., 103:1715 (1956).
24. R. E. Halsted and B. Segall, Phys. Rev. Lett., 10:392 (1963).
25. M. R. Lorenz and H. H. Woodbury, Phys. Rev. Lett., 10:215 (1963).
26. L. S. Pedrotti and D. C. Reynolds, Phys. Rev., 120:1664 (1960).
27. E. F. Gross, B. S. Razbirin, and S. A. Permogorov, Fiz. Tverd. Tela, 7:558 (1965).
28. G. A. Condas and J. H. Yee, Appl. Phys. Lett., 9:188 (1966).
29. F. E. Williams, J. Phys. Chem. Solids, 12:265 (1960).
30. D. G. Thomas, M. Gershenzon, and F. A. Trumbore, Phys. Rev., 133:A269 (1964).
31. K. Colbow and D. W. Nyberg, Phys. Lett., A, 25:250 (1967).
32. D. G. Thomas, J. J. Hopfield, and K. Colbow, Proc. Seventh Intern. Conf. on Physics of Semiconductors, Paris, 1964, Vol. 4, Radiative Recombination in Semiconductors, publ. by Dunod, Paris; Academic Press, New York (1965), p. 67.
33. K. Colbow, Phys. Rev., 141:742 (1966).
34. D. W. Nyberg and K. Colbow, can J. Phys., 45:2833 (1967).
35. Colbow and D. W. Nyberg, J. Phys. Chem. Solids, 29:509 (1968).
36. O. Goede and E. Gutsche, Phys. Status Solidi, 17:911 (1966).
37. N. N. Gerasimenko, Thesis for Candidate's Degree [in Russian], Novosibirsk (1968).
38. I. B. Ermolovich, A. V. Lyubchenko, and M. K. Sheinkman, Fiz. Tekh. Poluprovodn., 2:1639 (1968).
39. D. Curie and J. Prener, "Deep center luminescence," (ed. by M. Aven and J. S. Prener), North-Holland, Amsterdam (1967), p. 435.
40. A. A. Gippius, Zh. R. Panosyan, and G. A. Ivanov, Prib. Tekh. Eksp., No. 2, 212 (1970).
41. Yu. A. Vodakov, G. A. Lomakina, G. P. Naumov, and Yu. P. Maslakovets, Fiz. Tverd. Tela, 2:15 (1960).
42. V. S. Vavilov, A. F. Plotnikov, and A. A. Sokolova, in: Cadmium Telluride [in Russian], Nauka, Moscow (1968), p. 59.
43. E. S. Artobolevskaya, E. A. Afanas'eva, L. K. Vodop'yanov, and V. I. Sushkov, in: Cadmium Telluride [in Russian], Nauka, Moscow (1968), p. 97.
44. A. I. Zaslavskii, V. M. Sergeeva, and I. A. Smirnov, Fiz. Tverd. Tela, 2:2885 (1960).
45. P. Racquinot, Rev. Opt., 33:653 (1954).
46. S. A. Medvedev and Yu. V. Klevkov, in: Cadmium Telluride [in Russian], Nauka, Moscow (1968), p. 7.

47. S. A. Medvedev and Yu. V. Klevkov, Izv. Akad. Nauk SSSR, Neorg. Mater., 7:753 (1971).
48. S. N. Maksimovskii, Thesis for Candidate's Degree [in Russian], Lebedev Physics Institute, Academy of Sciences of the USSR, Moscow (1969).
49. S. N. Maksimovskii, S. A. Medvedev, and V. A. Rukavishnikov, in: Cadmium Telluride [in Russian], Nauka, Moscow (1968), p. 25.
50. B. M. Vul and V. A. Chapnin, in: Cadmium Telluride [in Russian], Nauka, Moscow (1968), p. 32.
51. L. K. Vodop'yanov and A. A. Abramov, in: Cadmium Telluride [in Russian], Nauka, Moscow (1968), p. 122.
52. B. Segall, M. R. Lorenz, and R. E. Halsted, Phys. Rev., 129:2471 (1963).
53. R. E. Halsted and M. Aven, Phys. Rev. Lett., 14:64 (1965).
54. V. A. Chapnin, Thesis for Candidate's Degree [in Russian], Lebedev Physics Institute, Academy of Sciences of the USSR, Moscow (1969).
55. E. F. Gross, S. A. Permogorov, and B. S. Razbirin, Fiz. Tverd. Tela, 8:1483 (1966).
56. J. R. Haynes, M. Lax, and W. F. Flood, Proc. Fifth Intern. Conf. on Physics of Semiconductors, Prague, 1960, publ. by Academic Press, New York (1961), p. 423.
57. P. Fisher and H. Y. Fan, Bull Am. Phys. Soc., 4:409 (1959).
58. D. T. F. Marple, J. Appl. Phys., 35:539 (1964).
59. M. Cardona and D. L. Greenaway, Phys. Rev., 131:98 (1963).
60. B. Segall and D. T. F. Marple, "Intrinsic exciton absorption," in: Physics and Chemistry of II-VI Compounds (ed. by M. Aven and J. S. Prener), North-Holland, Amsterdam (1967), p. 319.
61. V. S. Vavilov, S. G. Dzhioeva, and V. B. Stopachinskii, Fiz. Tekh. Poluprovodn., 3:727 (1969).
62. G. D. Mahan, J. Phys. Chem. Solids, 26:751 (1965).
63. D. T. F. Marple, Phys. Rev., 150:728 (1966); B. Segall, Phys. Rev., 150:734 (1966).
64. B. Segall, Proc. Ninth Intern. Conf. on Physics of Semiconductors, Moscow, 1968, Vol. 1, publ. by Nauka, Leningrad (1968), p. 425.
65. W. van Roosbroeck and W. Shockley, Phys. Rev., 94:1558 (1954).
66. J. R. Haynes and N. G. Nilsson, Proc. Seventh Intern. Conf. on Physics of Semi-conductors, Paris, 1964, Vol. 4, Radiative Recombination in Semiconductors, publ. by Dunod, Paris; Academic Press, New York (1965), p. 21.
67. Y. P. Varshni, Phys. Status Solidi, 19:459 (1967).
68. J. S. Blakemore, Semiconductor Statistics, Pergamon Press, Oxford (1962).
69. V. M. Agranovich, Theory of Excitons [in Russian], Nauka, Moscow (1968).
70. R. D. Cowan and G. H. Dieke, Rev. Mod. Phys., 20:428 (1948).
71. V. S. Vavilov, A. A. Gippius, and Zh. R. Panosyan (J. R. Panossian), in: II-VI Semi-conducting Compounds (Proc. Intern. Conf., Providence, R. I., 1967), ed. by D. G. Thomas, Benjamin, New York (1967), p. 743.
72. V. S. Vavilov, A. A. Gippius, and Zh. R. Panosyan, in: Cadmium Telluride [in Russian], Nauka, Moscow (1968), p. 103.
73. J. J. Hopfield and D. G. Thomas, Phys. Rev., 132:563 (1963).
74. G. D. Mahan and J. J. Hopfield, Phys. Rev., 135:A428 (1964).
75. J. J. Hopfield and D. G. Thomas, Phys. Rev., 122:35 (1961).
76. Zh. R. Panosyan (J. R. Panossian), A. A. Gippius, and V. S. Vavilov, Phys. Status Solidi, 35:1069 (1969).
77. R. Triboulet, C. R. Acad. Sci. B, 267:420 (1968).
78. M. R. Lorenz and B. Segall, Phys. Lett., 7:18 (1963).
79. G. Mandel, Phys. Rev., 134:A1073 (1964).
80. F. A. Kröger, J. Phys. Chem. Solids, 26:1707 (1965).
81. F. J. Bryant, A. F. J. Cox, and E. Webster, J. Phys. C, 1:1737 (1968).
82. W. Kohn, Solid State Phys., 5:257 (1957).

83. V. A. Chapnin, Fiz. Tekh. Poluprovodn., 3:566 (1969).

84. F. J. Bryant and W. Webster, J. Phys. D, 1:965 (1968).

85. K. Huang and A. Rhys, Proc. R. Soc. A, 204:406 (1950).

86. L. V. Keldysh, Zh. Eksp. Teor. Fiz. 45:364 (1963).

87. K. K. Rebane, Elementary Theory of the Vibrational Structure of the Spectra of Impurity Centers in Crystals[in Russian], Nauka, Moscow (1968).

88. V. S. Vavilov, É. L. Nolle, V. D. Egorov, and S. I. Vintovkin, Fiz. Tverd. Tela, 6:1406 (1964).

89. S. Yamada, J. Phys. Soc. Jap., 17:645 (1962).

90. J. P. Noblanc, J. Loudette, and G. Duraffourg, Phys. Status Solidi, 32:281 (1969).

91. F. J. Bryant and E. Webster, Phys. Status Solidi, 21:315 (1967).

92. M. Balkanski, R. Beserman, and L. K. Vodopianov, in: Localized Excitations in Solids (Proc. First Intern. Conf., University of California, Irvine, 1967), ed. by R. F Wallis, Plenum Press, New York (1968), p. 154.

93. W. Hayes and A. R. L. Spray, J. Phys. C, 2:1129, 1132 (1969).

94. G. S. Zavt, Fiz. Tverd. Tela, 7:2109 (1965).

INVESTIGATION OF THE ENERGY BAND STRUCTURE OF SEMICONDUCTORS BY DIFFERENTIAL OPTICAL METHODS

S. G. Dzhioeva

The electroreflection and thermoreflection spectra of GaAs single crystals were investigated in the 1.00-4.00 eV range at various temperatures. The results were used to determine the gaps between the energy bands and the values of the spin – orbit splitting of the valence band at the symmetry points Γ, L, and Λ of the Brillouin zone. Single crystals of CdTe were studied and it was found that the Coulomb interaction between electrons and holes should be allowed for in the electroreflection studies. The ground (n = 1) and the first excited (n = 2) exciton states were observed at liquid nitrogen temperature.

INTRODUCTION

All the important properties of semiconductors and other solids can, in principle, be understood and explained if the energy band structure is known. Therefore, the energy band structure is justly regarded as one of the central problems in experimental and theoretical solid-state physics. Investigations of the energy band structure reduce to the determination of the energy and wave function of an electron at each point in the wave-vector (or momentum) space in the Brillouin zone.

The energy spectrum of an electron in a crystal consists of bands of allowed energies separated by gaps known as the forbidden bands. In an ideal crystal an electron cannot have an energy which falls within a forbidden band. The structure and positions of the allowed and forbidden bands depend on the symmetry of the crystal structure, atomic spacings, and the nature of the atoms producing the crystal field. Consequently, each crystal has its own energy band model. There are some theoretical methods which make it possible to obtain a quite satisfactory qualitative picture of the energy band structure of many crystals in the range 10-20 eV. However, the precision of the theoretical energy band calculations is frequently insufficient and this leads to considerable errors in the positions of energy levels. Details of the band structure can be determined only when experimental data are used.

A considerable success has been achieved recently in studies of the energy band structure of semiconductors and the new information has been obtained mainly from wide-ranging experimental investigations of the fundamental optical properties and particularly of the absorption and reflection spectra. A comparison of the results obtained in these investigations with the theoretical calculations of the energy band structure has made it possible to determine reliably the characteristics of the energy bands in the k space for many crystals and to find the gaps in the vicinity of the principal points and directions in the Brillouin zone. (Descriptions and interpretations of the optical spectra from the point of view of the energy band theory are given in the reviews of Phillips [1] and Tauc [2].)

The most important and characteristic optical property of a semiconductor is the fundamental absorption edge E_0. The structure of this edge reflects the structure of the conduction and valence bands in the region where the energy gap between them is smallest. Studies of the fundamental absorption edge of various semiconductors have been reported in a very large number of experimental and theoretical papers.

Only relatively recently attention has been paid to the optical properties of semiconductors in the region extending from the absorption edge in the direction of higher energies. These studies have helped to understand better the electron structure of semiconductors. The knowledge of the energy-band structure in this range is necessary for the solution of many problems such as "hot" electrons, impact ionization, impact recombination, quantum efficiency of the internal photoeffect, external electron emission, optical properties, etc.

The aim of the investigation described below was to obtain information on the structure of semiconductors with the aid of differential optical methods. Such methods are justly regarded as the most important step forward in the experimental technology in recent years and this is the view shared by experimentalists and theoreticians working on the energy band structure of solids.

A common feature of these methods is the use of an external periodic perturbation (which can be an electric field, pressure, etc.) to alter the optical properties of a sample. These periodic changes are greatest near the critical points in the Brillouin zone, which are defined [3] as the points at which the "combined" density-of-states function has a singularity. Depending on the nature of this singularity (which can be a minimum, a maximum, or a point of inflection), we can distinguish four types of critical points [3], which are denoted by M_0, M_3, M_1, and M_2, respectively. Van Hove [4] was the first to draw attention to the importance of these points. The condition of the existence of these points [3] is $|\mathrm{grad}_\mathbf{k} (E_f - E_i)| = 0$, where E_i is the energy of an electron in its initial state; E_f is the energy in the final state; $\mathbf{k}$ is the wave vector.

The calculations carried out by Phillips [5] applying the group theory show that in the majority of cases the critical points lie at the points of symmetry or on the lines and planes of symmetry in the Brillouin zone. Figure 1 shows the Brillouin zone for crystals with the diamond or zinc-blende structure. Some symmetry points are identified by capital letters. A comparison with Fig. 2, which shows the energy bands of GaAs and the location and nature of the critical points important in interband transitions, confirms the validity of the results reported by Phillips.

It follows that the determination of the critical points, their nature, and energy positions can give considerable information on the energy band structure of a solid, namely on the energies of the band gaps and features of the structure of the bands in the k space. Differential optical spectra reveal the structure at energies corresponding to interband transitions in the

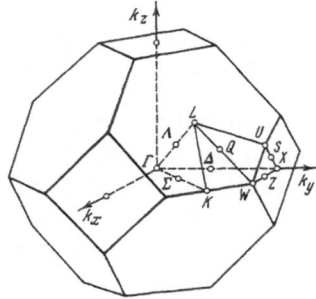

Fig. 1. Brillouin zone of a semiconductor with
a diamond or zinc-blende (sphalerite) structure.

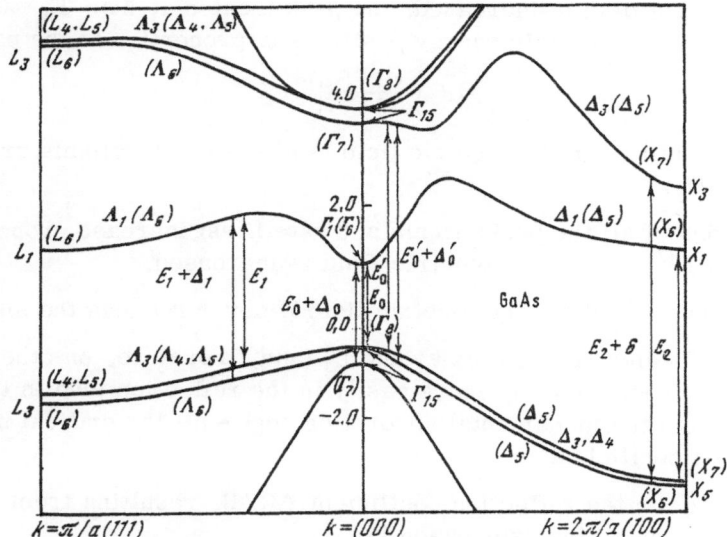

Fig. 2. Energy band structure of GaAs.

vicinity of the critical points, whereas elsewhere there is practically no differential signal. This increases the sensitivity and resolution by at least an order of magnitude compared with the conventional absorption and reflection methods. Moreover, it is possible to identify the critical points more accurately because the spectra obtained for different points are different.

The present author used two differential methods — electroreflection and thermoreflection — in the studies reported below. In order to interpret the experimental results correctly, it was necessary to know how the electric field (in electroreflection) or an increase in the temperature of a crystal (in thermoreflection) affected the optical properties of a solid and, consequently, its energy band structure.

In an electric field a steep fundamental absorption edge becomes broader and shifts in the direction of lower frequencies. This effect was predicted theoretically by Keldysh [6] and Franz [7] and is known an the Franz — Keldysh effect. The effect arises because an electron can be transferred from the valence to the conduction band by absorbing a photon of energy $\hbar\omega < E_g$, where E_g is the forbidden band width, and the deficit $E_g - \hbar\omega$, can be made up by acquiring energy from the electric field.

The theory developed by Keldysh, Franz, and other workers for the fundamental absorption edge and the critical points of type M_0 has now been extended to the critical points of all other types [8-10]. The results of these calculations are presented as expressions for $\Delta\varepsilon_2$ and $\Delta\varepsilon_1$, where $\Delta\varepsilon_2$ is the change in the imaginary component of the permittivity and $\Delta\varepsilon_1$ is the change in the real component. In the case of the four different types of critical points there are [9] six different forms of the expressions for $\Delta\varepsilon_2 (\omega, E)$ and $\Delta\varepsilon_1 (\omega, E)$.

We shall now list the main results that follow from analytic and graphical representations of the functions $\Delta\varepsilon_1 (\omega, E)$ and $\Delta\varepsilon_2 (\omega, E)$ near the critical points.

1. The functions $\Delta\varepsilon_1 (\omega, E)$ and $\Delta\varepsilon_2 (\omega, E)$ at $\hbar\omega = E_g$ (E_g is the energy corresponding to a critical point) have an asymmetric peak with satellites. The satellites may occur before or after E_g, depending on the nature of the critical point and the direction of the electric field. Away from E_g the amplitudes of the satellites decrease.

2. The main peak, corresponding to a critical point, need not be the strongest; in fact, it may be weaker than the satellites.

3. When the electric field is increased, the peak located at $\hbar\omega = E_g$ is not shifted, whereas all its satellites change their energy positions in proportion to the parameter

$$\Theta = \left| \frac{e^2 |E|^2}{2\hbar m^*} \right|^{1/3}.$$

The absence of a shift when the electric field is altered is a reliable criterion of the peak corresponding to E_g.

4. The amplitudes of all the peaks (main and satellites) increase proportionally to $\Theta^{1/2}$, i.e., proportionally to $E^{1/3}$, when the electric field is increased.

An increase in the intensity of the electric field E never reduces the signals $\Delta\varepsilon_1$ and $\Delta\varepsilon_2$.

The influence of the electric field on ε_1 and ε_2 (and, therefore, on other optical constants) can be observed experimentally more easily in the reflection than in the absorption spectra since — apart from the fundamental absorption edge — all the critical points lie at high values of the absorption coefficient.

The relative change in the reflection coefficient $\Delta R/R$, resulting from the application of an electric field, is given by the expression [8]

$$\frac{\Delta R}{R}(\omega, \mathbf{E}) = \alpha(\omega)\,\Delta\varepsilon_1(\omega, \mathbf{E}) + \beta(\omega)\,\Delta\varepsilon_2(\omega, \mathbf{E}).$$

Here, $\alpha(\omega)$ and $\beta(\omega)$ are the fractional coefficients which determine the contributions of $\Delta\varepsilon_1$ and $\Delta\varepsilon_2$ to $\Delta R/R$ in different spectral regions. They depend on the optical constants of the investigated material and on the angle of incidence of light on a crystal [8].

However, it should be pointed out that the experimental results can be compared only qualitatively with the theory because the inhomogeneity of the electric field, exciton effects, and various other interactions (for example, those between electrons and phonons) are ignored in the theory.

The changes in the optical properties of a crystal due to a change in its temperature are manifested [11-12] by the shift of the edge corresponding to a particular critical point and broadening of the edge.

Quantitative changes in the optical properties are manifested (as in the electroreflection case) by changes in the optical constants, which are the imaginary ε_2 and real ε_1 components of the permittivity. Approximate expressions for $\Delta\varepsilon_1$ and $\Delta\varepsilon_2$ are obtained in [13]. It is found that at all critical points the results may be expressed in terms of a function F(X), which has a maximum near a critical point and falls away from such a point.

The experimentally determined quantity, which is the relative change in the reflection coefficient $\Delta R/R$, is governed — as in the electroreflection case — by the quantities $\Delta\varepsilon_1$ and $\Delta\varepsilon_2$. Expressions for $\Delta R/R$ near critical points of all four types are given in [13]. It follows from these expressions that the thermoreflection signal increases with the depth of modulation ΔT and with the temperature coefficient dE_0/dT.

CHAPTER I

INVESTIGATIONS OF THE SINGULARITIES OF THE COMBINED DENSITY OF ELECTRON STATES IN SEMICONDUCTORS BY THE ELECTROREFLECTION METHOD

§1. Experimental Method

The most important procedure in the electroreflection measurements is the modulation of the reflection by an electric field.

It is known that light is reflected not from the whole crystal but from a surface layer whose depth is comparable with the wavelength of the incident light. Since the critical points usually correspond to wavelengths of the order of $0.1-1\,\mu$, the depth of penetration of light is usually about $10^{-6}-10^{-4}$ cm. For example, in the case of Ge the depth of penetration of 0.8 eV (corresponding to a critical point of type M_0 at the center of the Brillouin zone) and 2.1 eV photons (corresponding to a critical point of type M_1 on the [111] symmetry axis) is 5×10^{-4} and 5×10^{-5} cm, respectively [14].

In order to observe electroreflection it is sufficient to apply an alternating electric field penetrating a sample to a depth of the order of the depth of penetration of light. In semiconductors such a field can easily be applied by the parallel-plate capacitor method, which is known as the sandwich method. The application of an alternating electric field to such a capacitor, which is filled with the investigated material, modulates the surface potential barrier in a crystal and this is accompanied by changes in the reflection.

The construction of a sandwich is worth discussion, and, therefore, we shall describe it in detail. Our capacitor consisted of a crystal with a metal contact, a dielectric, and a transparent electrode. The dielectric was a polyethylene terephthalate (Mylar) film, $10\,\mu$ thick. The transmission of this film in the visible part of the spectrum was of the order of 90% but it was limited on the short-wavelength side to 4.5 eV (a Saran polymer film could be used to extend this range further in the ultraviolet direction). The refractive index of our dielectric film was n = 1.6 and its breakdown strength was $\sim 10^5$ V/cm.

A thin film of SnO_2, ~ 200 Å thick and evaporated on a quartz plate, was used as the transparent electrode. The surface resistance of the SnO_2 film was less than 80 Ω/cm^2 at room temperature. This relatively low resistance of the SnO_2 film was due to the high concentration of excess tin, which acted as a shallow donor. The doping with tin occurred during the evaporation of the SnO_2 film. When the temperature was lowered, the resistance of SnO_2 remained fairly low because this material was a degenerate n-type semiconductor. The SnO_2 electrode was transparent up to 4.5 eV and in the spectral range where the experiments were carried out there was no significant structure in the absorption of this electrode. Unfortunately, there was no conductor which would be more transparent in the ultraviolet range than SnO_2. The refractive index of the SnO_2 film was n= 1.7.

The modulating electric field was applied at right angles to the surface of the crystal. The capacitor described above acted also as an optical interferometer. Even when the mechanical contact between the components of the sandwich was very good, there was some interference at the signal frequency and this hindered considerably the recording of the electroreflection spectra. The undesirable interference could be avoided by bonding the crystal, dielectric, and transparent electrode by a viscous substance with a refractive index almost equal to that of the dielectric and SnO_2. In this case light would be reflected only from the surface of the crystal. We used silicone oil or its mixture with Canada balsam. The surface properties of the crystal were not affected by the presence of these substances but the interference was suppressed almost completely. Silicone oil and its mixture with Canada balsam were transparent up to about 4.7 eV; they were also good dielectrics and remained viscous right down to liquid nitrogen temperature. The use of these substances also avoided various breakdown effects which could occur at the high fields applied to the capacitor.

Figure 3 shows schematically (but not to scale) the sandwich described above. We applied a sinusoidal voltage, which modulated the electric field in the potential barrier on the surface of the investigated crystal. Our studies and those carried out by other workers indicated that the internal electric fields existing on the surface would be sufficient for the observation of the electroreflection effects. For example, it was reported in [15] that in the case of GaAs this field was of the order of 10^4V/cm. Therefore, in many cases there was no need to apply an additional static voltage, particularly as it was screened to a considerable

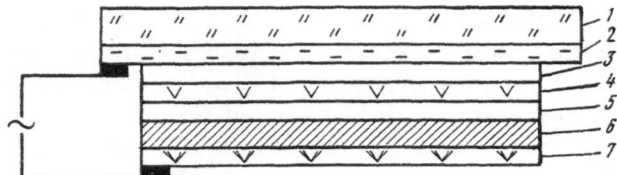

Fig. 3. Cross section through a sandwich structure: 1) quartz plate; 2) SnO_2 film; 3), 5) layer of
silicone oil or of its mixture with Canada balsam;
4) dielectric film; 6) investigated crystal; 7) metal
electrode (silver plate).

extent by the "slow" surface states. The alternating voltage in our experiments could be
varied between 5 and 1400 V. Voltages from 5 to 100 V were applied directly from an acoustic-
frequency oscillator of the GZ-33 type, whereas voltages in excess of 100 V were generated
by the same oscillator in conjunction with a high-voltage transformer of the NOM-10 type.

We made no attempt to determine the electric field in the surface barrier on a crystal.
This would be a fairly difficult task, as known from the physics of semiconductor surfaces.
This problem was encountered by most of the workers who studied the electrooptic effects.
Since the field was not known, it was not possible to carry out a quantitative comparison of
the experimental results with the available theory and, consequently, it was not possible to use
this comparison in the determination of certain important characteristic parameters such as
the effective masses of carriers. However, it was possible to determine the type and energy
positions of the critical points in the Brillouin zone without the knowledge of the modulating
electric field. This was the purpose of our experiments. The attainment of this purpose
yielded valuable information on the energy band structure of the investigated crystals.

The electroreflection signal depended on the frequency of the alternating voltage. This
was probably associated with the relaxation of the "fast" surface states because it was these
states that reacted to the electric field [16]. The measurements were carried out at the frequency which produced the strongest signal.

The amplitude of the signal depended also on the state of the reflecting surface of the
investigated crystal. We used mechanically polished samples which were subjected to a subsequent chemical etching and also freshly cleaved surfaces.

The experiments were carried out at two temperatures: 300 and 77.3°K. The sandwich
was placed in a special holder and at 77.3°K this holder was inmersed in a Dewar flask (with
optical windows), which was filled with liquid nitrogen.

Figure 4 shows a block diagram of the apparatus in which the electroreflection measurements were carried out.

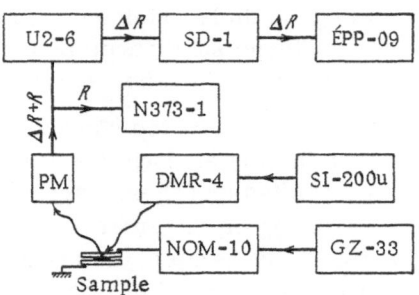

Fig. 4. Block diagram of the apparatus used
in electroreflection measurements.

Light from a ribbon incandescent lamp of the SI-200u type with a Uviol window (or light from a SI-300 lamp with an ordinary window) passed through a monochromator and was reflected from a sample at a particular angle (~22°). An attempt was made to carry out measurements in polarized light incident at the Brewster angle, but no increase in the electroreflection signal (predicted in [17]) was observed. This was evidently due to the fact that the Mylar film was itself a polarizer and the conditions for the effect predicted in [17] were not available.

The reflected light, modulated by the applied electric field, reached a photomultiplier denoted by PM in Fig. 4 (FÉU-28 or FÉU-18a, depending on the spectral region to be investigated). An important feature of the method employed was the simultaneous recording of two signals by the photomultiplier: one signal was constant and proportional to the reflection coefficient R of the sample and the other signal alternated at the frequency of the external field and it was proportional to the modulation ΔR. Dividing the alternating signal ΔR by the constant signal R, we obtain the value of $\Delta R/R$ which was independent of the incident light intensity. This avoided cumbersome and difficult measurements of this intensity, which would be necessary in direct measurements of the reflection. Moreover, this procedure automatically eliminated the influence of the instability of the light source. Several methods have been developed [18] for direct recording of the ratio $\Delta R/R$. We measured the electroreflection in narrow spectral regions where R was practically constant, and we determined R and ΔR separately. The constant signal was measured with an N-373-1 potentiometer. The alternating component of the signal was amplified with a U2-6 narrow-band amplifier, passed to an SD-1 synchronous detector, and recorded automatically by a ÉPP-09 potentiometer. The ratio $\Delta R/R$ was determined at separate points. Since the reflection coefficient R varied only slightly with the wavelength, the dependence $(\Delta R/R)(\hbar\omega)$ was practically indistinguishable from $\Delta R(\hbar\omega)$. Our system made it possible to determine the required ratio in the range $\Delta R/R \gtrsim 10^{-5}$.

A prism monochromator DMR-4 and a ZMR-3 monochromator with a diffraction grating (600 lines/mm) were used in measurements of the electroreflection spectra. In the former case the resolving power was of the order of 5×10^{-3} eV and in the latter case it was 1.5×10^{-3} eV.

The method employed made it possible to determine quite accurately the energy positions and the type of the critical point in the Brillouin zone. However, the method suffered from an important disadvantage in that the energy interval was limited by the transmission of the components of the sandwich (SnO_2 film, silicone oil, and dielectric). Thus, it was not possible to carry out measurements at energies exceeding 4.5 eV, although such measurements would have been desirable in a determination of the energy band structure.

The difficulty could be partly mitigated by using the electrolyte method [19], which would make it possible to extend the measurements to 6-7 eV. In this case the reflection could be modulated by the electric field at the semiconductor—electrolyte interface. However, even then there were some disadvantages: the spectral interval was limited on the long-wavelength side (~1 μ) because of the absorption of light in the electrolyte layer; low-temperature measurements were impossible (because of the lack of electrolytes which would not freeze); the electrolyte reacted chemically with the investigated sample.

One could also use other electroreflection methods [20, 21], which had not found wide acceptance. However, none of these methods was completely satisfactory in all respects and, therefore, we shall not consider them in detail but simply direct the interested reader to the literature. However, it should be pointed out that the electroreflection methods taken as a whole should make it possible to carry out measurements in a wide range of wavelengths and temperatures.

§ 2.   Electroreflection of n-Type GaAs

The experimental method described above was applied to n-type GaAs single crystals with a free-carrier density of the order of 10^{18} cm^{-3} (at room temperature). The samples

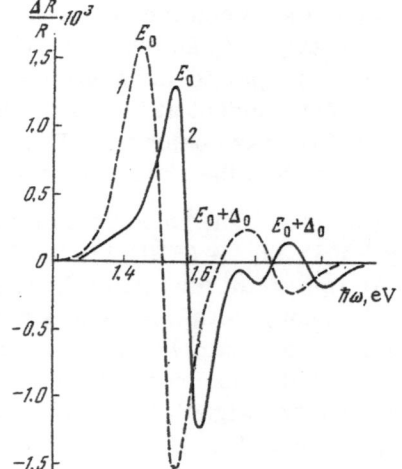

Fig. 5. Electroreflection spectrum of GaAs with a free-carrier density n ~ 10^{18} cm^{-3}, recorded in the region of the fundamental absorption edge at 300°K (1) and 77.3°K (2) using the same modulating voltage.

were plane-parallel plates, ~100-250 μ thick, and with linear dimensions of 5 × 10 mm. The reflecting surface, oriented in some cases parallel to the (111) plane, was mechanically polished and then etched for 10-15 sec in a polishing solution of the composition HF:H_2O:HNO$_3$ (1:2:3). An ohmic contact between the sample and a grounded copper holder was made by a silver plate. A capacitor containing a sample of GaAs was subjected to an alternating voltage of 200-1200 V amplitude and 775 Hz frequency.

The measurements were carried out at 300 and 77.3°K in the photon energy range 1.20-4.00 eV. Throughout this spectral range the resolution was of the order of 5 × 10^{-3} eV.

We shall compare our results with the earlier studies of the electroreflection of GaAs [19, 22].

A. Fundamental Absorption Edge. Figure 5 shows the electroreflection spectra GaAs, obtained at room and liquid nitrogen temperatures, in the range 1.20-2.20 eV, using the same modulating voltage. In agreement with many theoretical [23] calculations of the energy band structure of gallium arsenide and with experimental studies of this structure [19, 22, 24], we can attribute the observed spectrum to interband transitions $\Gamma_{15} \rightarrow \Gamma_1$ at the fundamental absorption edge. A positive peak at 1.45 eV (300°K) corresponds to the absorption edge (E_0 in the notation adopted in [19]) and it has a negative satellite at 1.56 eV. A weaker structure on the short-wavelength side of these peaks is due to transitions from the valence sub-band split by the spin — orbit interaction ($E_0 + \Delta_0$). It is worth noting that direct measurements of the reflection fail to reveal the transitions with the energy $E_0 + \Delta_0$ [24].

This identification of the spectrum can be justified by the following reasoning.

1. The fundamental absorption edge is a critical point of the M_0 type, whose satellite should be located on the short-wavelength side of the main peak. Since the positive peak at 1.45 eV has the longest wavelength, it should be regarded as the main peak corresponding to the fundamental absorption edge.

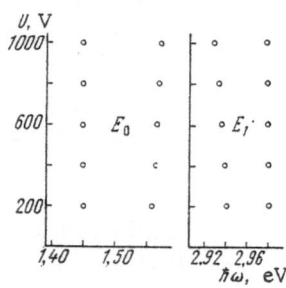

Fig. 6. Dependences of the energy position of the main peak and its satellites for critical points of type M_0 (E_0) and M_1 (E_1) on the voltage applied at 300°K.

2. The position of the main peak does not vary with the electric field in the sample, whereas the satellite shifts in the direction of shorter wavelengths. It is clear from Fig. 6, which gives the dependence of the peak positions on the applied voltage, that in the structure between 1.20 and 2.20 eV the main peak is located at 1.45 eV and the peak at 1.56 eV is the satellite of the main peak. Contrary to theoretical calculations of the electrooptic effects [9, 10], an investigation of the electroreflection of GaAs revealed some shift of the peak corresponding to E_0 with the applied voltage. This was evidently due to broadening effects which resulted in a shift of the position of the asymmetric peak E_0. However, the error in the determination of the energy of the fundamental edge was less than the width of the peak.

3. It follows from the many investigations of the fundamental absorption edge of GaAs that the energy position (1.45 eV) of the positive peak corresponds better to E_0 than the position of the negative peak.

Thus, at room temperature the fundamental absorption edge in the electroreflection spectrum of n-type GaAs (n ~ 10^{18} cm^{-3}) is located at 1.45 eV.

The structure due to transitions from the split-off valence sub-band to the conduction band can be interpreted in a similar manner. Here, $E_0 + \Delta_0$ corresponds to a peak located at 1.80 eV at room temperature. Hence, it follows that the spin − orbit splitting of the valence band of gallium arsenide at the symmetry point Γ in the Brillouin zone is $\Delta_0 = 0.35$ eV. This value is in agreement with the results of other workers [19, 22, 24].

However, it should be noted that both peaks, E_0 and $E_0 + \Delta_0$, are located at somewhat higher energies than those reported in [19, 22, 24]. Obviously, this is due to the Moss-Burstein effect which can shift the fundamental absorption edge of GaAs with a free carrier density of ~10^{18} cm^{-3} by about 10^{-2} eV. If we assume [22] that the fundamental edge is located at 1.43 eV at room temperature, the Moss-Burstein shift is 2×10^{-2} eV. This shift arises because, at high densities, free electrons (holes) occupy lower (higher) states in the conduction (valence) band and thus prevent the participation of these states in otical transitions.

The high density of free carriers also results in some broadening. The broadening effects are frequently the cause of errors in the determination of the fundamental absorption edge. Since the width of a peak is governed by the electric field in a crystal, we can say that in investigated samples the internal fields were high because of the presence of large numbers of carriers.

As pointed out in the Introduction, the electroreflection signal should depend in a particular way ($\infty E^{1/3}$) on the electric field in the investigated material. Since the electric field in the sample was not known, we simply determined the amplitude of the signal as a function of the modulating electric voltage. Figure 7 shows this dependence on a logarithmic scale in

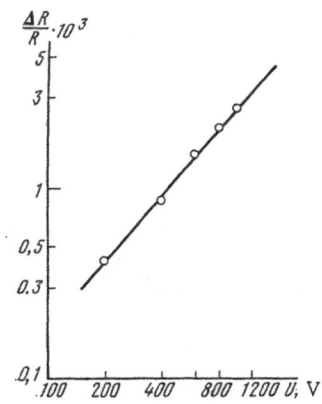

Fig. 7. Dependence of the electroreflection signal on the modulating voltage for the main peak E_0 at 300°K.

the vicinity of the fundamental absorption edge. Within the limits of the experimental error, the dependence $(\Delta R/R)(U)$ is the same for the main peak and the satellite.

Cooling from room to liquid nitrogen temperature shifts the whole structure near the fundamental absorption edge in the direction of higher energies (Fig. 5). The temperature coefficient of the shift is the same for E_0 and $E_0 + \Delta_0$, and amounts to $\sim 5 \times 10^{-4}$ eV/deg. In other words, the spin—orbit splitting is independent of temperature. This is in good agreement with the results obtained from other experiments [22]. Cooling also reduces somewhat the electroreflection signal. This is due to the same circumstances as found for Ge [25] in a simultaneous investigation of the electroreflection and surface properties. It is found that the height of the potential barrier on the surface decreases with temperature. This should naturally be accompanied by a reduction of the electroreflection signal.

B. Interband Optical Transitions at Higher Energies. Calculations of the energy band structure of gallium arsenide [23] have shown that between 1.29 and 4.00 eV one should observe not only transitions at the point Γ at the fundamental absorption edge, but also interband transitions at the point L ($L_{3'} \rightarrow L_1$) on the boundary of the Brillouin zone and at the point Λ ($\Lambda_3 \rightarrow \Lambda_1$) on the [111] symmetry axis (see Fig. 2 in the Introduction). The $L_{3'} \rightarrow L_1$ transitions can be assigned to a critical point of type M_0, and the $\Lambda_3 \rightarrow \Lambda_1$ transitions to a saddle critical point of type M_1. In the conventional reflection spectrum [24] the latter transitions appear as sharp peaks, whereas the $L_{3'} \rightarrow L_1$ transitions appear as a barely distinguishable weak structure. In spite of the higher sensitivity and resolution of the electroreflection spectrum, with the conventional reflection measurements, the signal which could be attributed to transitions at the point L is not observed and this is true of other investigations [19, 22]. This cannot be explained by shortcomings of the experimental method because these transitions should be separated [23] from the transitions at the point Λ by an energy gap much larger (by a factor of at least 50) than the spectral resolution which can be achieved in the electroreflection method. Consequently, it is difficult to accept, as has been done in the case of Ge [25], that the $L_{3'} \rightarrow L_1$ transitions are overlapped by the $\Lambda_3 \rightarrow \Lambda_1$ transitions.

The $\Lambda_3 \rightarrow \Lambda_1$ transitions are attributed, in agreement with the results of other experiments [19, 22], to a structure observed in the electroreflections spectrum of GaAs in the energy interval from 2.80 to 3.40 eV. Figure 8 shows this spectrum at room temperature. A negative peak located at 2.98 eV is the main peak (E_1 in the notation adopted in [19]) and a positive peak which precedes it is a satellite. Thus, a negative peak $E_1 + \Delta_1$ at 3.21 eV is due to optical transitions from the split-off (by the spin—orbit interaction) valence sub-band to the conduction band.

This interpretation is in agreement with the calculations of the energy band structure of GaAs and it is based on the principal conclusions which follow from the theory of the elec-

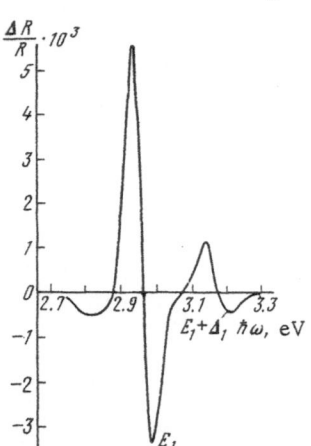

Fig. 8. Electroreflection spectrum of GaAs with $n \sim 10^{18}$ cm^{-3}, recorded in the 2.7–3.4 eV range at room temperature.

trooptic effects. According to this theory [9], the profile of the electroreflection signal for a saddle critical point of type M_1 depends on the orientation of the electric field relative to the principal symmetry axis, corresponding to the constant-energy surface $E_f = E_i = \hbar\omega$. Here, we must distinguish two cases, in which the direction of the field is closer to the parallel or perpendicular orientation with respect to this axis.

The analytic profile of the signal is governed [9, 10] by the quantitative relationship between the components of the reduce effective mass

$$\frac{1}{m^*} = \frac{1}{|E|^2}\left[\frac{E_x^2}{m_x^*} + \frac{E_y^2}{m_y^*} + \frac{E_z^2}{m_z^*}\right]$$

or by the quantity $\theta^3 = \theta_x^3 + \theta_y^3 + \theta_z^3$ ($\theta^3 = e^2E^2/2\hbar m_i^*$; i = x, y, z), which occurs in the argument of the Airy function [9]. Here, E is the electric field intensity; m_i^* represents the reduced effective masses. For a point of type M_1, we have $m_x^* > 0$, $m_y^* > 0$, and $m_z^* < 0$; the principal symmetry axis is the one which corresponds to the mass m_z^*. Depending on the field orientation relative to this axis, we can distinguish the following cases:

1) if $\theta_z^3 < \theta_x^3 + \theta_y^3 \equiv \theta_{xy}^3$ (the direction of the field is closer to the orientation which is perpendicular to the principal axis), the oscillations are located on the short-wavelength side of the peak corresponding to the edge E_g;

2) if $\theta_z^3 > \theta_{xy}^3$ (the direction of the field is closer to the orientation parallel to the principal axis), the satellites appear on the long-wavelength side of E_g.

If the electric field is exactly parallel to the principal axis ($\theta_{xy} = 0$), the Phillips theorem [26] applies. According to this theorem, under certain conditions the electric-field-induced change in the optical constants corresponding to a saddle critical point of type M_1 has an opposite sign to the change associated with a parabolic edge of type M_0. This should be manifested experimentally by a reversal, between points of type M_0 and M_1, of the relative positions of the main peak and its satellites, signs of $\Delta R/R$ of the main peaks, and — the most important point — the direction of shift of the structure with increasing electric field. Strictly speaking, in interpreting the experimental results we must allow for the contributions of different equivalent critical points. However, investigations carried out by the present author and by Seraphin [22] have shown that this theoretical representation is very useful in the interpretation of the electroreflection spectra.

In our experiments the electric field was applied at right angles to the reflecting surface of a GaAs crystal, which was oriented parallel to the (111) symmetry plane. In the case of interband transitions at the point Λ, located on the [111] symmetry axis in the Brillouin zone, the electroreflection structure shifts opposite to the direction of shift of the structure near the fundamental absorption edge associated with a critical point of type M_0 (Fig. 6). This anti-Franz—Keldysh effect, taken in combination with several other factors, allows us to attribute the transitions in question to a critical point of type M_1.

In fact, when the field is increased, the energy position of the short-wavelength negative peak remains unchanged, whereas the long-wavelength positive peak is displaced. Consequently, the former is the main peak corresponding to the absorption edge, and the latter is its satellite; at a critical point of type M_0 the main peak corresponds to positive values of $\Delta R/R$, whereas in this case these values are negative; the structure of the fundamental absorption edge includes oscillations on the short-wavelength side of E_g, whereas in the present case they are located on the long-wavelength side. Thus, we may conclude that the part of the electroreflection spectrum of GaAs observed in the energy range 2.80-3.40 eV corresponds to a critical point of type M_1.

The dependence of the electroreflection signal on the modulating voltage at a point of type M_1 is the same as at a point of type M_0.

Cooling from room to liquid nitrogen temperature shifts the spectrum in the direction of shorter wavelengths and, as in the case of the absorption edge, we have $dE_1/dT = d(E_1 + \Delta_1)/dT$, i.e., the spin−orbit splitting of the valence band at the point Λ is independent of temperature. According to our results, $\Delta_1 = 0.23$ eV. It should be noted that this value gives the spin−orbit splitting of the valence band at the point L on the boundary of the Brillouin zone. In fact, according to theoretical estimates [27], the splitting of a doubly degenerate valence level $L_3^!$ should be two-thirds of the corresponding splitting at Γ_{15} of the triply degenerate top of the valence band corresponding to k = 0. Since $\Delta_0 = 0.35$ eV at the point Γ and $\Delta_1 = 0.23$ eV at the point Λ, and since $\Delta_1/\Delta_0 \approx 2/3$, it follows that the spin−orbit splitting of the valence band at Λ is the same as at L.

The Moss−Burstein effect is not observed for the $\Lambda_3 \rightarrow \Lambda_1$ transitions but the broadening effects still occur though to a lesser degree.

§ 3. Characteristics of the Reflection of GaAs Crystals with Low Free-Carrier Densities

Investigations of the electroreflection of GaAs crystals with carrier densities of $\sim 10^{18}$ cm^{-3} have shown that the experimental results are in good qualitative agreement with the existing theory. The electroreflection spectrum has a pronounced structure at the energies of interband transitions near the Van Hove critical points. We can determine not only the position but also the type of the critical point using particular characteristics. In fact, the determination of the type and position of critical points is the main purpose of the electroreflection method. However, in many cases the electroreflection spectra are complicated by secondary effects. For example, it is reported in [19, 22, 28] that there is a peak on the long-wavelength side of the fundamental absorption edge E_0 and in the case of GaAs this peak is attributed to impurities [19, 22], whereas in the case of CdS and CdSe it is attributed to modulated light reflected from the rear face of the investigated crystal [28]. This modulation is due to electroabsorption in both charged surface layers of a crystal.

Secondary effects not associated with the energy band structure of a semiconductor were also found in the present study when the electroreflection spectra were recorded for n-type GaAs crystals with free-carrier densities of $\sim 10^{15}$ cm^{-3} (300 °K). The measurements were carried out at room temperture in accordance with the method described in §1 and using plane-parallel samples 100-150 μ thick. These measurements were made in the vicinity of the fundamental absorption edge.

Figure 9 shows the dependence $(\Delta R/R)(\hbar\omega)$, which is the dependence of the relative change in the reflection coefficient in the presence of an electric field on the photon energy,

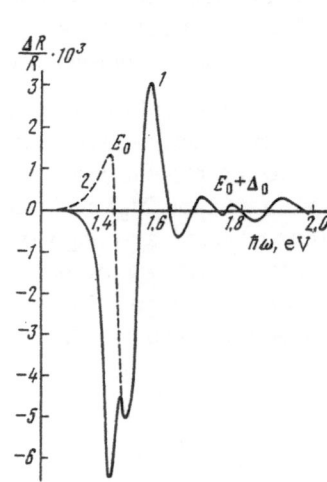

Fig. 9. Electroreflection spectrum of GaAs (1) with n $\sim 10^{15}$ cm^{-3}, recorded in the fundamental absorption region at T = 300°K. The dashed curve (2) represents a possible profile of the peak corresponding to the edge E_0.

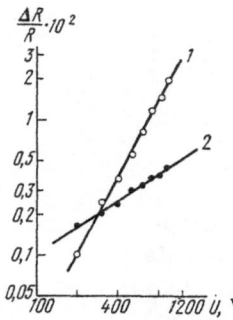

Fig. 10. Dependences of the amplitudes of the long-wavelength (1) and short-wavelength (2) negative electroreflection peaks on the modulating voltage applied to GaAs with n ~ 10^{15} cm^{-3}.

in the spectral range from 1.20 to 2.00 eV. This spectrum differs strongly from the corresponding electroreflection spectrum of GaAs crystals with n ~ 10^{18} cm^{-3} (see Fig. 5 in § 2).

Two negative peaks are observed at 1.417 and 1.470 eV and neither of them can be attributed to the fundamental absorption edge E_0.

The peak $E_0 + \Delta_0$, corresponding to transitions from the split-off valence sub-band, can be identified from the known dependences, as described in § 2. We can see from Fig. 9 that this peak is positive and, consequently, E_0 should correspond to positive values of $\Delta R/R$.

The amplitude of the long-wavelength negative peak (Fig. 10) varies too rapidly (approximately in accordance with the law $\propto U^2$) with the modulating voltage. Moreover, it is important to note that when the voltage is increased, this peak behaves differently from the neighboring short-wavelength negative peak. It follows the two peaks are associated with different features of the energy band structure.

The dependence $(\Delta R/R)(U)$ for the short-wavelength negative peak is approximately the same (Fig. 10) as in the case of E_0 of GaAs crystals with n ~ 10^{18} cm^{-3} (see § 2). However, the position of this peak varies with the field and its short-wavelength tail intersects the energy axis at different points when the voltage is varied (Fig. 11). Consequently, this peak is a satellite.

It follows from our discussion that the main peak E_0, which should be located [22] between the two negative peaks discussed above, does not appear in this spectrum. Its possible form is represented by a dashed curve in Fig. 9.

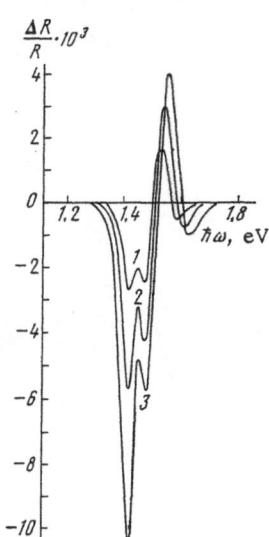

Fig. 11. Dependence $(\Delta R/R)(h\omega)$ recorded at T = 300°K in the range 1.2-1.8 eV using different values of the modulating voltage ($U_1 < U_2 < U_3$) applied to GaAs with n ~ 10^{15} cm^{-3}.

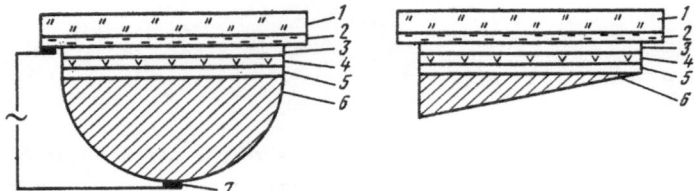

Fig. 12. Shematic representation of a sandwich structure in which the investigated crystal has a deliberately
selected shape of a half-cylinder or a wedge (this shape
prevents the light reflected from the rear face of the
crystal from reaching the photomultiplier). The notation is the same as in Fig. 3.

The long-wavelength negative peak is not associated with impurities. It is estimated
that the electroreflection signal associated with impurities in the investigated crystals should
be of the order of 10^{-4}, whereas the peak under consideration is characterized by a value of
$\Delta R/R$ of the order of several percent. Moreover, since — according to [22] — the room-temperature fundamental absorption edge of the investigated material is located at 1.430 eV, the
ionization energy of the postulated impurities wuold be 0.013 eV. The literature on GaAs
[22, 29] does not include any reference to impurities with this ionization energy.

It is possible that the long-wavelength negative peak is due to the electroabsorption in
the rear face of the crystal.

We determined the relative transmission$(\Delta I/I)(\hbar\omega)$ using a particular value of the voltage.
The amplitude and energy position of the signal obtained in this way suggested that the electroreflection peak on the long-wavelength side of the absorption edge may be due to electroabsorption.

In order to confirm that the peak in question was indeed due to the reflection of the incident monochromatic light from the rear face of the crystal, we selected a particular geometry
known to exclude the stray signal due to such reflection from the photomultiplier. The geometry
(a wedge or a half-cylinder, as shown in Fig. 12) was such that the light fell on the rear surface
only at right-angles to it and, consequently, it was reflected also at right angles, whereas the
angle of reflection of light from the front face was of the order of 22°, as mentioned in § 1.
Under these conditions a photomultiplier records only that part of the light which is reflected
from the front face of the crystal. It follows from Fig. 13, which shows the dependence $(\Delta R/R)$
$(\hbar\omega)$ obtained at room temperature for these samples of special geometry, that the peak in

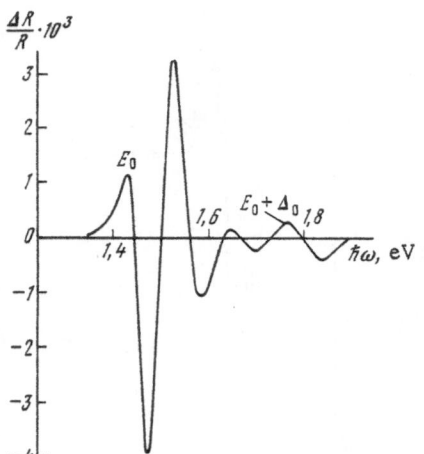

Fig. 13. Electroreflection spectrum of GaAs
with $n \sim 10^{15}$ cm^{-3}, recorded in the region of
the fundamental absorption edge at T = 300°K
using a sample of geometry preventing the light
reflected from the rear face from reaching the
photomultiplier.

questions is no longer observed. The same conclusion follows also from the dependence of
the electroreflection signal on the modulating voltage of the structure shown in Fig. 13. This
dependence does not differ from the results obtained for pure electroreflection.

The fact that the main peak corresponding to the edge is weaker than the satellites does
not contradict the theory of the electrooptic effects (see Introduction). In fact, the conditions
under which the changes in the real and imaginary components of the permittivity in an electric
field are given by the expressions

$$\Delta\varepsilon_1\,(E) = \varepsilon_1\,(E) - \varepsilon_1\,(0),$$
$$\Delta\varepsilon_2\,(E) = \varepsilon_2\,(E) - \varepsilon_2\,(0),$$

do not correspond to reality. Usually, the electric field causing modulation of the optical
constants does not rise from zero. In other words, we have

$$\Delta\varepsilon_1 = \varepsilon_1\,(E_{max}) - \varepsilon_1\,(E_{min}),$$
$$\Delta\varepsilon_2 = \varepsilon_2\,(E_{max}) - \varepsilon_2\,(E_{min}),$$

where E_{max} and E_{min} are the average maximum and minimum values of the electric field
applied to the sample in question. If $(E_{max} - E_{min}) \ll E_{max}$, we find, as pointed out in
[19], that the modulation is proportional to the derivatives $d\varepsilon_1/dE$ and $d\varepsilon_2/dE$. By way of
illustration, the spectral dependences of the functions

$$\frac{d\varepsilon_1}{dE} \cdot \frac{\omega E}{C\theta^{1/2}} \quad \text{and} \quad \Delta\varepsilon_1\frac{\omega}{C\theta^{1/2}}$$

were calculated for a critical point of type M_0. Here, c is a constant and Θ is a quantity
mentioned earlier.

$$\Theta = \left|\frac{e^2\,|E|^2}{2\hbar m^*}\right|^{1/3}.$$

Calculation showed that both these functions oscillated but, whereas in the case of $\Delta\varepsilon_1$ the
signal was strongest near the edge E_0, the amplitude of the oscillations of the derivative
$d\varepsilon_1/dE$ increased away from E_0. This indicated that the strongest peak in the experimentally
obtained electroreflection spectrum need not correspond to the fundamental edge. True, the
dependence $(\Delta R/R)(\hbar\omega)$ never follows exactly the behavior of the derivatives $d\varepsilon_1/dE$ and
$d\varepsilon_2/dE$, i.e., a continuous rise of the oscillation amplitude on increase of the energy separation from the edge is not observed. This is probably due to the broadening effects resulting
from various interactions and from the inhomogeneity of the electric field (ignored in the
Franz—Keldysh theory), which "quench" the oscillations further away from the edge. Nevertheless, the experimentally obtained spectrum should depend on the quantitative relationship
between E_{max} and E_{min}. Therefore, the main peak may be weaker than its satellites. This
is the situation observed, for example, in the electroreflection spectrum of ZnSe [30].

Generalizing the foregoing discussion, we may conclude that in the case of plane-parallel
n-type GaAs crystals with free-carrier densities of ~10^{15} cm^{-3}, the electroreflection peak
located on the long-wavelength side of E_0 is due to the modulation of the light reflected from
the rear face of the investigated crystal.

We can easily explain why this peak is not observed for GaAs with n ~10^{18} cm^{-3} investigated under analogous conditions. The depth of penetration of the electric field into a crystal is proportional now to n$^{-1/2}$ [19], where n is the free-carrier density. Consequently, compared with a sample with n ~ 10^{15} cm^{-3}, the depth of penetration is 30 times lower in a sample with n ~ 10^{18} cm^{-3}. This means that the electroabsorption is weaker and, therefore, the signal produced by it is much smaller.

It should be pointed out that this secondary effect may distort considerably the electroreflection spectrum near the fundamental absorption edge and it should occur in all plane-parallel GaAs crystals with free-carrier densities of ~10^{15} cm^{-3}. Usually, the rear face is left in the rough state (it is not polished) in order to avoid a significant reflection but this is not sufficient because after etching [22] the rear face becomes smooth and this favors the appearance of the effect in question.

Therefore, it is possible that the interpretation of the characteristic features of the electroreflection spectra of gallium arsenide with this free-carrier density, suggested in [19], is in error. This seems to be highly likely because all the main dependences obtained in our case for the peak due to electroabsorption are fully analogous to the dependences reported by other workers, who have adopted a similar interpretation.

§ 4. Exciton Electroreflection of CdTe

In the period (over a decade) since the publication of the Franz – Keldysh effect, there have beeeen many theoretical papers [9, 10, 31] on the influence of the electric field on the optical properties of solids. All the calculations have been made on the assumption that the interaction between electrons and holes can be ignored. It has been shown that the electric field alters the imaginary ε_2 and real ε_1 components of the permittivity and, therefore, all the other optical constants of the material. As pointed out earlier, the greatest changes occur near the critical points in the Brillouin zone. In the light of these results, the experimental data (electroreflection and electroabsorption spectra) are usually attributed [8, 9] to optical transitions near various critical points. By definition, these points represent the locations of the singularities of the combined density of states in the Brillouin zone. As mentioned earlier, the condition for the existence of such points is

$$\mathrm{grad}_k\, E_c(\mathbf{k}) = \mathrm{grad}_k E_v(\mathbf{k}), \tag{1}$$

where $E_c(\mathbf{k})$ is the energy of an electron in its final states (in the conduction band) and $E_v(\mathbf{k})$ is the energy in the initial state (in the valence band).

It should be noted that exciton states are also singularities of the combined density of states. In fact, excitons may be formed from electron and hole states with equal group velocities, when an electron and hole move together. We must assume that

$$[\mathrm{grad}_k\, E_c(\mathbf{k})]_{k_e} = [\mathrm{grad}_k\, E_v(\mathbf{k})]_{k_h}. \tag{2}$$

In the case of "direct" excitons, i.e., those for which $k_e = k_h$, the condition (2) is analogous to the condition (1). Consequently, direct exciton states may exist only at the critical points of the function $E(\mathbf{k}) = E_c(\mathbf{k}) - E_v(\mathbf{k})$. Hence, we may expect the application of an electric field to cause the greatest changes in the optical constants not only at the energies of interband transitions but also in the region of exciton energies. Investigations of the electrooptic effects carried out on Ge [32] have shown that the exciton effects may play an important role and they must be allowed for in the interpretation of the electroabsorption and electroreflection spectra.

The question of the influence of the electric field on the exciton absorption was considered in [33, 34]. The same problem was discussed in greater detail by Duke and Alferieff [35].

The reported calculations show that the electric field causes splitting (Stark effect), displacement, considerable broadening, and even disappearance of exciton peaks. We must also remember that a consequence of the Coulomb interaction between an electron and a hole is not only the appearance of the discrete levels in the forbidden band but also a change in the density of states in the allowed range, which should change the modulation of the permittivity $\Delta \varepsilon$ due to the electric field. However, it is possible to compare only qualitatively the experimental results with the conclusions reached in these papers because the theoretical treatments ignore the finite width of an exciton line in the absence of an electric field.

Excitons appear in the differential electroabsorption and electroreflection spectra as structures with a characteristic form even when they are not observed at all in the conventional absorption and reflection spectra. For example, the exciton peak is hardly visible in the room-temperature absorption spectrum of GaAs [36], whereas a strong exciton signal $\Delta I/I \sim 4 \times 10^{-2}$ (I is the transmission) is observed in the electroabsorption spectrum. The binding energy of excitons is reported in [37] to be $\Delta E_{ex} = 3.4$ meV. Moreover, the electrooptic effect can be used to study the excited states of excitons [38], bound states of excitons [39], exciton transitions with phonon participation [30], etc. In all these investigations the spectra were recorded near the fundamental absorption edge, i.e, the information was obtained on parabolic excitons. It should be noted, however, that the exciton effects may also play a considerable role near other critical points located at higher energies [16].

Thus, the Coulomb interaction between electrons and holes must be allowed for in investigations of the electroabsorption and electroreflection spectra.

We determined the electroreflection near the fundamental absorption edge of cadmium telluride. The electroreflection of CdTe had been reported in [19], where the structure of the peaks was attributed to interband transitions at the appropriate critical points.

We shall show that the positions of the observed peaks near the fundamental absorption edge cannot be identified with the forbidden band width at the point k = 0. Even in a qualitative description of the experimental data on the basis of the Franz – Keldysh effect, we must allow for the Coulomb interaction between electrons and holes.

We selected CdTe because it was available in the form of cubic crystals with a large spin – orbit splitting of the valence band ($\Delta_0 = 0.9$ eV [40]), for which the existence of excitons has been thoroughly established in many experiments [41-44].

Our measurements were carried out on freshly cleaved or freshly etched [for 10 min in a polishing etchant $K_2Cr_2O_7 : HNO_3 : H_2O$ (4:10:20)] n-type CdTe single crystals with a free-carrier density of $\sim 10^{15}$ cm^{-3}. The reflection from the rear face of the crystal was suppressed by choosing a suitable geometry. The measurements were carried out in the energy range 1.550-1.630 eV at liquid nitrogen temperature. The resolution of the spectroscopic system was ~1.5 meV.

A crystal was subjected to an alternating sinusoidal voltage of 775 Hz frequency. The amplitude of the voltage was varied within a wide range from 5 to 1200 V.

Figure 14 shows the electroreflection spectrum of CdTe obtained at a voltage of 100 V.

According to the theory [23], which ignores the Coulomb interaction between electrons and holes, the position of the edge (a critical point of type M_0) should correspond to the first (long-wavelength) positive peak. However, according to the results published in [42], the fundamental absorption of CdTe at T = 77.3°K is located at 1.595 eV, which is closer to the second positive peak.

Moreover, the observed structure differs in its form from that expected on the basis of the theory of the Franz – Keldysh effect (the negative peak is considerably narrower than the positive peaks).

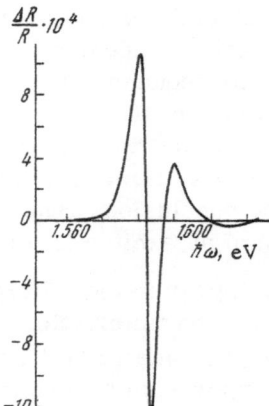

Fig. 14. Electroreflection spectrum of CdTe, recorded in the fundamental absorption edge at T = 77.3°K under a voltage of 100 V.

Moreover, if interband transitions take place, all the peaks should broaden with increasing electric field, the position of the first positive peak should be unaffected, and its satellites should be shifted in the direction of shorter wavelengths. In fact, it is found that an increase in the applied voltage hardly alters the position and width of the negative peak, whereas the positive peaks are broadened and slightly displace.

This disagreement between the experimental results and the theory which ignores the Coulomb interaction between the carriers becomes obvious also when we consider the dependence of the electroreflection signal on the applied voltage. According to this theory, a signal representing interband transitions near a critical point should always increase with the electric field and the nature of the increase should be the same for different peaks (i.e., for the main peak and its satellites). It is clear from Fig. 15, which shows the logarithmic dependence $(\Delta R/R)(U)$ for the first positive and negative peaks, that this does not occur. At low values of the voltage the amplitudes of both peaks increase with rising U but the dependence $(\Delta R/R) \cdot (U)$ is steeper for the negative peak than for the positive peak. Beginning from about 300 V (this "critical" voltage varies from crystal to crystal in the range 200-400 V), a further increase in the voltage reduces both peaks and the fall of $\Delta R/R$ with rising U is faster for the negative peak than for the positive one. (It should be noted that the short- and long-wavelength positive peaks behave in the same way when the voltage is raised.)

It thus follows from our discussion that the electroreflection spectrum of CdTe shown in Fig. 14 cannot be attributed to interband transitions at a critical point of type M_0 (at the fundamental absorption edge).

The form of the structure obtained as well as the energy positions and behavior of the various peaks can be explained quite easily in a qualitative manner if we assume that the spectrum represents the exciton electroreflection. It is convenient and illuminating to consider

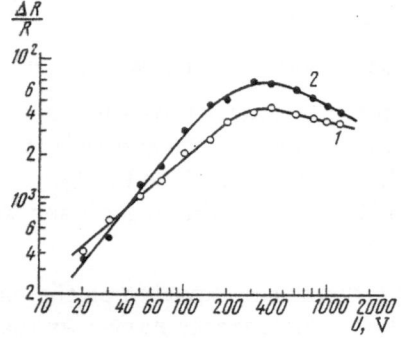

Fig. 15. Dependences (on a logarithmic scale) of the amplitudes of the long-wavelength positive (1) and negative (2) peaks on the voltage applied to CdTe.

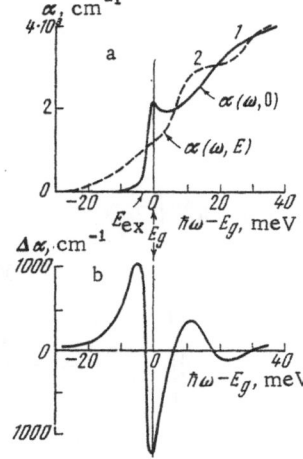

Fig. 16. Schematic representation of the electroabsorption near the fundamental edge in the presence of excitons: a) absorption coefficient in the absence (1) and presence (2) of a field E; b) change in α under the influence of the field.

the influence of the electric field on the absorption spectrum with an exciton peak. The electric field alters the optical absorption. The exciton electroabsorption near the fundamental edge of Ge is explained in [45] by a model shown schematically in Fig. 16. Here, we can see the dependence of the absorption coefficient on the photon energy in the absence of an electric field $\alpha(\omega, 0)$ and the behavior of $\alpha(\omega, E)$, which is the absorption coefficient modified by an electric field E. The field broadens the exciton peak, which gradually disappears as the field is increased, merging with the continuum. The corresponding differential spectrum is shown in the same figure as the dependence $\Delta\alpha(\omega, E)$. According to this model, the negative peak is due to the annihilation of an exciton and the positive peaks should be regarded as a consequence of the broadening of an exciton line by the field. The negative peak in the spectrum of $\Delta\alpha(\omega, E)$ has the same energy as the exciton absorption peak in the absence of the field. Its width is also approximately equal to the width of the exciton line in $E = 0$. This follows also from theoretical calculations give in [35]. The width and energy position of the negative peak should be hardly affected when the field is increased.

Since the optical constants are related to one another by the Kramers – Kronig relationship, it follows that if we know $\Delta\alpha(\omega, E)$, we can calculate the corresponding change in the reflection coefficient caused by the field. Such a calculation is reported in [32] for Ge and it shows that the dependences $\Delta\alpha(\omega, E)$ and $(\Delta R/R)(\omega, E)$ are identical. Generalizing this fact, we may conclude that all the main features characterizing the exciton electroabsorption should be observed also in the electroreflection spectra.

Our experimental electroreflection spectra of CdTe can be understood on the basis of the model illustrated in Fig. 16. This model explains not only the form of the signal but also its dependence on the applied voltage (Fig. 15). In weak fields an exciton is annihilated and the amplitude of the negative peak rises strongly. In this case the spectrum is of purely exciton nature, i.e., the electric field is far too low to alter the interband states. As the field rises (beginning from about $U \geq 300$ V in our case), a positive contribution to $\Delta\alpha$ is made by a continuous distribution of the states adjoining the exciton energy. In very strong fields this contribution may be greater than the contribution of the exciton peak and, together with the broadening effects, this reduces the amplitudes of all the peaks.

Similar results were obtained in [32, 39, 46]. It was concluded in these papers that the observed electroreflection (or electroabsorption) spectra were more likely to be due to exciton than to interband transitions.

The exciton peak in our electroreflection spectra of cadmium telluride (Fig. 14) is located at 1.584 eV at liquid nitrogen temperature. Since at this temperature the forbidden

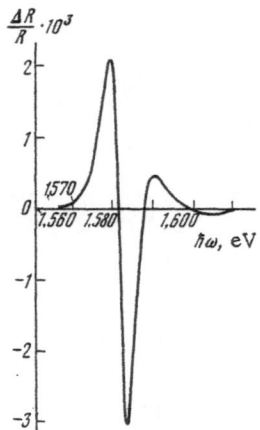

Fig. 17. Electroreflection spectrum of CdTe, recorded in the fundamental absorption region T = 77.3°K using a voltage of 50 V.

band width of cadmium telluride is 1.595 eV [42], the exciton binding energy is 0.011 eV, which is in agreement with the results of other investigations [42, 44].

Apart from the ground (n = 1) exciton state in CdTe, it is also possible to observe excited states since this material is a cubic crystal with a large spin – orbit splitting of the valence band. With this point in mind, we measured the electroreflection at the minimum possible voltages because it was known that weaker electric fields were neeeded to annihilate a state with n = 2 than a state with n = 1. Thus, whereas an application of 50 V (Fig. 17) produced a spectrum which was entirely due to the ground exciton state, the situation was quite different at 20 V (Fig. 18). On the short-wavelength side of the strong (n = 1) negative peak we observed, at a distance of the order of 0.009 eV, an additional weaker negative peak, which we attributed to the n = 2 excited exciton state. Unfortunately, we were unable to study the dependence of the amplitude and behavior of this peak on the applied voltage because the lowest voltage which we could apply (see § 1) was limited to 5 V (this was related to the sensitivity limit of the system). Thus, the range from 5 to 20 V was too narrow to establish any definite relationships. Therefore, we were unable to compare the influence of the electric field on the ground (n = 1) and excited (n = 2) states of an exciton in order to obtain evidence which would leave no doubt as to the nature of the short-wavelength negative peak.

Two exciton peaks were observed in the electroabsorption spectrum of Ge [45] and the short-wavelength peak was attributed to mechanical stresses resulting from the difference

Fig. 18. Electroreflection spectrum of CdTe, recorded at T = 77.3°K under a voltage of 20 V (n = 1 is the ground exciton state and n = 2 is the first excited state).

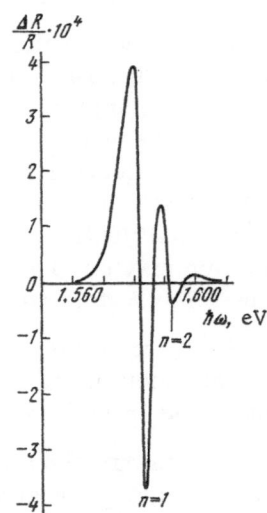

between the compressibilities, at T = 77.3°K, of the crystal and the glass substrate to which the crystal was bonded. In our case a sample was in an unstressed state because all the air gaps in the sandwich were filled with the viscous silicone oil or its mixture with Canada balsam, and the whole capacitor was located in a Teflon holder by an elastic spring.

The electroreflection spectrum observed in the region of the n = 2 peak could not be due to the stresses caused by the electric field. These should shift the fundamental absorption edge and exciton lines, which should be manifested by changes in the absorption (reflection) similar to those observed in the electrooptic effects. Even in hexagonal CdS, which is a good piezoelectric, the changes caused by mechanical stresses were negligible [39], compared with those caused by the electric field. Consequently, in the case of cadmium telluride the electroreflection spectrum could only be due to the electric field and not due to piezoelectric stresses.

In direct measurements of the reflection the exciton peak corresponding to the n = 2 excited state was observed only below 22°K [43]. According to the results reported in [43], the energy separation between the ground (n = 1) and the first excited (n = 2) exciton states was 7.5 meV at 2°K and the corresponding exciton binding energy was 10 meV. These values differed somewhat from those deduced by us from the electroreflection spectra. However, the difference was of the order of the resolution of our spectroscopic system.

Thus, the electroreflection spectrum of CdTe crystals observed at 77.3°K was primarily due to exciton transitions. As pointed out above, the electric field destroyed excitons, but there also other possibilities. The field of the potential barrier on the surface could sometimes be sufficient to destroy excitons without any external electric field. The subsequent application of a voltage of a polarity tending to reduce the surface-barrier field should be accompanied by the appearance of excitons and, consequently, of a corresponding signal in the electroreflection (electroabsorption) spectrum.

Some CdTe crystals did indeed exhibit an electroreflection spectrum which was basically in agreement with the proposed model. It should be pointed out that the model was suitable for the description of the processes occurring in crystals with a low exciton binding energy, such as GaAs. According to [15], the observed electroreflection spectrum of GaAs was due to the neutralization of the electric field in the surface potential barrier by free carriers generated by intense He–Ne laser radiation. The results of this investigation, in which the spectrum was interpreted on the basis of the exciton effects, confirmed the proposed model. It should be pointed out that a characteristic feature of the spectrum in the case where excitons appeared due to the application of an external field was the sign of the modulation of the absorption ($\Delta\alpha/\alpha$) or reflection ($\Delta R/R$). Since the exciton annihilation corresponded to negative values of $\Delta\alpha/\alpha$ and $\Delta R/R$, the exciton appearance should be characterized by positive peaks in the electroabsorption and electroreflection spectra. It is interesting to note also that in the first case the exciton peak and the peak corresponding to the interband transition energy should be directed in opposite ways, whereas in the second case they should be directed in the same ways because (see the Introduction) the fundamental absorption edge always corresponds to positive values of $\Delta\alpha/\alpha$ and $\Delta R/R$.

If at some definite value of the electric voltage the spectrum exhibits simultaneously an exciton peak and a peak corresponding to E_g or to an excited state of an exciton, it should be possible to determine directly the binding energy of the exciton with a much higher sensitivity and resolution than by other methods; moreover, in the other methods one has to carry out experiments at very low temperaturs. In general, it must be stressed that differential electrooptic measurements provide additional extensive information on exciton states in semiconductors.

CHAPTER II

INVESTIGATION OF THE THERMOREFLECTION OF GaAs

It is pointed out in the Introduction that various external perturbations (electric field, pressure, temperature, magnetic field, etc.) act in different ways on a crystal and alter its optical properties. The changes in these properties are strongest near the critical points in the Brillouin zone and can be used to determine the energy positions and nature of these points, which is one of the principal tasks in the studies of the energy band structures of solids. Clearly, in order to perform this task as well as possible it is desirable to carry out investigations using various differential methods, because each of them may provide different information. Therefore, a study was made of the thermoreflection of GaAs in order to supplement the information deduced from the electroreflection spectra. This material was selected for reasons given earlier and for the following additional reason. According to [12], the thermal modulation increases strongly the sensitivity and resolution only in the case of the direct transitions. Since direct transitions occur in GaAs, one may expect the thermoreflection method to give information on the energy band structure of gallium arsenide.

§ 1. Experimental Method

Investigations of the thermoreflection were carried out on crystals with a free-carrier density $n \sim 10^{16}$–10^{17} cm^{-3} (at room temperature). The samples were plane-parallel crystals of arbitrary dimensions and 100–150 μ thick. The reflecting surface was polished mechanically and then the crystals were etched for 10–15 sec in a polishing solution of the composition $HF:H_2O:HNO_3$ (1:2:3).

The thermal modulation in all the thermoreflection investigations [12, 13] was provided by passing a current of a selected frequency either directly through the sample or through its substrate which was in good thermal contact with the sample. In our experiments we heated a sample using a CO_2 gas laser ($\lambda = 10.6$ μ) whose radiation was modulated at a frequency of 20 Hz using a mechanical modulator. Therefore, there was no need to provide electrical contacts (this avoided carrier injection) or a thermal contact with the heater (normally necessary in the case of high-resistivity samples).

In the direct heating by a current a sample would have to be sufficiently thin [11] to ensure a thermal modulation at reasonable frequencies, i.e., to avoid the use of very low frequencies which would cause experimental difficulties. In our case there was no need to restrict the thickness and we could use samples of any thickness and shape because the heating was local.

The absorption of $\lambda = 10.6$ μ laser radiation by free carriers resulted in the heating of a crystal located in a copper holder. The temperature of the sample heated in this way varied at the modulation frequency of the laser radiation (20 Hz). The constant component of the temperature rise was deduced from the shift of the fundamental absorption edge (the temperature coefficient dE_0/dT was assumed to be constant and equal to 5×10^{-4} eV/deg [47] in the investigated temperature range). The amplitude of the temperature oscillations was difficult to determine. However, it could be estimated knowing the value of $\Delta R/R$. Assuming that $R = [(1 - n)/(1 + n)]^2$, where R is the reflection coefficient and n is the refractive index of the investigated substance, we found that

$$\frac{\Delta R}{R} = \frac{4n}{n^2 - 1} \cdot \frac{\Delta n}{n}.$$

The values of $\Delta n/n\Delta T$ for gallium arsenide were given in [48]. We found ΔT from $\Delta n/\Delta T$ given in [48] and we used Δn calculated from the thermoreflection spectrum employing $\Delta R/R$. These estimates yielded the depth of modulation $\Delta T \sim 15$ deg.

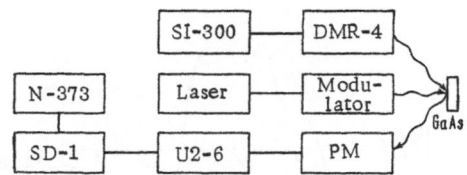

Fig. 19. Block diagram of the apparatus used
in thermoreflection measurements.

However, it should be pointed out that the method was not free of shortcomings. In particular, it was difficult to investigate the temperature dependence of the thermoreflection effect because it was difficult to regulate the constant temperature when a laser was used.

Figure 19 shows a block diagram of the apparatus. Light from a ribbon incandescent lamp of the SI-300/10 or SI-200u type passed through a DMR-4 monochromator and reached a GaAs sample at right-angles to the reflecting surface or at the Brewster angle (~74°). A laser beam, passed through a modulator and a germanium lens-filter, reached the same part of the crystal. The monochromatic light reflected from the crystal was received by a photomultiplier (PM in Fig. 19) of the FÉU-28 or FÉU-18a type, depending on the spectral region being investigated. As in the electroreflection case, the photomultiplier current had a constant component proportional to the reflection coefficient of the sample and an alternating one, which was proportional to the change in the reflection coefficient ΔR. This change was due to the thermal modulation.

In contrast to the electroreflection spectra, the thermoreflection peaks were quite wide, so that we were unable to carry out measurements in narrow energy intervals and we had to allow for the spectral dependence of the reflection coefficient R. In other words, in the thermoreflection case we could not assume (as was done in Chap. I, § 1) that the spectrum $(\Delta R/R)$ $(\hbar\omega)$ did not differ from the dependence $\Delta R(\hbar\omega)$. Hence, we had to abandon the method of determining separately ΔR and R and then calculating $\Delta R/R$. The ratio $\Delta R/R$ was measured by a method described in [19]. The essence of this method was that a constant signal proportional to R was maintained throughout the investigated spectral range. The constancy of the signal was ensured by varying the high voltage supplied by a VSV-2 rectifier, which was applied to the photomultiplier. The alternating signal, amplified by a U2-6 narrow-band amplifier, was passed to a SD-1 synchronous detector and recorded with an N-373 or ÉPP-09 potentiometer. When the signal proportional to R was kept constant, the alternating signal was proportional to $\Delta R/R$.

The thermoreflection spectra of GaAs were investigated above room temperature in the photon energy range from 1.10 to 4.00 eV.

The resolution of the spectroscopic system was of the order of 5×10^{-3} eV.

2. Results and Discussion

Figure 20 shows the thermoreflection spectrum of GaAs obtained at a temperature of the order of 500°K in the photon energy range 1.10-4.00 eV.

A strong negative peak at 1.29 eV is due to the light reflected from the rear face of the investigated plane-parallel crystal and it is a consequence of the thermal modulation of the absorption in the bulk of the crystal. This is indicated by the amplitude and energy position of the relative transmission signal $(\Delta I/I)(\hbar\omega)$. A similar peak, located on the long-wavelength side of the fundamental absorption edge, is observed in the thermoreflection spectra of Ge [13] and CdS [49], and in the electroreflection spectra of GaAs, CdS, and CdSe [28]. For some crystals this signal is weaker or altogether absent probably because of the uneven etching which destroyed the parallel orientation of the surfaces.

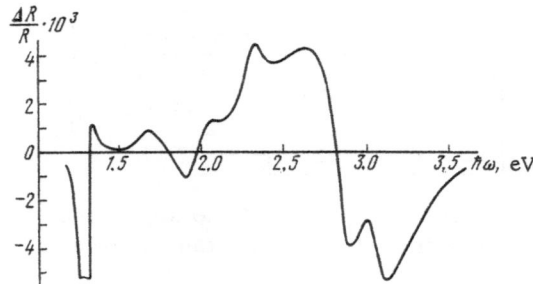

Fig. 20. Thermoreflection spectrum of GaAs at
~500°K recorded in the photon energy range 1.1-4.0
eV.

Positive peaks E_0 and $E_0 + \Delta_0$ at energies of 1.33 and 1.68 eV are the results of the
$\Gamma_{15} \rightarrow \Gamma_1$ transitions at the fundamental absorption edge. The separation between these peaks,
$\Delta_0 = 0.35$ eV, is in agreement with the splitting of the valence band by the spin–orbit interaction [23].

The range from 1.80 to 2.50 eV is of special interest. According to theoretical calculations of the energy band structure of GaAs [23], this region should correspond to the $L_3 \rightarrow L_1$,
transitions associated with the critical point of type M_0 on the boundary of the Brillouin zone.
At this point the spin–orbit splitting of the valence band should be 1.5 times smaller then
the corresponding splitting at the point Γ [27], where the top of the triply degenerate valence
band is located at k = 0.

It follows from our discussion that the structure observed in this range can be attributed
to the $L_{3'} \rightarrow L_1$ transitions. The separation between the positive peaks e_1 and $e_1 + \Delta e_1$, located
at 2.08 and 2.32 eV, obeys the rule

$$\Delta_{s-o}(L) = \frac{2}{3}\Delta_{s-o}(\Gamma).$$

According to our data, the spin–orbit splitting is $\Delta_{s-o}(\Gamma) = \Delta_0 = 0.35$ eV and $\Delta_{s-o}(L) = \Delta e_1 =$
0.24 eV. A negative peak at 1.90 eV is due to the special form of the thermoreflection signal
[12], which manifests the influence of temperature on the imaginary and real components of
the permittivity.

Direct measurements of the reflection of GaAs in liquid nitrogen [24] revealed a very
weak structure which was attributed to the same transitions. However, these transitions
were not manifested in the spectra obtained by other differential methods (electroreflection,
piezoreflection). At first sight it seemed strange because the differential methods were known
to have a higher sensitivity and resolution than the conventional reflection method. It should
be pointed out that these $L_{3'} \rightarrow L_1$ transitions were observed more or less reliably not only
in GaAs but also in all semiconductors investigated by the differential methods. However, a
study of the electroreflection of Ge [50] yielded barely distinguishable peaks at 2.05 and 2.24
eV, which were attributed to the $L_{3'} \rightarrow L_1$ transitions at room temperature.

These energies differed very strongly from Potter's results [51], who determined the
imaginary ε_2 and real ε_1 components of the permittivity of Ge without recourse to the absolute
reflectivity measurements. Potter used a polarimetric method and, having determined the
Brewster angle to within 5', he first observed experimentally the thresholds of the $L_{3'} \rightarrow L_1$
transitions. According to his results, these transitions occurred at 1.74 and 1.94 eV at room
temperature. Thus, the difference between these values and those obtained from the electrore-

flection spectra in [50] was considerable. In order to resolve any possible doubts, other investigators [52] carried out a careful study of the electroreflection of Ge in a wide temperature range (300-14°K). However, the $L_{3'} \rightarrow L_1$ transitions were not observed in the spectra reported in [52].

The same was also true of GaAs. Investigations of the electroreflection spectra carried out by the present author (see Chap. I, § 2) and other workers [19, 22] failed to reveal the $L_{3'} \rightarrow L_1$ interband transitions on the boundary of the Brillouin zone in the temperature range from 77.3 to 300°K.

One of the possible explanations of these discrepancies could be as follows. The electroreflection signal depends [9] on the reduced effective mass of carriers and the greater this mass the weaker is the signal. If we assume that the $L_{3'} \rightarrow L_1$ transitions correspond to relatively high values of m^*, we can see why the electroreflection method is insufficiently sensitive to reveal these transitions.

In the thermoreflection case the signal depends on the temperature coefficient of the shift of the fundamental absorption edge ($\Delta R/R \propto dE_0/dT$ [13]). The $L_{3'} \rightarrow L_1$ transitions correspond to a critical point of type M_0, i.e., of the same type as the fundamental absorption edge. Extensive experimental data show [53] that in the case of III-V compounds, which include GaAs, the temperature coefficient of the shift of the fundamental absorption edge increases with the reduced effective mass of carriers (Fig. 21). Hence, it follows that even if m^* is large for the $L_{3'} \rightarrow L_1$ transitions, the value of $\Delta R/R$ should be considerable in the thermoreflection spectrum.

Moreover, the depth of modulation ΔT was considerable in our experiments and this also tended to increase the thermoreflection signal because, according to [13], $\Delta R/R \propto \Delta T$. The transitions under discussion were not observed in the thermoreflection spectrum of Ge at $T \sim 400°K$ [13]. This could probably be explained by the fact that the transitions in Ge at the points L and Λ were fairly close in respect of the energy $(\Lambda_1 - \Lambda_3) - (L_1 - L_{3'}) = 0.25$ eV [54], i.e., the separation was several times smaller than in the case of GaAs [23]. Since the thermoreflection peaks were fairly wide, the pattern at the point L was masked by the transitions at Λ.

The short-wavelength part of the thermoreflection spectrum of GaAs (Fig. 20) resembled, in the 2.50-3.50 eV range, the thermoreflection spectrum of Ge in the 1.70-2.70 eV range [13]. The structure found in this range for Ge was attributed in [13] to the $\Lambda_3 \rightarrow \Lambda_1$ transitions at a saddle critical point of type M_1. The thermoreflection structure of Ge corresponding to the $\Lambda_3 \rightarrow \Lambda_1$ transitions was calculated in [13]. It was found that the experimental spectrum was in good agreement with the calculations.

In view of the fact that in the case of GaAs the dependence $(\Delta R/R)(\hbar\omega)$ in the 2.50-3.50 eV range was identical with the thermoreflection of Ge in the 1.70-2.70 eV range (which, as

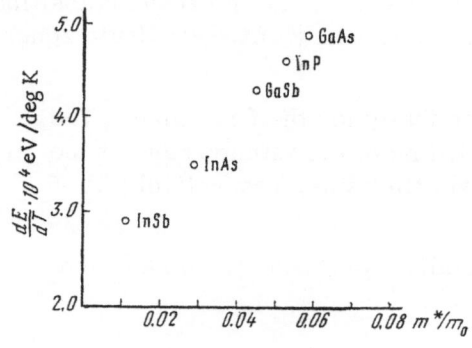

Fig. 21. Dependence of the temperature coefficient of the shift of the fundamental absorption edge on the reduced effective mass of carriers in III-V semiconductors.

shown convincingly in [13], corresponded to the $\Lambda_3 \rightarrow \Lambda_1$ transitions) and in view of the numerous experimental [19, 22, 24] and theoretical [23] results, which demonstrated that in this region only the transitions at the point Λ in the Brillouin zone could occur, it was concluded that the structure in question was due to the $\Lambda_3 \rightarrow \Lambda_1$ interband transitions. The structure consisted of negative peaks E_1 and $E_1 + \Delta_1$ at energies of 2.91 and 3.14 eV. The separation between these peaks, $\Delta_1 = 0.23$ eV, corresponded to the spin–orbit splitting of the valence band. The spin–orbit splitting of the valence band was practically the same at the points Λ and L. This was in agreement with the theoretical calculations [23, 27].

CONCLUSIONS

1. An investigation was made of the energy band structure of semiconductors by the electroreflection and thermoreflection methods. In the electroreflection case the optical properties of the investigated crystal were varied by modulating the electric field in the potential barrier on the surface of a sample placed in a capacitor. The thermal modulation was provided by a CO_2 gas laser ($\lambda = 10.6$ μ), whose radiation intensity was varied at selected frequency. The absorption of laser radiation by free carriers resulted in an alternating heating of the investigated crystal and a consequent change in the optical constants. The interpretation of the results was hindered by some secondary effects. These included the interference of light in measurements of the electroreflection or the reflection of light from the rear surface of a crystal. The latter was particularly important because when it was ignored, the experimental results could be interpreted erroneously. These secondary effects could be eliminated by a suitable selection of the experimental conditions: the interference of light was eliminated by a special construction of the capacitor and the reflection of light by a suitable geometry of the sample.

2. Measurements were made of the electroreflection of n-type GaAs crystals in the photon energy range 1.20-4.00 eV at two different temperatures (T = 300 and 77.3°K).

The experimental results were found to be in good qualitative agreement with the current theory of the electrooptic effects and with the calculations of the energy band structure of GaAs. In the investigated range of energies the electroreflection spectrum included two groups of peaks which were attributed to the $\Gamma_{15} \rightarrow \Gamma_1$ interband transitions at the center of the Brillouin zone and to the $\Lambda_3 \rightarrow \Lambda_1$ transitions at the point Λ on [111] symmetry axis. The values of the spin–orbit splitting of the valence band of GaAs were determined at the symmetry points Γ and Λ in the Brillouin zone. The splitting was, respectively, $\Delta_0 = 35$ eV and $\Delta_1 = 0.23$ eV; within the limits of the experimental error, the splitting was independent of temperature in the range 77.3-300°K.

3. The thermoreflection spectra of n-type GaAs single crystals were determined in the photon energy range 1.10-4.00 eV. The thermoreflection spectrum of GaAs exhibited a peak structure which was attributed, in accordance with the calculations of the energy band structure and with theoretical discussions of the thermoreflection effect, to the interband transitions $\Gamma_{15} \rightarrow \Gamma_1$ (at the center of the Brillouin zone) $L_{3'} \rightarrow L_1$ (on the boundary of the Brillouin zone), and $\Lambda_3 \rightarrow \Lambda_1$ (on the [111] symmetry axis).

The $L_{3'} \rightarrow L_1$ transitions in GaAs were observed practically for the first time. The thermoreflection measurements yielded the spin–orbit splitting of the valence band at the symmetry points Γ, L, and Λ in the Brillouin zone. The splitting was, respectively, $\Delta_0 = 0.35$ eV, $\Delta_1 = 0.24$ eV, and $\Delta_1 = 0.23$ eV.

The rule $\Delta_{s-o}(\Gamma) = (3/2)\Delta_{s-o}(L)$, predicted theoretically, was found to be satisfied exactly.

4. In the case of CdTe single crystals it was found that the interpretation of the experimental electroreflection spectra was sometimes impossible without allowance for the Coulomb interaction between electrons and holes, i. e., without allowance for the exciton effects. Moreover, the electroreflection method, which ensured a high sensitivity and resolution, provided a powerful tool for investigating the exciton states in semiconductors. At liquid nitrogen temperature the electroreflection spectrum of n-type CdTe included, in the region of the fundamental absorption edge, a peak due to the ground (n = 1) exciton state located at 1.584 eV. The binding energy of the excitons was estimated to be 0.011 eV. At low voltages, the first excited (n = 2) state of excitons appeared at 1.593 eV. This exciton state appeared in the reported conventional reflection spectra only at temperatures below 20°K.

LITERATURE CITED

1. J. P. Phillips, Spectra-Structure Correlation, Academic Press, New York (1964).
2. J. Tauc (ed.), The Optical Properties of Solids (Proc. E. Fermi Intern. Physics School Varenna, 1965), Academic-Press, New York (1966).
3. D. Brust, Phys. Rev., 134:A1337 (1964).
4. L. Van Hove, Phys. Rev., 89:1189 (1953).
5. J. C. Phillips, Phys. Rev., 104:1263 (1956).
6. L. V. Keldysh, Zh. Eksp. Teor. Fiz., 34:1138 (1958).
7. W. Franz, Z. Naturforsch., 13a:484 (1958).
8. B. O. Seraphin and N. Bottka, Phys. Rev., 145:628 (1966).
9. D. E. Aspnes, Phys. Rev., 147:554 (1966); 153:972 (1967).
10. D. E. Aspnes, P. Handler, and D. F. Blossey, Phys. Rev., 166:921 (1968).
11. M. Cardona, Proc. Ninth Intern. Conf. on Physics of Semiconductors, Moscow, 1968, Vol. 1, publ. by Nauka, Leningrad (1968), p. 365; G. S. Krinchik, Usp. Fiz. Nauk, 94:143 (1968).
12. E. Matatagui, A. G. Thompson, and M. Cardona, Phys. Rev., 176:950 (1968).
13. B. Batz, Solid State Commun., 5:985 (1967).
14. H. R. Philipp and E. A. Taft, Phys. Rev., 113:1002 (1959).
15. R. E. Nahory and J. L. Shay, Phys. Rev. Lett., 21:1569 (1968).
16. B. O. Seraphin, Phys. Rev., 140:A1716 (1965).
17. J. E. Fischer and B. O. Seraphin, Solid State Comm., 5:973 (1967).
18. A. Frova and P. Handler, Phys. Rev. Lett., 14:178 (1965); W. E. Engeler, H. Fritzsche, M. Garfinkel, and J. J. Tiemann, Phys. Rev. Lett., 14:1069 (1965); B. O. Seraphin and N. Bottka, Phys. Rev. Lett., 15:104 (1965).
19. M. Cardona, K. L. Shaklee, and F. H. Pollak, Phys. Rev., 154:696 (1967).
20. V. Rehn and D. S. Kyser, Phys. Rev. Lett., 18:848 (1967).
21. Y. Hamakawa, P. Handler, and F. A. Germano, Phys. Lett. A, 25:617 (1967).
22. B. O. Seraphin, Proc. Phys. Soc., 87:239 (1966); J. Appl. Phys., 37:721 (1966).
23. M. L. Cohen and T. K. Bergstresser, Phys. Rev., 141:789 (1966); F. H. Pollak, C. W. Higginbotham, and M. Cardona, J. Phys. Soc. Jap., Suppl., 21:20 (1966).
24. D. L. Greenaway, Phys. Rev. Lett., 9:97 (1962).
25. B. O. Seraphin, R. B. Hess, and N. Bottka, J. Appl. Phys. 36:2242 (1965).
26. J. C. Phillips, in: The Optical Properties of Solids, (Proc. E. Fermi Intern. Physics School, Varenna, 1965, ed. by J. Tauc), Academic Press, New York (1966); Phys. Rev., 146:584 (1966).
27. L. M. Roth and B. Lax, Phys. Rev. Lett., 3:217 (1959).

28. E. Gutsche and H. Lange, Phys. Status Solidi, 22:229 (1967).

29. D. E. Hill, Phys. Rev., 133:A866 (1964).

30. K. Ikeda, Y. Hamakawa, H. Komiya, and S. Ibuki, Phys. Lett. A, 28:647 (1969).

31. D. S. Bulyanitsa, Zh. Eksp. Teor. Fiz., 38:1201 (1960); J. Callaway, Phys. Rev., 134:A998 (1964); K. Tharmalingam, Phys. Rev., 130:2204 (1963).

32. Y. Hamakawa, F. A. Germano, and P. Handler, Phys. Rev., 167:703, 709 (1968).

33. B. O. Seraphin, Proc. Seventh Intern. Conf. on Physics of Semiconductors, Paris, 1964, Vol. 1, Physics of Semiconductors, publ. by Dunod, Paris; Academic Press, New York (1964), p. 165.

34. F. Seitz, Phys. Rev., 76:1376 (1949).

35. C. B. Duke, Phys. Rev. Lett., 15:625 (1965); C. B. Duke and M. E. Alferieff, Phys. Rev., 145:583 (1966).

36. M. D. Sturge, Phys. Rev., 127:768 (1962).

37. Q. H. F. Vrehen, J. Phys. Chem. Solids, 29:129 (1968).

38. R. A. Forman and M. Cardona, in: II-VI Semiconducting Compounds (Proc. Intern. Conf., Providence, R. I., 1967, ed. by D. G. Thomas), Benjamin, New York (1967), p. 100.

39. B. B. Snavely, Phys. Rev., 167:730 (1968).

40. M. Cardona and D. L. Greenaway, Phys. Rev., 131:98 (1963).

41. D. T. F. Marple, Phys. Rev., 150:728 (1966); B. Segall, Phys. Rev., 150:734 (1966); V. S. Vavilov, A. A. Gippius, Zh. R. Panosyan, in: Cadmium Telluride [in Russian], Nauka, Moscow (1968), p. 103.

42. D. G. Thomas, J. Appl. Phys. Suppl., 32:2298 (1961).

43. B. Segall and D. T. F. Marple, in: Physics and Chemistry of II-VI Compounds (ed. by M. Aven and J. S. Prener), North-Holland, Amsterdam (1967), Chap. 7, pp. 319-381.

44. R. E. Halsted, M. R. Lorenz, and B. Segall, J. Phys. Chem. Solids, 22:109 (1961); V. S. Vavilov, A. F. Plotnikov, and A. A. Sokolova, in: Cadmium Telluride [in Russian], Nauka, Moscow (1968), pp. 59, 69.

45. Q. H. F. Vrehen, Phys. Rev., 145:675 (1966).

46. V. K. Subashiev and G. A. Chalikyan, Fiz Tverd. Tela, 10:1343 (1968); Proc. Ninth Intern. Conf. on Physics Semiconductors, Moscow, 1968, Vol. 1, publ. by Nauka, Leningrad (1968), p. 397; P. I. Perov, L. A. Avdeev, M. I. Elinson, and G. V. Stepanov, Radiotekh. Elektron., 13:1721 (1968).

47. H. Welker and H. Weiss, Solid State Phys., 3:1 (1956) [see p. 51].

48. M. Cardona, Proc. Fifth Intern. Conf. on Physics of Semiconductors, Prague, 1960, publ. by Academic Press, New York (1961), p. 388; R. Weil, Bull. Am. Phys. Soc., 13:1657 (1968).

49. H. Lange and W. Henrion, Phys. Status Solidi, 23:K67 (1967).

50. A. K. Ghosh, Phys. Rev., 165:888 (1968).

51. R. F. Potter, Phys. Rev., 150:562 (1966).

52. Y. Hamakawa, T. Nishino, and J. Yamaguchi, Proc. Ninth Intern. Conf. on Physics of Semiconductors, Moscow, 1968, Vol. 1, publ. by Nauka, Leningrad (1968) p. 384.

53. C. Hilsum and A. C. Rose-Innes, Semiconducting III-V Compounds, Pergamon Press, Oxford (1961).

54. D. Brust, J. C. Phillips, and F. Bassani, Phys. Rev. Lett., 9:94 (1962).

LUMINESCENCE DELAY AND DETERMINATION OF NONRADIATIVE RELAXATION TIMES

É. L. Nolle

An analysis is made of the kinetics of the luminescence of a three-level system in which a transition takes place from a lower metastable state (lifetime τ_1) under the influence of excitation pulses of duration t_0 to an upper state (lifetime $\tau_2 < \tau_1$) from which the system relaxes back to the metastable state. It is shown that if $t_0 \lesssim \tau$, the luminescence is delayed after the end of excitation and this delay can be used to calculate the nonradiative relaxation time τ_2. The luminescence delay was observed experimentally when $Y_3Al_5O_{12}$:Nd^{3+} was excited with electrons pulses of $t_0 \lesssim 3$ μsec duration (radiative transitions of Nd^{3+} from the $^4F_{3/2}$ level due to radiative relaxation from an S level of the neodymium ions) and $t_0 = 10$ nsec duration (transitions from an S level due to the transfer of energy from the host lattice to Nd^{3+}). It was found that the transfer time was $\tau_2 = 40$ nsec. The delay of the luminescence emitted from electron—hole drops in Ge at 4.2°K was used to determine the drop formation time, which was 200 nsec.

In many cases the excitation of semiconductors is, in contrast to the resonance case, a multistage process. Initially the system goes into a relatively short-lived excited state from which it relaxes, frequently in a nonradiative manner, to a metastable state characterized by a finite radiative transition probability. Examples of such processes are the intrinsic radiative recombination in semiconductors due to the excitation with pulses of electrons or photons of energies greater than the forbidden band width; other such processes are the binding of excitons into complexes or electron—hole drops followed by the emission of light, transfer of energy in crystal phosphors from the host substance to the activator, and so on. Usually, particularly in the case of semiconductors, the nonradiative relaxation time is short (according to [1], the slowing-down time of carriers in Ge is $\tau_s \approx 5 \times 10^{-10}$ sec at 300°K and $\tau_s \approx 2 \times 10^{-9}$ sec at 4.2°K) and it is considerably shorter than the duration of the excitation pulses. Therefore, a stationary short-lived state is established during the excitation pulse but this has practically no influence on the luminescence kinetics on transition from a metastable to the ground state, at least during the early stages of luminescence. For some laws of relaxation in electron transitions between states, for example, in the case of the hyperbolic law for a state with a short lifetime or the exponential law for a metastable state, the existence of a short-lived state may result in the delay of luminescence in later stages [2]. The relaxation of a system in accordance with such laws has been considered for instantaneous excitation and it was observed in phosphors [3]. Since the duration of excitation was considerably less than the lifetimes (10^{-4}-10^{-3} sec), the luminescence intensity increased by a factor of 1.5 after the end of excitation.

However, in many situations (particularly in the case of semiconductors) the relaxation of a system from a short-lived state is exponential with a very short time constant comparable with the duration of excitation pulses. Therefore, it would be interesting to consider the

227

kinetics of such processes as a function of the duration of excitation and thus show that it should be possible to determine the nonradiative relaxation times, which is the aim of the present investigation.

1. Relaxation and Luminescence Kinetics

Let us consider a three-level semiconductor (Fig. 1) with its upper energy level 2 excited at a rate G. This excitation is followed by the relaxation in a time τ_2 to a state with an energy level 1 and then in a time τ_1 to the ground state 0, accompanied by the emission of a photon. For the sake of simplicity we shall assume that the energy gaps between the levels are considerably greater than kT so that the thermal transitions between the levels can be ignored. Then, the change in the number of excited particles at the level 2, denoted by Δm, and the level 1, denoted by Δn, is described by the system of rate equations

$$\frac{d(\Delta m)}{dt} = G - \frac{\Delta m}{\tau_2}, \tag{1a}$$

$$\frac{d(\Delta n)}{dt} = \frac{\Delta m}{\tau_2} - \frac{\Delta n}{\tau_1}. \tag{1b}$$

Excitation alters Δm because $\Delta m = G\tau_2(1 - e^{-t/\tau_2})$, so that substituting the expression in Eq. (1b) and bearing in mind that $\Delta n = 0$ at $t = 0$, we find the change Δn with time under excitation conditions

$$\Delta n = G\frac{\tau_1^2}{\tau_1 - \tau_2}(1 - e^{-\frac{t}{\tau_1}}) - G\frac{\tau_1\tau_2}{\tau_1 - \tau_2}(1 - e^{-\frac{t}{\tau_2}}). \tag{2}$$

After the end of excitation we have $\Delta m = G\tau_2(1 - e^{-t_0/\tau_2})e^{-t/\tau_2}$, where t_0 is the duration of the excitation pulse. Substituting Δm into Eq. (1b) and bearing in mind that at $t = 0$ the value of Δn is given by Eq. (2) modified by the substitution $t = t_0$, we obtain

$$\Delta n = G\frac{\tau_1^2}{\tau_1 - \tau_2}(1 - e^{-\frac{t_0}{\tau_1}})e^{-\frac{t}{\tau_1}} - G\frac{\tau_1\tau_2}{\tau_1 - \tau_2}(1 - e^{-\frac{t_0}{\tau_2}})e^{-\frac{t}{\tau_2}}. \tag{3}$$

It is clear from Eq. (3) that after the end of excitation the number of particles at the level 1 and, consequently, the luminescence intensity $I \propto \Delta n/\tau_1$ increase with time reaching their maximum values at

$$t_{\max} = \frac{\tau_1\tau_2}{\tau_1 - \tau_2}\ln\frac{1 - \exp\left(-\frac{t_0}{\tau_2}\right)}{1 - \exp\left(-\frac{t_0}{\tau_1}\right)}. \tag{4}$$

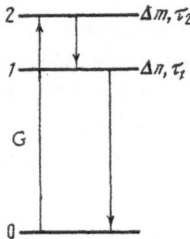

Fig. 1. Electron transition scheme.

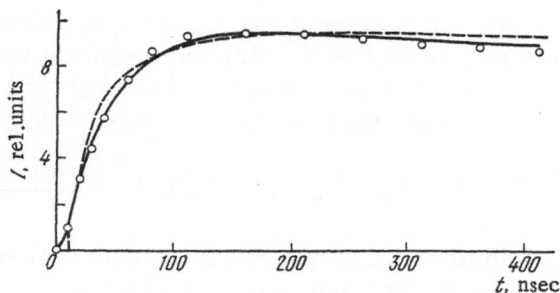

Fig. 2. Experimental points and calculated curves [the continuous curve is based on Eq. (3) and the dashed one on Eq. (6)] demonstrating the kinetics of the S-level luminescence of $Y_3Al_5O_{12}:Nd^{3+}$ in the case of excitation with pulses of $t_0 = 10$ nsec duration, which is marked by a dashed vertical line near the origin.

If the conditions $\tau_2 \ll \tau_1$ and $t_0 \ll \tau_1$ are satisfied, the ratio of the maximum luminescence intensity to the intensity after the end of excitation is

$$\frac{I_{max}}{I_0} = \frac{t_0}{t_0 - \tau_2\left[1 - \exp\left(-\frac{t_0}{\tau_2}\right)\right]}. \tag{5}$$

In the case of longer excitation pulses when $t_0 \gg \tau_2$, it follows from Eq. (5) that $I_{max} \approx I_0$ and from Eq. (4) that $t_{max} \ll t_0$, so that the last term in Eq. (3) can be ignored, i.e., after the end of excitation the luminescence decays. Therefore, in order to determine the time τ_2 from Eqs. (3), (4), or (5), we must satisfy the conditions $t_0 \leq \tau_2$.

The nonradiative relaxation time τ_2 can be determined conveniently from the measured luminescence delay using a graphical solution of Eq. (4) and the experimentally determined parameters t_0, t_{max}, and τ_1. This approach can be applied also in the case of nonexponential relaxation provided $\tau_2 = f(t)$. The value of τ_2 obtained in this case represents the instantaneous lifetime when a given degree of excitation Gt_0 is reached. For example, in the case of the hyperbolic relaxation law $d(\Delta m)/dt = G - \gamma(\Delta m)^2$ and instantaneous excitation ($\Delta n = 0$ at $t = 0$),* the number of particles at the level 1 after the end of excitation depends on t in accordance with the expression

$$\Delta n = \text{th}^2\left(t_0\sqrt{G\gamma}\right)\sqrt{\frac{G}{\gamma}}\left(e^{-\frac{t}{\tau}} - \frac{1}{t\sqrt{G\gamma}+1}\right) - \frac{\text{th}^2\left(t_0\sqrt{G\gamma}\right)}{\tau} \times$$
$$\times e^{-\frac{1}{\tau\sqrt{G\gamma}}}\left[E_i\left(\frac{1}{\tau\sqrt{G\gamma}}\right) - E_i\left(\frac{t\sqrt{G\gamma}+1}{\tau\sqrt{G\gamma}}\right)\right]e^{-\frac{t}{\tau}}. \tag{6}$$

Figure 2 shows the dependences $I \propto \Delta n(t)$ calculated from Eq. (3) for $t_0 = 10^{-8}$ sec, $\tau_1 = 3.2 \times 10^{-6}$ sec, $\tau_2 = 4.2 \times 10^{-8}$ sec, as well as the functions calculated from Eq. (6) for $t_0 = 10^{-8}$ sec, $\tau = 3.2 \times 10^{-6}$ sec, $\sqrt{G\gamma} = 5 \times 10^7$ sec^{-1}. We can see that the dependences $\Delta n(t)$ for the relaxation laws corresponding to Eqs. (3) and (6) by not more than 20%. This makes it somewhat difficult to find the law of relaxation of a system to a metastable radiative state but

---

*In general, the solution of the system (1) for the hyperbolic relaxation law cannot be expressed in terms of simple functions.

it permits us to use a simple expression, corresponding to the exponential law, in the determination of the "instantaneous" relaxation times. The nature of the relaxation can be deduced from an investigation of the dependence of the luminescence delay time t_{max} on the excitation rate because — according to Eq. (4) — the value of t_{max} is proportional to τ_2.

2. Luminescence Delay and Energy Transfer Time from Host to Activator in $Y_3Al_5O_{12}:Nd^{3+}$

The luminescence of neodymium ions in yttrium aluminum garnet (YAG) single crystals excited by x-ray [4] or electron-beam [5] bombardment was found to be due to radiative transitions from an upper level S and from the level $^4F_{3/2}$ located approximately 3.35 below the S level. The quantum efficiency of the luminescence of $Y_3Al_5O_{12}:Nd^{3+}$ was approximately the same for the transitions from these two levels and it amounted to about 0.7. The lifetime at the level S was 3.5 ± 0.3 μsec, which was 50 times shorter than the lifetime at the level $^4F_{3/2}$. Therefore, we concluded that the electron-beam excitation was transferred from the host substance first to the S level of the neodymium ions and then the level $^4F_{3/2}$ was excited. In this case the excitation with pulses of $t_0 \lesssim 3.5$ μsec duration should generate a delayed luminescence from the $^4F_{3/2}$ level. Since the transition from the S level was radiative, we could determine the lifetime at this level τ from the decay of its luminescence and from the growth of the luminescence of the $^4F_{3/2}$ level. This should make it possible to determine the nature of the excitation of Nd^{3+} and to check the relationships (2)-(5) as a function of the duration of excitation.

The luminescence of samples of $Y_3Al_5O_{12}:Nd^{3+}$ (0.3 wt.%), grown by V. V. Osiko and M. I. Timoshechkin by the Czochralski method, were bombarded with pulses of 90 keV electrons (current density 5×10^{-3}-0.5 A/cm^2) at room temperature. The pulse duration could be varied smoothly from 10^{-8} to 10^{-5} sec. The edges of the pulses were 3-5 nsec long (Figs. 3b and 3d). The luminescence was detected with an FÉU-39 or an FÉU-28 photomultiplier.

The decay of the luminescence of the S level was exponential with a time constant equal to the lifetime $\tau_S = 3.2 \pm 0.3$ μsec (Fig. 3c and curve 1 in Fig. 4) and the decay of the lu-

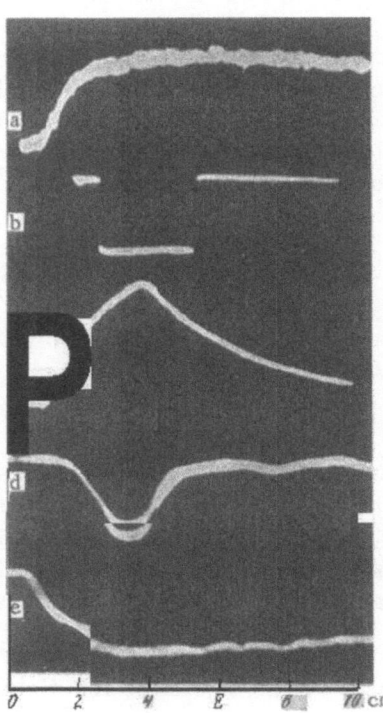

Fig. 3. Oscillograms of the excitation and luminescence pulses obtained for a crystal of $Y_3Al_5O_{12}:Nd^{3+}$: a) luminescence due to transitions from the $^4F_{3/2}$ level (scale 2.5 μsec/cm) excited by a pulse of $t_0 = 1$ μsec duration; b) electron excitation pulse (1 μsec/cm); c) luminescence due to transitions from the S level (1 μsec/cm), $t_0 = 3$ μsec; d) electron excitation pulse (5 nsec/cm); e) S-level luminescence (50 nsec/cm), $t_0 = 10$ nsec.

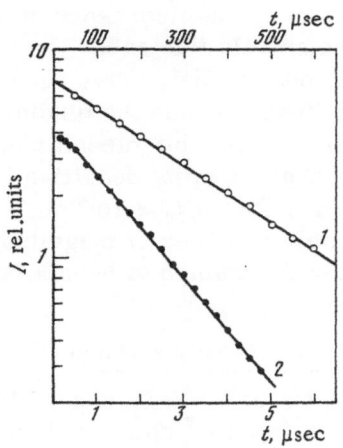

Fig. 4. Decay of the luminescence of $Y_3Al_5O_{12}$: Nd^{3+} due to transitions from the S level (curve 1, lower abscissa) and from the $^4F_{3/2}$ level (curve 2, upper abscissa).

minescence of the $^4F_{3/2}$ level was also exponential with $\tau_F = 170 \pm 10$ μsec (curve 2 in Fig. 4). At the end of an excitation pulse of $t_0 \lesssim 3$ μsec duration we observed a considerable increase in the intensity of the $^4F_{3/2}$ luminescence, as shown in Fig. 3a. Figure 3 shows also the end of an excitation pulse of $t_0 = 1$ μsec duration (the pulse was recorded because of the induced signal in the photomultiplier). It is clear from Fig. 5 that the experimental growth of the luminescence was in agreement with the values calculated from Eq. (3) and this was true for three different durations of the excitation pulses $t_0 = 1$, 2, and 3 μsec. The time constant $\tau_1 = \tau_F = 1.70$ μsec was determined using the decay of the $^4F_{3/2}$ luminescence at a time $t > \tau_2 = \tau_S$, whereas the constant deduced from Eq. (4) and the growth of the $^4F_{3/2}$ luminescence after the end of excitation was $\tau_2 = \tau_S = 2.7 \pm 0.7$ μsec, which was close to the value obtained from the decay of the S-level luminescence.

These results indicated that in the case of excitation with an electron beam the energy was transferred from the host substance (garnet) first to the upper level (S) of the neodymium ions and the delay of the luminescence excited by short pulses could be used to determine the nonradiative relaxation time.

When pulses of several microsecond duration were used (Fig. 3b) the S-level luminescence was not delayed (Fig. 3c). However, when the duration of the excitation pulses was reduced to 10 nsec (Fig. 3d) the intensity of the S-level luminescence continued to rise after the end of excitation and reached its maximum in 200 nsec (Fig. 3e). The time constant τ_2, deduced in this case from the luminescence growth (Fig. 2), was $\tau_2 = \tau_t = 40$ nsec and it was independent of the rate of excitation which could change by about two orders of magnitude. This constant obviously represented the time for the transfer of energy from the host substance to the neodymium ions because the electron-beam excitation initially generated electron−hole pairs in the allowed bands of the garnet.

Fig. 5. Experimental points and calculated curves [based on Eq. (3)] showing the kinetics of the luminescence of $Y_3Al_5O_{12}$:Nd^{3+} due to transitions from the $^4F_{3/2}$ level, excited with pulses of different duration (μsec): 1) 1; 2) 2; 3) 3. The end of each pulse is denoted by a vertical dashed line.

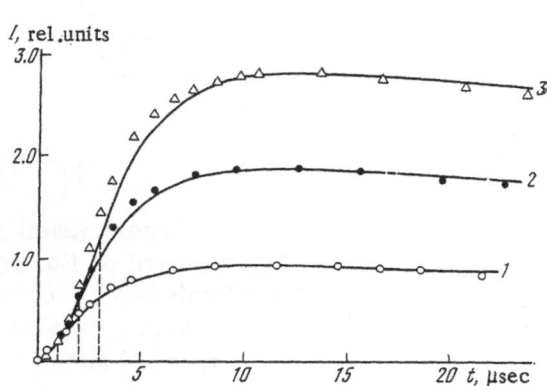

It was pointed out in [5] that the high quantum efficiency of the luminescence of $Y_3Al_5O_{12}$: Nd^{3+} was evidently due to the exciton energy transfer mechanism. In this case, the transfer time constant τ_t could represent the time of diffusion of excitons to Nd^{3+}. This time could not be governed by the binding of electrons and holes into excitons because the binding time should depend on the rate of excitation (the formation of excitons was a bimolecular process). Moreover, the exciton formation time at the lowest electron-beam current densities $j \approx 10^{-2}$ A/cm^2, which corresponded to a density of electron—hole pairs $\Delta n = Gt_0 \approx 10^{15}$ cm^{-3} in the case of pulses of $t_0 = 10^{-8}$ sec duration, should be approximately an order of magnitude smaller than the measured value τ_t if the exciton binding coefficient was assumed to have a fairly small value $\gamma_{ex} \approx 10^{-7}$ cm^3/sec.

3. Formation Time of Electron — Hole Drops in Germanium

The short-wavelength luminescence emitted from pure Ge at 77°K is due to free excitons [6]. Since the radius of an exciton in Ge is large ($a_0 = 1.2 \times 10^{-6}$ cm, which corresponds $\gamma_{ex} \approx 10^{-5}$ cm^3/sec), the exciton binding time is about 10^{-9} sec for $\Delta n \approx 10^{14}$ cm^{-3}. Therefore, in the case of excitation pulses of $t_0 \approx 10^{-8}$ sec duration there should be no delay of the exciton luminescence due to the binding of electrons and holes into excitons or due to their slowing down in the allowed bands.

However, according to [7, 8], excitons in Ge are condensed into electron—hole drops at T < 10°K or, according to [9], they are bound into biexcitons. These processes can occur in times much shorter than 10^{-8} sec.

In order to obtain more information on this subject, we studied samples of pure Ge. Such samples were bonded to a helium bath of a cryostat in such a way that they remained at the bath temperature. The samples were excited with pulses of 90 keV electrons. A fast response germanium photodiode was used as a radiation detector.

At 77°K the luminescence spectrum was due to free excitons [6] and there was no delay of the luminescence after the end of an excitation pulse (curve 2 in Fig. 6). The luminescence decayed exponentially with a time constant $\tau_1 = \tau_{ex} = 370$ nsec, representing the exciton lifetime. The relatively small value of τ_{ex} was evidently due to the influence of the surface recombination because the 90 keV electrons penetrated into Ge to a depth of about 20 μ.

At 4.2°K the excitation with electron-beam current densities from 0.2 to 4 A/cm^2 produced a luminescence spectrum which corresponded to electron—hole drops. The free-exciton luminescence was not observed because the excitation rate was too high and the drop luminescence intensity was proportional to the cube of the exciton concentration [8]. The current density $j = 1$ A/cm^2 corresponded, in the case of $t_0 = 50$ nsec pulses, to an average (in the excitation

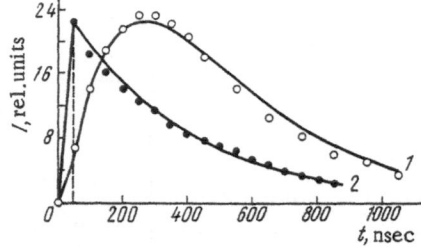

Fig. 6. Experimental points and calculated curves [based on Eq. (3)] showing the kinetics of the luminescence of Ge: 1) 4.2°K, $t_0 = 50$ nsec, $\tau_1 = 300$ nsec, $\tau_2 = 200$ nsec; 2) 77°K, $t_0 = 50$ nsec, $\tau_1 = 370$ nsec, $\tau_2 = 0$.

region) exciton concentration $\Delta n_{ex} \approx 2 \times 10^{17}$ cm^{-3}. The volume of the excitation region was governed by the diffusion length of excitons which was approximately 200 μ for $\tau_1 = \tau_{ex} = 300$ nsec [8]. In the j $\gtrsim$ 1 A/cm^2 range the luminescence spectrum broadened somewhat and shifted slightly toward longer wavelengths. This could be explained by the interaction between electrons and holes because the electron − hole drops filled completely the excitation region.

A study of the kinetics of the luminescence of electron − hole drops revealed a delay due to the drop formation time $\tau_2 = \tau_0$ (curve 1 in Fig. 6). This study yielded $\tau_2 = \tau_0 = 200$ nsec and the drop lifetime $\tau_d = 300$ nsec, which were independent of the excitation rate in the range j = 0.2-4 A/cm^2. These results indicated that the 4.2°K luminescence was not due to the formation of biexcitons, since such formation was a bimolecular process and τ_0 should have decreased with the rate of excitation.

If we assumed that the observed electron − hole drops condensed at nuclei, which could be impurities or defects present in a concentration N [8], the time constant τ_0 should represent the time needed for the capture of first excitons by the nuclei because the capture cross section should increase with the number of capture excitons n_c as $n_c^{2/3}$. The later stages of the drop formation could be ignored by considering only the time during which the capture cross section of a nucleus increased by a factor of 10, which should occur after the capture of 30 excitons. The capture time could then be expressed in the form of the series given below and then used to find the electron − hole drop formation time:

$$\tau_0 = \sum_{l=1}^{L} \frac{1}{l^{1/3} \sigma v N} = \frac{S}{\sigma v N} , \qquad (7)$$

where v is the thermal velocity of excitons; σ is the cross section for the capture of the first exciton by a nucleus, which is governed by the exciton radius a_0 and is equal to πa_0^2; S is the sum of L terms of the above series, which is 7 for L = 30. Substituting the measured value $\tau_0 = 2 \times 10^{-7}$ sec, we obtained − depending on the coefficient S − a value N $\approx 5 \times 10^{12}$ cm^{-3}, which was approximately equal to the concentration of residual impurities in Ge.

The reported investigation demonstrated the possibility of determination of the nonradiative relaxation time from the delay of the luminescence observed after the end of excitation. This can be done if the duration of the excitation pulses t_0 does not exceed the relaxation time τ_2. Usually, the duration of the luminescence from a metastable state is much longer than τ_2 and when $t_0 = \tau_2$ the ratio of the maximum luminescence intensity to the intensity at the end of the excitation pulse is e [Eq. (5)], which makes it possible to determine quite accurately the value of τ_2.

Mode-locked lasers can emit light pulses of up to 10^{11} W power and of duration shorter than 10^{-12} sec [10]. However, at present the lowest limit of the relaxation times that can be measured is set by the response time of photodetectors, which is of the order of 10^{-11}-10^{-10} sec in the case of germanium photodiodes and vacuum photocells. These detectors can be used to measure directly the slowing-down time of carriers in such semiconductors as germanium and silicon and to determine the formation time of exciton complexes and possibly also the binding time of electrons and holes in excitons.

The author is grateful to V. S. Vavilov, V. V. Osiko for the discussion of the results, and to A. Fazilov for his help in this investigation.

LITERATURE CITED

1. O. N. Krokhin and Yu. M. Popov, Zh. Eksp. Teor. Fiz., 38:1589 (1960).
2. V. A. Yastrebov, Dokl. Akad. Nauk SSSR, 90:1015 (1953).

3. M. V. Fok, Opt. Spektrosk., 2:127 (1957);T. P. Belikova, Opt. Spektrosk., 16:862 (1964).

4. Yu. K. Voron'ko, B. I. Denker, V. V. Osiko, A. M. Prokhorov, and M. I. Timoshechkin, Dokl. Akad. Nauk SSSR, 188:1258 (1969).

5. Yu. K. Voron'ko, É. L. Nolle, V. V. Osiko, and M. I. Timoshechkin, Zh. Eksp. Teor. Fiz., 13:125 (1971).

6. J. R. Haynes and N. G. Nilsson, Proc. Seventh Intern. Conf. on Physics of Semiconductors, Paris, 1964, Vol. 4, Radiative Recombination in Semiconductors, publ. by Dunod, Paris; Academic Press, New York (1965), p. 21.

7. V. M. Asnin and A. A. Rogachev, ZhETF Pis'ma Red., 9:415 (1969).

8. Ya. E. Pokrovskii and K. I. Svistunova, ZhETF Pis'ma Red., 9:435 (1969); Fiz. Tekh. Poluprovodn., 4:491 (1970).

9. C. Benoit à la Guillaume, F. Salvan, and M. Voos, J. Lumin., 1:315 (1970).

10. N. Bloembergen, Comments Solid State Phys., 1:37 (1968).